Genetics and Breeding of Sugar Beet

Genetics and Breeding of Sugar Beet

Editors

Enrico Biancardi
Istituto Sperimentale Per Le Colture Industriali
Rovigo, Italy

Larry Campbell
USDA Agricultural Research Station
Fargo, ND
USA

Marco De Biaggi
Istituto Sperimentale Per Le Colture Industriali
Rovigo, Italy

George N. Skaracis
Hellenic Sugar Industries
Thessaloniki
Greece

Science Publishers, Inc.

Enfield (NH), USA Plymouth, UK

SCIENCE PUBLISHERS, INC.
Post Office Box 699
Enfield, New Hampshire 03748
United States of America

Internet site: *http://www.scipub.net*

sales@scipub.net (marketing department)
editor@scipub.net (editorial department)
info@scipub.net (for all other enquiries)

Library of Congress Cataloging-in-Publication Data

Genetics and breeding of sugar beet/editors, Enrico Biancardi ... [et al.].
p.cm.
Includes bibliographical references and index.
ISBN 1-57808-366-4
1. Sugar beet—Breeding. 2. Sugar beet—Genetics. I. Biancardi, Enrico.
SB221.G365 2004
633.6'32—dc22

2004057821

ISBN 1-57808-366-4

Published by Science Publishers Inc., Enfield, NH, USA

Printed in India

Foreword

It was a great honour to be asked by the editors to write a foreword to their substantial book. Comprehensive publications on the improvement and genetics of sugar beet are few indeed. The most recent is the chapter by Professor Bosemark in "The Sugar Beet Crop" edited by Cooke and Scott in 1993. This field is developing quickly and this update is well timed.

Sugar beet is one of the most recent plants cultivated on a large scale. Cultivation of beet species is ancient but its industrial use is recent. It was first used as a source of sucrose at the beginning of the nineteenth century, after the German Achard built the first sugar factory in Cunern in Silesia, with the help of the King of Prussia (1802). Because of the continental blockade, Napoleon gave the sugar beet industry the resources necessary for its early development.

Given its recent origin, it is quite easy to follow the evolution of the sugar beet compared to most cultivated plants that were domesticated before the discovery of writing and recording of history. From the beginning of its industrial development, sugar beet has benefited from an impressive improvement effort, for three reasons:

— First: its fodder origin. The "*Weisse schlesische Rübe*" used by Achard had a sugar content of only 7%. The nineteenth century was devoted to increasing sugar content and hence the breeders were often called "chemists".

— Second: sugar beet is a biennial which requires specific conditions for reproduction, in contrast to cereals, legumes, and crucifers for which farmers can readily reproduce the seed.

— Third: industrial customers were not able initially to easily control the sugar content, of prime necessity for the profitability of their production. So they entrusted the task of good and stable sugar content to specialized breeders.

Sugar beet is the first scientifically bred crop and breeders, confident of their sales, have invested considerable resources in its improvement. Thanks to the initial improvement efforts, techniques and breeding methods have continually improved during the nineteenth and twentieth century. Therefore, sugar beet can be considered a model in plant breeding.

Among other subjects, this book gives considerable emphasis to selection for resistance to pests and diseases in relation to employment of new technologies. Host plant resistance research became essential in plant breeding in response to decreasing agricultural prices and a corresponding reduction in inputs in order to decrease costs. A reduction in pesticide use also became absolutely necessary for environmental protection.

My congratulations to the several authors for their contributions covering numerous breeding objectives of considerable scientific interest.

Michel Desprez

Templeuve, December 1, 2003

Preface

The incentive for this book was the realization that, unlike other important crops, a monograph entirely dedicated to sugar beet genetics and breeding was not published. Pertinent information is available only in chapters of several books devoted either to the breeding of different crops or to various aspects of sugar beet production. Contributions by E. Knapp (1958), K.H. Barocka (1985), both in German, G.A. Smith (1987), R.J. Hecker and R.H. Helmerick (1985), and L.G. Campell (2002) belong to the former group, while those of J.S. McFarlane (1971) and N.O. Bosemark (1993) belong to the latter. These basic publications have comprehensively documented the current state of the art and proved beneficial to those engaged in the theoretical and applied aspects of sugar beet genetic improvement.

The book comes at a time of rapid expansion in molecular technology–based selection approaches destined to modify or supplement conventional breeding methodology. These new technologies will allow genetic and physiological factors influencing sugar yield and quality to be assessed in great detail and then manipulated. These novel techniques are also expected to facilitate reduction in reliance of the sugar beet crop on chemical pesticides and fertilizers, by exploiting unique and improved resistance mechanisms against the various abiotic stresses and diseases and by producing varieties that use soil resources more efficiently. An entire chapter is devoted to current information on the development of these new techniques and their integration into sugar beet breeding.

As evidenced by the numerous references cited, it has been necessary to synthesize an impressive volume of scientific information. For all the subjects addressed, the authors attempted to maintain an appropriate balance between relevant theoretical considerations and application to sugar beet genetic improvement, albeit the latter is often complicated by local considerations. The contents of this book have been updated with the most recent information on breeding activities. Several researchers at public institutes, universities and private seed companies with vast experience in the numerous subjects have collaborated in this work. Some authors, directly involved in sugar beet breeding, have released lines and varieties of notable importance from both a commercial and historical perspective. Their experience is partially acknowledged in the respective sections.

The editors are grateful to the essential contributions that collaborators provided in the preparation of this volume. The critical review of some chapters by John Macefield was particularly appreciated.

The Editors

About the Cover Illustration

Reproduction of the first known illustration of a beet dates back to the manuscript *Materia Medica*, written by Pedanius Dioscurides. The author was a physician living in Rome during the reign of Emperor Tiberius Claudius Nero (around 60 AD). During the Middle Ages, the book was thought to be one of the most important manuals of medicine in the ancient world. This color miniature was found at the Biblioteca Marciana of Venice, in the codex named *Diuscurides Neapolitanus*. The Herbarium included in the manuscript, probably reproduced in Constantinople or Ravenna, is the earliest illustrated source of botanical knowledge and natural therapeutic practises. Some historians speculated that the Herbarium could date back to Crateuas, physician of Mithridates VI king of Pontus (120-63 BC). At least the white form of the plants described in the original caption seems to refer to the sea beet (*Beta vulgaris* ssp. *maritima*), used at that time for food and often as a drug.

ϹΕΥΤΛΟΝ ΜΕΛΑΝ ΑΓΡΙΟΝ

ΡΩΜΑΙΟΙ ΒΙΤΑ ϹΙΛΒΑΤΙΚΑ.

ΔΙΤΤΟΝ. ΤΟ ΜΕΝ ΜΕΛΑΝ ϹΤΑΛΤΙΚΩΤΕΡΟΝ ΚΟΙΛΙΑϹ ϹΥΝ ΦΑΚΗ ΕΨΗ-
ΘΕΝ, ΚΑΙ ΜΑΛΛΟΝ Η ΡΙΖΑ, ΤΟ ΔΕ ΛΕΥΚΟΝ ΕΥΚΟΙΛΙΟΝ. ΑΜΦΟΤΕΡΑ ΔΕ
ΚΑΚΟΧΥΛΑ ΔΙΑ ΤΟ ΕΝ ΑΥΤΟΙϹ ΝΙΤΡΩΔΕϹ, ΟΘΕΝ Ο ΧΥΛΟϹ ΑΥΤΩΝ
ΚΑΘΑΙΡΕΙ ΚΕΦΑΛΗΝ ΡΕΙΝΕΓ'ΧΥΤΟΥΜΕΝΟϹ ΜΕΤΑ ΜΕΛΙΤΟϹ ΚΑΙ
ΩΤΑΛΓΙΑϹ ΩΦΕΛΕΙ. ΤΟ ΔΕ ΤΩΝ ΡΙΖΩΝ Κ(ΑΙ) ΤΩΝ ΦΥΛΛΩΝ ΑΦΕ-
ΨΗΜΑ ΠΙΤΥΡΩΝ ΚΑΙ ΚΟΝΙΔΩΝ ϹΜΗΚΤΙΚΟΝ ΚΑΙ ΧΕΜΕΘΛΩΝ ΠΡΑ-
ΥΝΤΙΚΟΝ ΚΑΤΑΝΤΛΟΥΜΕΝΟΝ. ΟΜΟΙΩϹ ΔΕ ΤΟΙϹ ΦΥΛΛΟΙϹ ΚΑΤΑ-
ΠΛΑΤΤΕΙΝ ΑΛΦΟΥϹ ΠΡΟΕΚ'ΝΙΤΡΩΘΕΝΤΑϹ ΚΑΙ ΑΛΩΠΕΚΙΑϹ ΚΑΤΑ-
ΚΝΙϹΘΕΙϹΑϹ ΚΑΙ ΤΑ ΝΕΜΟΜΕΝΑ ΕΛΚΗ. ΕΦΘΗ ΔΕ ΕΞΑΝΘΗΜΑΤΑ ΚΑΙ
ΠΥΡΙΚΑΥΤΑ ΚΑΙ ΕΡΥϹΙΠΕΛΑΤΑ ΪΩΝΤΑΙ : ›››

Translation of the Greek caption (translated by Prof. George Gialouris, Panorama High School, Thessalonica, Greece): Wild black beet. According to Romans: *Sylbatica* (wild beet of the forest). The beet is of two kinds: one is black, systolic for the abdomen when cooked (especially the root) with lentils; the other kind is white and is a laxative. The gruel of both has a bad taste because it contains nitrous salts and this is why it clears the head. When infused together with honey in the nose, it comforts from ear pain. The juice of roots and leaves fights off dandruff and nits, while a footbath mollifies chilblains. In addition, using leaves as cataplasm on previously soda-washed areas heals leprosy, the itching from alopecia and extended skin sores. It also anticipates pimples and heals blisters from burning and erysipelas.

Contents

Contributors

Detlef Bartsch
Federal Office of Consumer Protection and Food Safety (BVL)
Dept. Genetechnologies
Taubenstrasse 42/43, 1077 Berlin, Germany
E-mail: detlef.bartsch@bvl.bund.de

Enrico Biancardi
Istituto Sperimentale per le Colture Industriali, Sezione di Rovigo
Viale Amendola 82, 45100 Rovigo, Italy
E-mail: e.biancardi@isci.it

Larry G. Campbell
USDA, Agricultural Research Service, Northern Crop Science Laboratory
P.O. Box 5677, Fargo, ND 58105, USA
E-mail: campbell@fargo.ars.usda.gov

Marco De Biaggi
Istituto Sperimentale per le Colture Industriali, Sezione di Rovigo
Viale Amendola 82, 45100 Rovigo, Italy
E-mail: isciro@tecna.it

Brian V. Ford-Lloyd
School of Biosciences, University of Birmingham, Edgbaston, Birmingham, B15 2TT, UK
E-mail: b.ford-lloyd@bham.ac.uk

Karen Klotz
USDA, Agricultural Research Service, Northern Crop Science Laboratory
P.O. Box 5677, Fargo, ND 58105, USA
E-mail: klotzk@fargo.ars.usda.gov

Robert T. Lewellen
USDA, Agricultural Research Service, 1636 East Alisal Street, Salinas, California 93905, USA
E-mail: rlewellen@pw.ars.usda.gov

Mitchell J. McGrath
USDA, Agricultural Research Service, 494D Plant and Soil Science Building, Michigan State University, East Lansing, MI 48824-1325, USA
E-mail: mitchmcg@msu.edu

Leonard W. Panella
USDA, Agricultural Research Service, Crop Research Laboratory, 1701 Center Avenue, Fort Collins, CO 80526, USA
E-mail: lpanella@lamar.colostate.edu

George N. Skaracis
Hellenic Sugar Industries S.A., Mitropoleos Street 34, 54110 Thessaloniki, Greece
E-mail: skarebz@mail.otenet.gr

Gerhard Steinrücken
Syngenta Seeds AB, P.O. Box 302, 26123 Landskrona, Sweden
E-mail: gerhard.steinrucken@syngenta.com

Piergiorgio Stevanato
Istituto Sperimentale per le Colture Industriali, Sezione di Rovigo
Viale Amendola 82, 45100 Rovigo, Italy
E-mail: p.stevanato@isci.it

Ming H. Yu
Formerly, USDA, Agricultural Research Service, 1636 East Alisal Street, Salinas, CA 93905, USA
E-mail: myu243@yahoo.com

Introduction

E. Biancardi, L.G. Campbell, G.N. Skaracis, and M. De Biaggi

Success in plant breeding has always depended on the ability to recognize and create genetic variation and proficiency in selecting superior genotypes for environments of interest.

Natural genomic plasticity was the sole source of variation exploited in early breeding efforts. Observable differences within a species were the result of unpredictable genetic changes in chromosome and DNA structure. With the advances in genetics and related scientific fields, along with the systematic adaptation of new technologies, plant breeding has become a scientifically based endeavor that continuously and consciously stirs the hereditary material and effects planned changes in the genetic constitution of plants for societal benefit. Human interventions have impacted all levels of organization from the whole plant down to the subcellular and molecular levels.

A retrospective view of developments beneficial to the breeder over the last 100 years reveals that the contribution of quantitative genetics, coupled with the application of statistical theory to field plot techniques, has been of paramount importance because it added scientific precision to a profession previously regarded, more or less, as an art, and increased breeder efficiency in synthesizing and manipulating populations to be improved. Concomitantly, it is worth noting that several so-called revolutionary approaches based on newly acquired knowledge have appeared almost every 20 years, promising monumental increases in efficiency. Interspecific hybridization, induced polyploidy and artificial mutagenesis are examples. Although none of these individual "breakthroughs" justified the initial enthusiasm, they have all contributed significantly to plant breeding.

Sugar beet is recognized as the first crop whose development is based solely on a modern understanding of genetics. Its present status is the result of scientifically based selection during the nineteenth and twentieth centuries, an endeavor that has capitalized on most of the aforementioned scientific advances. According to M. Desprez, an enumeration of the milestones of progress definitely should include developments in: i) mass selection (Achard, 1790); ii) pedigree selection and estimation of sugar concentration by root density (de Vilmorin, 1837 to 1859); iii) optic measurement of sugar by polarization (d'Arago and Biot, 1856); iv) use of cytoplasmic male sterility (Owen, 1945) allowing for production of pure hybrid seed; v) employment of the male sterility gene (Owen, 1952; Doggett, 1970; Bosemark, 1972); vi) change in seed morphology with the discovery of the monogerm character, a development that safeguarded the future of sugar beet cultivation (Bordonos and Savitsky, 1934); vii) selection for resistance to diseases, initially by USDA researchers; and viii) selection for industrial qualities, juice purity, low dirt tare.

Successes in sugar beet breeding resulting from the above and other advances over the years can only be described as remarkable; yet, there is much more to be accomplished. The current and future need for a quantitatively and qualitatively improved sugar beet crop produced in the framework of an agriculture characterized by sustainability, high precision and adaptability to rapidly changing agroclimatic conditions and market demands, will challenge breeders. Their work will become even more demanding, but exciting developments in molecular biology and biotechnology, proceeding at an unprecedented pace in the last 2-3 decades, have already, and will continue to supply sugar beet breeders with fascinating new technologies to solve old and new problems.

Within the unlimited possibilities created by the emergence of a global economy, it is certain that breeding will continue to exert a substantial positive influence on sugar beet production and agriculture.

1

History and Basic Biology

Sugar beet and sugar cane provide most of the sugar (sucrose) produced in the world today. The history of the former spans about 200 years, during which time it benefited from continual progress in the fields of genetics, agronomy, and factory processing. Sugar beet was literally invented and constructed by man (Coons, 1936), making it unique in the history of agriculture.

1.1 Brief History of Sugar and Sugar-producing Plants (E. Biancardi)

Sugars are natural substances produced by plants that provide a significant food source for man and animals. They consist of carbon, hydrogen, and oxygen, containing the latter two in a 2:1 ratio. A monosaccharide contains 5 or 6 carbon atoms and forms a molecule that is difficult to hydrolyze. A pentose, such as L-arabinose, is formed by 5 carbon atoms, while monosaccharides, such as glucose and fructose (hexoses), possess 6 carbon atoms. Based on the number of monosaccharides that make up a molecule, it is possible to have monosaccharides, disaccharides, trisaccharides and polysaccharides. A molecule of sucrose (sugar), the most economically important disaccharide, is composed of two monosaccharides, glucose and fructose **(Fig. 1.1)**.

Chemical properties of saccharides depend on their molecular weight. Simple sugars (glucose, fructose, sucrose, galactose, lactose, etc.) are highly soluble and reactive, in contrast to more complex polysaccharides such as starch, cellulose, and lignin. Monosaccharides and disaccharides readily provide energy, while polysaccharides are able to provide it only after they have been transformed into simple sugars (McGinnis, 1982).

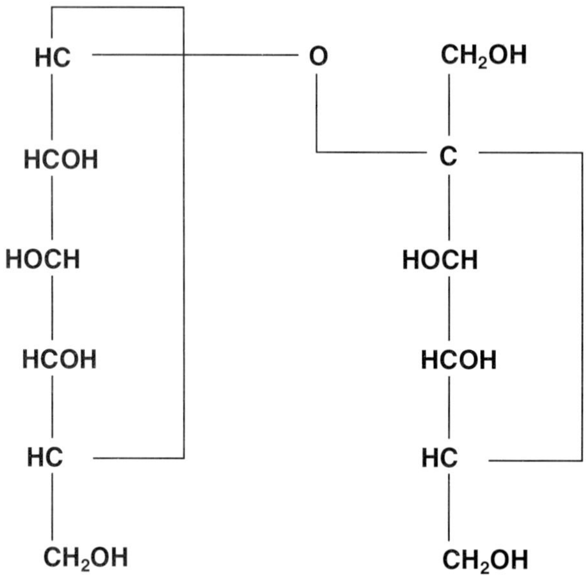

Fig. 1.1 Structural diagram of sucrose.

High sugar concentrations may occur in fruits, roots, tubers, rhizomes, stems etc. Sugar cane (*Saccharum officinarum*) and sugar beet (classified by Lange et al., 1999, as *Beta vulgaris* L. ssp. *vulgaris* Sugar Beet Group) are essentially the only plants utilized for the industrial extraction of sucrose.

Sugars have a pleasant taste and even in the earliest civilizations were widely used to sweeten foods and drinks to make them more appetizing. Bee honey or concentrated fruit syrups were used, but these altered the flavor of the foods to which they were added (Olbrich, 1989). Use of crystallized sugar in food became widespread after the discovery of the Americas. It was actually used much earlier as a medicinal substance, or was presumed as such. As for the origin of the term "sugar", which is similar in all modern languages except Chinese, von Lippmann (1929) believed it derived from the Sanskrit "*Sakkara*" or "*Sukkârêî*", both of which indicate superior mercantile quality.

According to Surface (1918), the first plant from which sweet juice was extracted for food was bamboo (*Bambusa* spp.), not cane. The author quotes a sentence from Theophrastus (300 B.C.) that alludes to "bamboo honey" as well as to the "Indian salt" that can be obtained from it. Cane cultivation began near the mouth of the Ganges River around 300 B.C., spreading to China, Java, and then to Tibet and the island of Ceylon (Surface, 1918). In the same period, Indian salt appeared in Greece after

Alexander the Great's military expeditions into the Orient. Seneca wrote about a honey obtained from cane, while Pliny the Elder (77) reported, "*sugar is produced in Arabia, but the Indian one is more appreciated.*" Around the eighth century, there was significant sugar cane production between the Tigris and Euphrates Rivers (von Lippmann, 1929). Sugar, which required a long and complex purification process, was still limited to use in medicine. Cane was introduced by the Arabs into Egypt and Ethiopia, and later, toward the thirteenth century, into Sicily, Cyprus, Crete and Andalusia (von Lippmann, 1929).

Sugar, easily preserved and particularly expensive, began to be traded in the Republic of Venice alongside silk and spices. The process for refining the thick juice, partially crystallized in the zones where the cane was cultivated was developed in Venice. The refining process was considered a military secret, especially after the discovery of an innovative method around 1420 (Surface, 1918). The sugar monopoly began to waver once sugar cane spread to the Canary Islands and Madeira, and especially following the discovery of America. The Portuguese began growing sugar cane in the New World, first in Haiti in 1515 and then in Brazil and San Domingo (von Lippmann, 1929). From the outset, they produced higher yields than in Asia and Africa. Transatlantic transport became ever more intense, causing a sharp drop in sugar prices in the Old World and a dramatic increase in demand and consumption.

The center of world trade shifted from Venice to Lisbon, where raw sugar arrived from the Americas as did the refined product from Antwerp. Portuguese supremacy did not last long, as trade had shifted to London by the end of the sixteenth century. In the meantime, the British began growing sugar cane in Louisiana and Cuba (von Lippmann, 1929). Consumption in Europe had been further boosted by the spread of drinks such as tea, coffee and chocolate. Coming to the close of the eighteenth century, the market was firmly in the grip of British naval and commercial power (Winner, 1993).

1.2 Brief History of Sugar Beet Cultivation (E. Biancardi)

All cultivated beet species derive from sea beet (Coons, 1936). Even before confirmation obtained from DNA analysis (Hjerdin et al., 1994; Jung et al., 1993), this ancestry appeared the most probable. In fact, sea beet possesses sweet, tender, emerald green leaves that are readily used for food and not easily mistaken for leaves of other plants (Coons, 1936; Biancardi, 1999). These useful features are far less developed in other wild *Beta* species.

Sea beet **(Fig. 1.2)**, classified as *Beta vulgaris* L. ssp. *maritima* (L.) Arcang by Lange et al., (1999), is still common along the Mediterranean coastline and the central and northern Atlantic coasts bordering Europe and Africa (Ulbrich, 1934). Judging from the modern-day geographical spread of wild *Beta* and in particular of sea beet, the origin of genus *Beta* coincides with the area stretching from the British Isles and the North Atlantic coast across the Mediterranean and the Black Sea as far as the Persian Gulf and the mouth of the Indus River (Ulbrich, 1934). Coons (1936), on the other hand, concluded that *Beta* originated in Transcaucasia or Asia Minor.

It is believed that sea beet leaves were collected by man in prehistoric times and that it was domesticated around the Persian Gulf at nearly the same time as wheat and barley, about 12,000 years ago (Simmonds, 1976). Wild plants tend to be annual as this ensures greater seed production. Indeed, annuals are most prevalent among sea beet (Bartsch and Schmidt, 1997). However, the earliest growers selected and reproduced biennial individuals, i.e., those that flowered the year after seeding, allowing a longer time for leaf production, the only part used for food. Von Lippmann (1925, 1929) asserted that between 1000 and 850 B.C. beet was widely grown throughout Greece, Sicily, Syria, and Persia

Fig. 1.2 Flowering sea beet (After Munerati, 1909).

and disagreed with the assertion, attributed to Herodotus, that beet was listed two millennia earlier as being a food of the Egyptian pyramid builders.

The earliest references to beet and its usage date back to Aristophanes (445-385 B.C.) and Euripides (480-406 B.C.). From Greece, the plant spread toward modern-day Italy with no difficulty, where it was referred to by the Latin name *beta*, later used by Linnaeus to identify the taxonomic genus. Cato, Cicero, and Virgil made references to beet, while Columella and Pliny the Elder described its botanic characteristics, growing techniques, and nutritional and medicinal properties. The latter author wrote not only about the use of the leaves for food, as in Greece, but also the use of roots (Pliny, 77 A.D.). This was made possible by the appearance of suitable types, ancestors of the modern day garden beet, characterized by colored round roots that were sweet tasting and with low fiber content **(Plate 1.1)**.

Pliny mentioned the possibility of autumn or spring sowing and, probably in reference to sea beet, asserted that "*there is a wild beet that some call limonium and others neuroides.*" Along with the Roman Empire, the crop spread throughout Europe, producing uncountable biotypes with a retinue of names (Winner, 1993). As Roemer (1927) stated, descriptions of the plant are often confused because the distinction (persisting in common language today) between cultivated beet and turnip (*Brassica*

Plate 1.1 Root shape and colour of some cultivated beets, clockwise from the top left: garden, fodder, ethanol, sugar, and leaf beets (After Biancardi, 1999).

spp.) was not always clear. The first to describe the various types of beet known at that time was Cesalpino, 1519-1603.

During the Middle Ages, alongside the leaf and root types an additional one developed which was suitable for cattle fodder. Its large roots were primarily used for winter feed. The new application, which probably started in Spain or France (von Lippmann, 1925), spread to Holland and northern Germany (Magdeburg, Silesia), where the fourth and now most important use would later be created (Knapp, 1958).

The history of the sugar beet is often said to start at the turn of the seventeenth century, when the French agronomist Oliver de Serres described several types of beet "*just arrived from Italy,*" that after cooking produced "*a juice similar to sugar syrup, with its vermilion colour a beauty to behold*" (Dureau, 1886; von Lippmann, 1925). The next major development did not occur until 1747, in which year Andreas Siegmund Marggraf revealed to the Berlin Academy of Science that locally grown beets contained a crystalline substance very much like cane sugar. This discovery was given only a nod, probably because the yield of sugar extracted with alcohol was only 1.6% of the root weight (Marggraf, 1767; Knapp, 1958). The foregoing prompted Franz Carl Achard, a scholar of Marggraf, to reconsider the matter. It is highly probable that Achard used fodder beet to begin mass selection based on root morphology. This work led to the development of "White Silesian" (Weisse schlesische Rübe), the progenitor of all sugar beet. One difficulty in this breeding effort was the high variability within the populations (Achard, 1803; Coons, 1936; Heinisch, 1960; Fischer, 1989). Due to the inadequate means at his disposal, Achard sought an audience with the Prussian king Frederick Wilhelm II to request funding for additional experiments. This moment was of fundamental importance for sugar beet, as it is very unlikely that conditions favorable for its birth as an industrial crop would have arisen a second time (Biancardi, 1984).

Cane cultivation and trade in the sugar produced from it reached supernational dimensions, engendering enormous economical interests. By 1801 the British monopoly had become so politically irritating that Achard was granted funding to build the first sugar beet factory, in Cunern (Lower Silesia), which by the spring of 1802 began processing the beet crop harvested the previous autumn (Winner, 1993). Despite endless technical problems, due in part to the low concentration of sugar in the roots (Coons, 1936), the first year experience demonstrated that sugar production from beets was a practical possibility.

Achard encountered numerous difficulties but thanks to his vast knowledge of technical subjects combined with the honesty and obstinancy of a dedicated scientist he surmounted them. Subsequent to a fire at the sugar factory in Cunern in 1807, he collaborated more closely with the von Koppy barons who were reproducing "White Silesian" seed and who, two years earlier, had built a sugar factory in Krayn, near Cunern. With the benefit of the Cunern experience, the new factory, capable of processing 500 tons of roots per campaign, utilized an improved extraction process. Cultivation techniques were also improved (Briem, 1895).

Meanwhile Achard founded a school of sugar technology that operated until 1814 (Winner, 1993). In the 4 years it was open, the school was attended by technicians from numerous countries and the knowledge they gained was probably essential for survival of the new crop. After a few years, interest in the crop gradually waned, due to a political crisis in all beet sugar production areas. The Krayn factory was eventually closed (von Lippmann, 1929). Fortunately, Achard's ideas had found supporters and a small sugar factory had been built in Tula (Russia) in 1802, an initiative backed by the German chemist Theodor Lowitz (Anon., 1998).

Napoleon was instrumental in the crop's survival. Because of a naval blockade on British imports, he had to find an alternative to ensure normal supplies of sugar in France. Before 1802, Delessert had begun extraction experiments using beet from Germany. Due to the positive results obtained, Napoleon set up five schools which specialized in sugar technology (Geschwind and Sellier, 1902). After evaluating the previous French and German experiments, a government commission consented to cultivating sugar beet on a vast scale. With the decree of March 25, 1811, Napoleon authorized sowing 32,000 hectares, later reduced to 7,000 due to lack of seed, and the building of numerous factories with state funding. The following year, the production area was increased to 100,000 hectares and construction of 334 more sugar factories authorized. Half of these were in operation before the battle of Waterloo (Geschwind and Sellier, 1902).

In the following years, beet cultivation was cut back drastically. Sugar cane regained the ground it had lost and appeared to have eliminated its competitor. But sugar beet proved persistent in areas where it had been grown. Furthermore, it increased the productivity of other crops, primarily cereals, in the rotation and improved distribution of labor. Moreover, it did not just bring the industry to the countryside, but also new techniques, capital, and the initiative spirit characteristic of the industrial revolution, which was rapidly changing society (Biancardi, 1984).

Construction of sugar factories began once again in Germany around 1830, based on the improvements obtained in France. Furthermore, the yield of cultivated varieties increased thanks to the analytical and breeding systems developed in France by de Vilmorin (McFarlane, 1971). In Austria-Hungary in 1856, 108 factories were in operation, while in Bohemia and Czechoslovakia several companies specialized in building equipment for sugar factories. Sugar beet cultivation expanded rapidly in Russia, where 180 factories were producing 250,000 tons of sugar annually around 1870. Cultural practices were improved with the introduction of mineral fertilizer and steam tractors, which made possible deeper ploughing compared with traditional methods. The crop's yield improved further, thanks to the German and French seed companies that sprang up in the middle of the nineteenth century.

Toward the end of the century, other countries joined the great European sugar beet producers (France, Russia, Austria-Hungary and Germany). The first sugar factory was built in the United States in 1838, in Northampton, Massachusetts, but was closed after its first campaign (Ware, 1880). After several failed attempts in various states, the first regularly operating factory was built in Alvarado, California.

In the new century, sugar beet cultivation gradually spread to other countries: Italy, England, Ireland, Canada, Chile, Uruguay, Japan, Turkey, China, etc. In recent years, it has been introduced into warmer climates that are more suited to sugar cane cultivation, such as Egypt, Morocco, Tunisia and Algeria. In these cases sugar beet cultivation was encouraged to accelerate progress in regions where agricultural practices are relatively primitive.

The twentieth century was characterized by continuous progress. Agronomic advances resulted in reduced production costs and a steady yield increase, similar to that observed in other crops (Robertson-Scott, 1911). These benefits resulted, in large part from genetic research. Singling, which required about 100 working hours per hectare, was first reduced by planting mechanically processed seed (precision seed) and then completely eliminated following the introduction of genetic monogerm seed. More recent research has produced pelleted seed with fungicides (Leach and Bainer, 1942) and precision sowing systems.

Protection from pests has further improved yields. Since sugar beet responds favorably to many defensive operations, it was one of the first crops to be treated with arsenic, nicotine, sodium fluoride, sulfur, copper etc. It soon became clear that long rotation periods were required to reduce losses from some soilborne diseases (Schacht, 1859; Briem, 1895), while host plant resistance offered the only solution for other diseases. Sugar beet was one of the first crops to demonstrate the potential of chemical weed control. The first selective herbicide, Prophan, was used in the USA in 1950 (Winner, 1993).

Tillage techniques should not be forgotten for their part in the steady rise in productivity, nor should the techniques used in seedbed preparation, mechanical harvesting and conservation practices. Proper fertilizer management not only improved sugar yield, but also facilitated industrial extraction. Irrigation proved fundamental in dry areas. The most important advancements of the last fifty years include introduction of hybrid varieties, which began on a vast scale after 1960, rhizomania-resistant varieties and in the last twenty years, meristematic multiplication techniques and molecular analyses.

1.3 Anatomy and Physiology (K. Klotz)

Sugar beet is a herbaceous dicotyledon and a member of *Chenopodiaceae*, a family characterized morphologically by inconspicuous, radially symmetric, petalless flowers and non-fleshy fruit. The family is distinctive in that most members, including sugar beet, are halophytes (Watson and Dallwitz, 2000). Other agriculturally significant members of this family include the horticultural crops, red beet and Swiss chard (*Beta vulgaris*), spinach (*Spinacia oleracea*) and quinoa (*Chenopodium quinoa*), and the noxious weeds lamb's quarters (pigweed, *Chenopodium album*), Russian thistle (*Salsola kali*), and kochia (*Kochia scoparia*).

Although annual and biennial genotypes of beet exist, the cultivated sugar beet is a biennial (Krasochkin, 1959; Lexander, 1985). Sugar beet grows vegetatively its first year as a near-rosette plant and develops a large fleshy taproot that contains the food reserve for the second year of growth. In the second year, sugar beet becomes reproductive, induced to flower by the combined exposure to low temperatures from overwintering and long photoperiods during the second year of growth. The anatomy and major physiological events in the two-year life cycle of sugar beet are described below.

1.3.1 Vegetative Growth

1.3.1.1 Seed germination and early plant development

Sugar beet germinates from a dry, hard, irregular fruit, often referred to as a seed ball. In actuality, the fruit is an utricle, a fruit containing a single seed attached at a single point to the internal fruit wall (Copeland and McDonald, 2001). Nearly all sugar beet in culture today is derived from monogerm seed, a term that actually refers to the singleness of the fruit which contains the true seed. Early sugar beet varieties were multigerm with the propagule comprising two or more fruits fused at their bases into a single unit (Artschwager, 1927).

Seed germination is initiated by the absorption of water which triggers the return of the embryo from a quiescent state to active growth as the water content increases from 8 to 12% in the dry seed to 80 to 90% in the emerged seedling (Lexander, 1993). Water absorption is driven principally by the polar attraction between water and the proteins of the seed and requires no metabolic energy (Copeland and McDonald, 2001). Water imbibition rehydrates the proteins, pectin, and cellulose of the seed, reactivates stored enzymes, and initiates a return to active metabolism. The return to active metabolism is characterized by RNA and protein synthesis and an increase in cellular respiration. Metabolism during reactivation is fueled by the degradation of lipids stored in the cotyledons (Copeland and McDonald, 2001; Elamrani et al., 1992). Seed germination is complete with the emergence of the radicle from the sugar beet fruit body. The radicle elongates by cell division and elongation and emerges from the fruit by displacement of the operculum or fruit cap (Lawrence et al., 1990; Sliwinska et al., 1999).

Germination normally occurs three to five days after planting (Doney, 1979). The rate and extent of germination, however, are highly dependent on temperature, water availability, oxygen uptake, and concentration of endogenous chemical inhibitors within the fruit wall. Maximum germination occurs between 22 and 25°C, with little to no germination at temperatures less than 3°C, and reduced germination at temperatures in excess of 30 to 35°C (Campbell and Enz, 1991; Gummerson, 1986). Between 10 and 25°C, germination rate increases linearly with temperature (Campbell and Enz, 1991). An adequate supply of water is necessary for seed germination and an inverse linear relationship between germination rate and osmotic potential has been demonstrated (Gummerson, 1986). Water availability is determined by water supply, soil matrix potential, and the osmotic potential created by solutions of fertilizers near the seed. Oxygen supply to the seed also affects germination since oxygen is a requisite substrate for respiration (Capron et al., 2000; Santos and Pereira, 1989). Oxygen uptake is restricted by the fruit wall, requiring nearly all oxygen to enter through a pore at the base of the fruit where the fruit was attached to the mother plant (Perry and Harrison, 1974). Blockage of the fruit basal pore, such as by a film of water during wet conditions, severely restricts oxygen uptake and inhibits seed germination. Germination inhibition by compounds in the fruit wall has also been documented, although the identity of the compounds and the nature of inhibition have not been definitively determined (Lexander, 1978; Tolman and Stout, 1940). Compounds implicated in germination inhibition include potassium nitrate (Inoue and Yamamoto, 1975), potassium salts (Lexander, 1978), oxalate (Inoue and Yamamoto, 1974; Miyamoto, 1957), phenolic compounds (Battle and Whittington, 1969), *cis*-4-cyclohexene-1,2-dicarboximide (Mitchell and Tolbert, 1968), and ammonia (Stout and Tolman, 1941). These proposed germination inhibitors are all water-soluble and can be leached from the fruit wall.

Germination is epigeal, and after radicle emergence the seedling hypocotyl elongates and lifts the cotyledons above the soil surface. Emergence occurs one to two days after germination under favorable conditions, although the time from germination to emergence is dependent on temperature, water availability and planting depth (Doney, 1979; Durr and Boiffin, 1995). Postgermination, preemergent growth is fueled primarily by degradation of starch from the perisperm, the main nutritive tissue of the seed (Lawrence et al., 1990; Elamrani et al., 1992). The seedling dry weight at emergence is proportional to the initial weight of the seed (Durr and Boiffin, 1995). After emergence, the cotyledons become photosynthetically active. Subsequent seedling growth, based on dry matter accumulation, is exponential until the first true leaves appear, approximately five to seven days after emergence (Doney, 1979; Durr and Boiffin, 1995).

1.3.1.2 Shoot morphology and development

The vegetative shoot is composed of large elongate leaves arranged in a close spiral on the crown of the sugar beet taproot (Artschwager, 1926). The leaves are triangular with a conspicuous midrib and petiole, and a netted pattern of leaf venation. The leaf lamina is typically glabrous with an undulating margin. Stomata are present on both abaxial and adaxial leaf surfaces, but are more prevalent on the lower, abaxial leaf surface. Stomatal densities of 91 to 114 and 144 to 162 stomata per square millimeter have been reported for the adaxial and abaxial leaf surfaces respectively (Artschwager, 1926).

Leaves emerge continuously during the first season of growth, typically at a rate of two to four leaves per week. Leaf emergence rate is temperature dependent, however, and emergence rates of four to six leaves per week are possible under optimal conditions (Doney, 1979; Milford et al., 1985a,b; Terry, 1968). Leaf size, measured as the leaf area after full expansion, is variable (Milford et al., 1985b). Leaf size increases progressively from the first to approximately twelfth leaf, then progressively decreases with each subsequent leaf. The lowermost leaves of the rosette senesce during the growing season with leaves senescing in the order in which they were produced (Milford et al., 1985a). By the end of the vegetative growing season, the shoot comprises thirty to seventy leaves with the leaf number dependent on environmental and genetic factors (Wyse, 1982).

1.3.1.3 Root morphology and development

The sugar beet taproot (**Fig. 1.3**) is a creamy white, highly elongated, tapering, conical structure (Artschwager, 1926) composed of three regions, crown, neck and true root. The crown is the broadened apex of the

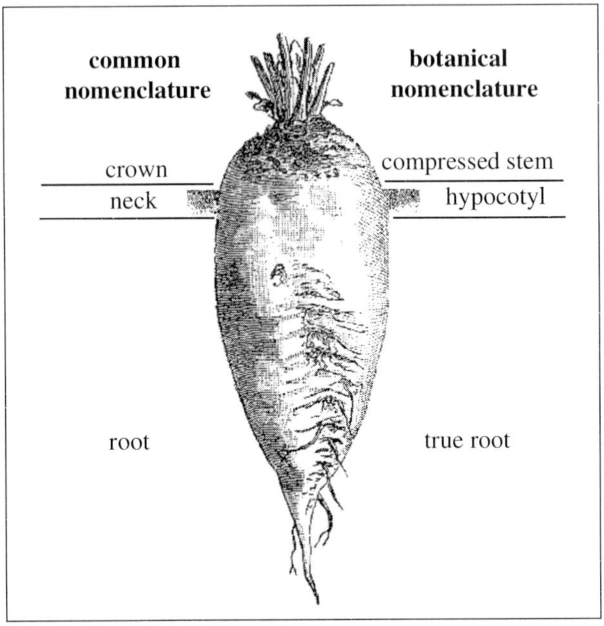

Fig. 1.3 Schematic diagram of a sugar beet taproot with common and botanical nomenclature for its main components (Adapted from Biancardi, 1999).

taproot that bears the rosette of leaves and leaf buds which comprise the shoot. The crown is actually a compressed stem and contains 6 to 17% of the fresh mass of the taproot (Milford and Houghton, 1999). The neck of the taproot is a narrow zone of tissue below the crown and is the broadest part of the taproot. The neck is a thickened hypocotyl devoid of lateral roots. The true root comprises the largest portion of the taproot. The true root extends downward from the neck, decreasing basipetally in width and commonly extends to soil depths of 1.5 meters (Artschwager, 1926; Martin, 1983). The circumference of the root is not symmetrical but slightly flattened on opposing sides. Each flattened surface contains a single groove that descends the length of the root in a shallow spiral. Numerous small, filiform lateral roots extend from the two grooves of the taproot (Artschwager, 1926).

The surface of the root is protected by periderm, a thin cork layer five to eight cells deep with suberized cell walls and a lignified middle lamella (Artschwager, 1926). The periderm aids in the defense of the root from pathogens and prevents desiccation. Internal to the periderm, the root comprises alternating concentric rings of vascular and parenchymal tissue surrounding a central star-shaped core (**Fig. 1.4**). The central core or vascular cylinder contains the vascular and ground tissue that developed

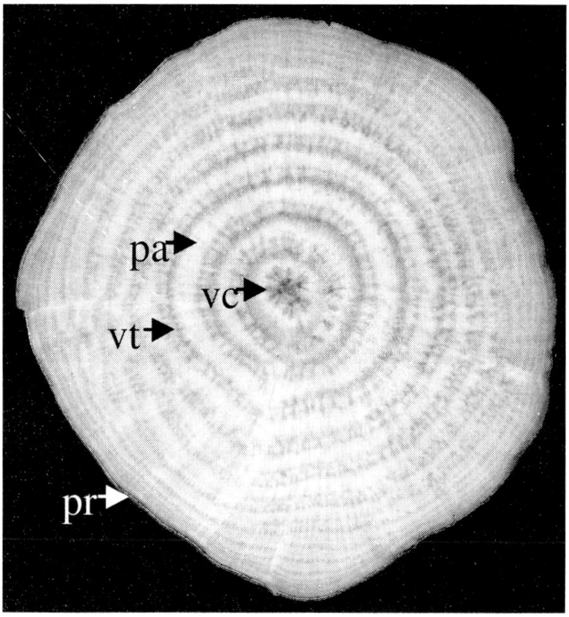

Fig. 1.4 Cross-section of a sugar beet taproot with alternating concentric rings of vascular and parenchymal tissue surrounding a central vascular cylinder (pr = periderm; pa = parenchymal tissue; vc = vascular cylinder; vt = vascular tissue).

mainly during root primary growth. Centrifugal to the central core are 8 to 13 vascular rings (Artschwager, 1926; Wyse, 1982). These rings contain bundles of vascular tissue separated by rays of parenchymal tissue (Artschwager, 1930) and are themselves separated by broad bands of parenchymal cells. These bands of parenchyma are widest in the central portion of the root, decrease in width centrifugally, and are very narrow at the outermost regions of the root. Due to this pattern of development, the innermost five to six rings of parenchyma account for 70 to 80% of the volume of the root (Wyse, 1982). The parenchymal cells are the sites for sucrose storage in the sugar beet root. Sucrose is stored in the parenchymal cell vacuole, which typically comprises about 95% of the cell volume (Wyse, 1982).

Development of the taproot from the seedling hypocotyl and primary root occurs rapidly. During the first six weeks of growth, the primary root increases in dry matter content by approximately 4000%, develops the structure of a mature taproot, and stores sucrose (Klotz, unpubl. data, 2002; Klotz and Finger, 2002; Milford, 1973). Most of the vascular rings present in the mature root develop when the root is less than one centimeter in diameter (Artschwager, 1926). Subsequent growth occurs simultaneously in all rings by cell division and expansion. Additional vascular rings develop during the growing season at the rate of approximately one new vascular ring per month, and the root grows in size and accumulates dry matter at nearly linear rates throughout the remainder of the growing season (Klotz and Finger, 2002; Milford, 1973).

1.3.2 Reproductive Growth

1.3.2.1 Transition to reproductive growth

Transition from vegetative to reproductive growth requires exposure to both low temperatures and long photoperiods; the term photothermal induction has been used to describe the environmental requirements for this transition (Owen et al., 1940). Exposure to cold, but nonfreezing temperatures initiates the transition from vegetative to reproductive growth. Vernalization acts upon the meristematic cells of the shoot apical meristem and young leaves, triggering a series of events that alters the competency of leaves to produce a floral stimulus in response to long photoperiods (Chrobozek, 1934; Crosthwaite and Jenkins, 1993). Photoperiodic induction, accompanying or following vernalization, leads to the production of a mobile floral stimulus in the leaves. The identity of the floral stimulus is not known. Measurement of photoperiod length is mediated by phytochrome which becomes physiologically activated upon absorption of the red wavelengths of white light. The light-induced floral stimulus is transported from the leaves to the shoot apex, completing the transition to reproductive growth.

Vernalization requires temperatures in the range of 4 to 10°C for a period of 10 to 14 weeks (Curth, 1962; Lexander, 1980). Vernalization requirements are highly variable among sugar beet genotypes, however, and can be complete in just two weeks for some early bolters (Lexander, 1980). The age of the plant also influences vernalization requirements and older plants generally require a shorter vernalization period than young plants (Lexander, 1980). Photoinduction is influenced primarily by length of the day period and light quality. Daylength requirements vary among genotypes, although a 16-hour daylength suffices to induce flowering in most genotypes. As a phytochrome response, photoinduction is influenced by light quality. Photoinduction is promoted by red light (wavelengths 600 to 700 nanometers) and inhibited by far-red light (wavelengths 700 to 760 nanometers), and light with a high ratio of red to far-red light is needed to promote photoinduction (Lane et al., 1965). Blue light (wavelengths 400 to 500 nanometers) can also promote photoinduction, although much less effectively than red light (Lexander, 1980). Light intensity is not a major factor; photoinduction can occur at very low light intensities.

1.3.2.2 Reproductive growth and floral anatomy

In the second year of growth, after vernalization during overwintering, leaves emerge from the taproot crown in a rosette similar in appearance to the vegetative growth of the previous growing season. Two to six weeks later, however, reproductive plant morphology becomes distinctly different from vegetative plant morphology. Newly formed leaves become progressively smaller and rapid cell division at the shoot apex and elongation of the cells immediately below the shoot apex result in formation of a stem (Artschwager, 1927; Lexander, 1980). The leaves that develop on the stem are significantly smaller than those of the basal rosette. In the axils of the primary stem leaves, secondary branches develop by rapid cell division and elongation, and tertiary branches develop in the leaf axils of the secondary branches. The primary stem and all branches from it terminate in an inflorescence. The inflorescence (**Fig. 1.5**) is composed of open and variably branched spikes bearing flowers (Artschwager, 1927). Flower opening begins at the base of the spike and

Fig. 1.5 Sugar beet inflorescence composed of branched spikes bearing small sessile flowers, each subtended by a bract.

Fig. 1.6 Sugar beet flower after anther dehiscence. Sepals are fully expanded, and four of the five anthers have dehisced and are pollen-covered. The two-lobed, tetrasporangiate anther structure can be seen in the unopened anther. The central stigma is three lobed and partially opened (an = anther; se = sepal; st = stigma).

proceeds acropetally. Flowering is indeterminate and continues for the remainder of the plant's life (Smith, 1987).

The sugar beet flower is sessile and subtended by a bract. Flowers are borne in the bract axil singly in monogerm genotypes or in clusters of two to seven flowers in multigerm genotypes (Artschwager, 1927; Smith, 1980). The small, inconspicuous flower is composed of five green, narrow, inward curving sepals surrounding five stamens and a central pistil (**Fig. 1.6**). Petals are absent. The bases of the sepals and stamens are positioned above the ovary of the pistil. The ovary is therefore inferior and the flower epigynous (Elliot and Weston, 1993; Esau, 1977).

The pistil is tricarpellate with a short style that terminates in a three-lobed, or sometimes four- or five-lobed stigma (Artschwager and Starrett, 1933). The epidermal cells of the stigma and the central canal of the style are covered with papillae, protuberances that become covered with a sticky secretion when the stigma and style reach maturity. The basal portion of the pistil, the ovary, encloses the ovule which is attached laterally to the ovary wall by a short stalk or funiculus (Artschwager, 1927). Contained within the ovule is the embryo sac, a seven-celled structure that contains the egg. The stamens extend above the pistil with the anthers facing inward toward the center of the flower. The anthers at maturity are yellow and tetrasporangiate, consisting of two lobes that each contain two pollen-filled locules (Artschwager, 1927). The pollen is spherical with a thick, sculptured outer wall punctuated with numerous thin-walled pores for the exit of pollen tubes (Artschwager and Starrett, 1933). The pollen is trinucleate and contains a vegetative cell and the two sperm cells needed to effect the double fertilization typical of angiosperm reproduction (Brewbaker, 1967; Esau, 1977).

1.3.2.3 Fertilization

Flower opening hallmarks the beginning of reproduction. Flowers open mainly in the morning, although some flower opening occurs at all times of the day (Smith, 1980). Flowers open by expansion of the sepals exposing the stamens that are held upright above the pistil. During sepal expansion, or shortly thereafter, the anthers dehisce, releasing pollen (Artschwager, 1927; Artschwager and Starrett, 1933). The anthers open spontaneously by the formation of a longitudinal break in each anther lobe. The pollen grains upon release initially adhere to each other. They quickly dry, however, and separate into individual pollen grains. Under natural conditions, the pollen is carried by the wind or occasionally insects to effect cross-pollination (Smith, 1980). At the time of anther dehiscence, the stigma lobes are typically closed. The stigma lobes open slowly, beginning approximately five to seven hours after flower opening and continuing over the next thirty hours (Artschwager and Starrett, 1933). The rate at which stigma lobes open and expand is highly variable among flowers.

The pollen that reaches the receptive surface of a stigma can germinate within two hours after anther dehiscence (Artschwager and Starrett, 1933). Germination occurs on both open and unopened stigmas, although pollen tube growth proceeds more rapidly on open stigmas. The pollen tube elongates between the papillae of the stigma surface, through the central canal of the style and through the cavity of the ovary to reach and penetrate the ovule and its enclosed embryo sac. The germinating pollen grain typically requires twenty hours to complete pollen tube elongation, although pollen tubes have been observed to penetrate the ovule in less than ten hours. Numerous pollen grains typically germinate on a stigma and many pollen tubes reach the ovule. Fertilization is effected by a single pollen grain, however.

Fertilization involves the fusion of the two sperm cells of the pollen grain with the egg and the central cell of the embryo sac. The sperm cells are released from the pollen grain by rupture of the pollen tube after its entry into the embryo sac (Esau, 1977). One sperm cell fuses with the egg to produce the zygote that develops into the embryo. The other sperm cell fuses with the binuclear embryo sac central cell to produce the primary endosperm nucleus that develops into the endosperm. The endosperm is triploid and is the nutritive tissue that supports the growth and development of the embryo during its initial growth. Fertilization occurs soon after the pollen tube reaches the embryo sac and the actual fusion of cells to produce the zygote and the primary endosperm nucleus is rapid (Artschwager and Starrett, 1933).

Sugar beet is largely cross-pollinated due to a high degree of self-incompatibility. Sugar beet exhibits gametophytic self-incompatibility under the control of possibly two (Maletsky and Weisman, 1978; Owen, 1942) or four complementary S-loci (Larsen, 1977; Lundquist et al., 1973). Each S-locus is polymorphic, with multiple alleles at each locus. An incompatible mating occurs when all S-loci alleles are the same in the pollen and the pistil. If one or more S-loci alleles differ in pollen and pistil, mating is compatible. In an incompatible mating, the pollen grain germinates on the stigma but pollen tube elongation terminates in the style (Brewbaker, 1967). Incompatibility between pollen and pistil may be due to S-loci encoded ribonucleases, synthesized in the style, which enter pollen tubes and specifically destroy RNA in incompatible pollen tubes. Such a mechanism is responsible for gametophytic self-incompatibility in Solanaceous species (Kao, 1992), although it is not known whether the same mechanism operates in sugar beet.

1.3.2.4 Seed morphology and development

Sugar beet seed is round to reniform, and small, typically weighing approximately three mg and measuring approximately three mm on its longest axis (Artschwager, 1927; Martin, 1983; Smith, 1980). The true seed contains an embryo surrounded by perisperm or nutritive tissue, and is enclosed in a seed coat. Surrounding the true seed is a hard, brown, heavily lignified fruit subtended by the remains of the flower sepals (Perry and Harrison, 1974). The fruit is small with an average weight of approximately 20 mg (Martin, 1983).

The true seed develops from the fertilized egg and its surrounding ovule. The first cell division of the embryo occurs 12 to 14 h after fertilization, and subsequent cell division leads to the formation of a mature embryo 11 to 13 d after fertilization (Artschwager and Starrett, 1933). The ovule grows rapidly after fertilization and develops to the size of the mature seed before the embryo matures. The fruit, which develops from the ovary wall of the pistil, also grows rapidly following fertilization and reaches its full size prior to embryo maturation. At the time of embryo maturation, the fruit is green, but subsequently dries to form the hard seed ball typically propagated in sugar beet culture (McFarlane, 1975). The flower sepals also enlarge after fertilization and curve upward over the developing fruit. The sepals dry after the embryo has matured and remain attached to the fruit.

1.4 Cytology and Cytogenetics (G.N. Skaracis)

Cytogenetic information in sugar beet, although recently enriched thanks to novel molecular approaches, is still very limited compared to other important crop species. Identification of its chromosomes by standard cytological techniques has been particularly difficult due to the small size and uniformity. Conventional cytogenetic analysis has also been hampered by lack of appropriately characterized stocks of structural chromosomal aberrations, such as translocations, inversions, etc. The need for chromosome and ploidy level manipulation required by polyploid breeding, as well as interest in transferring disease resistance through interspecific hybridization, provided the initial incentives for developments in cytology and cytogenetics of the crop and its related species. Apart from knowledge acquired through molecular cytogenetic studies initiated in the early 1990s, most scientific advances have been thoroughly documented in a review by Nakamura et al., (1991). Since some specific subjects are covered elsewhere in this book, only a few fundamental topics are highlighted here.

1.4.1 Chromosome Morphology and Structure

Sugar beet is a true diploid species with a basic haploid chromosome number of x = 9. Seven metacentric and two sub-metacentric somatic metaphase chromosomes, with an average length of 2.5 μm, were described in the first complete karyotype of a homozygous diploid sugar beet line (Bosemark and Bormotov, 1971). Except for the longest SAT-chromosome, it was not possible to identify individual chromosomes. Later this inability to differentiate chromosomes was verified using specific banding techniques (De Jong and De Bock, 1978; Loptien, 1985). A second somatic karyotype, based on classifying the sugar beet chromosomes of haploid plants into three distinct groups, was provided by Cistue et al. (1985). The average chromosome length was 2.39 μm and all nine chromosomes appeared identifiable on the basis of their total length and arm ratio. This karyotype was subsequently used for characterization of a complete trisomic series (Romagosa et al., 1986, 1987). Following preliminary work by Yu (1977), a standard chromosome karyotype was established at the pachytene stage of the meiotic prophase, including four metacentric and five submetacentric chromosomes (Nakamura and Tsuchiya, 1982b) with a second SAT-chromosome pair evident, while a third pair with nucleolar-organizing capacity was observed at diakinesis of the meiotic prophase (Nakamura and Tsuchiya, 1982a). Despite these as well as some more recent findings (e.g. Srivastava and Srivastava, 2000), cytological identification of sugar beet chromosomes continues to be an extremely difficult task.

In the framework of recent molecular cytogenetic studies, major types of repetitive DNA sequences present in *Beta* species genomes (i.e. satellite DNA, microsatellites and retrotransposons) have been used to investigate the large-scale organization of sugar beet chromosomes. This information contributed to the structural plant chromosome model proposed in Section 4.1.

1.4.2 Meiotic Course

Haploid, diploid, triploid, and tetraploid genotypes have been utilized in sugar beet breeding programs and their meiosis studied by several researchers. In particular, meiotic studies at the normal, diploid, level can provide baseline information to which abnormalities observed in materials of interest (i.e. diploid genotypes originating from monosomic alien addition lines) are compared. Also, since triploids are the source of primary trisomics, the study of their meiotic mechanism is useful in examining the relationship between chromosome assortment and distribution in the progeny after selfing or crossing with diploids. The occurrence of meiotic instabilities in autotetraploid sugar beet is related to the frequency of aneuploid individuals present in commercial triploid hybrids and, finally, the behavior of chromosomes during meiosis in haploid plants may provide an indication of the frequency of genomic duplications.

1.4.2.1 Haploids

Haploids sporadically occur at a low frequency in diploid populations, in the progeny of crosses between various ploidy levels and of interspecific crosses, as a result of the effect of nonfunctional pollen and through gynogenesis (see Sect. 4.3). As expected, meiosis in haploids proceeds in a highly abnormal manner with univalents, bivalents and trivalents as well as secondary associations of up to three chromosomes, resulting in a mean of 0.18 chiasmata per meiocyte occurring at the first metaphase. It was suggested that only two major duplications were present in the haploid sugar beet genome (Yu, 1980). Recent molecular information, indicating unique patterns of repetitive DNA distribution for each chromosome, also suggests that there is little duplication in the diploid sugar beet genome (Hallden et al., 1998; Schondelmaier et al., 1996). An examination of meiosis in haploids obtained from unfertilized ovule culture revealed a similarly high degree of abnormality, resulting in almost complete sterility of the plants (Szota and Goska, 1990).

1.4.2.2 Diploids

Following fragmental information from a few early relevant studies (for an extensive review, see Skaracis, 1980), the standard diploid meiotic process with its characteristic features and abnormalities was described in both inbred lines and broad genetic base open-pollinated germplasm. Meiosis was highly normal, although not synchronized within an individual anther. Mean chiasma frequency was 13.2 per meiocyte at diakinesis with a rare occurrence of paracentric inversion bridges, univalent formation, delayed separation of bivalents, and irregular chromosome distribution. These abnormalities were largely confined to the inbred lines as has been observed in other allogamous species. Nonhomologous chromosomes appeared interconnected with a chromatin structure at all stages, except metaphase I (Skaracis, 1980). Associations among the centromeres of nonhomologous chromosomes at the meiotic prophase were also reported by De Jong and Stam (1985). The normal meiotic process in sugar beet and related *Beta vulgaris* L. subspecies was recently summarized by Srivastava and Srivastava (1999).

1.4.2.3 Triploids

In a detailed study of triploid sugar beet meiosis, the average number of trivalents was 7.8, ranging from 3 to 9 per meiocyte, while chromosomes in general were randomly distributed, producing gametes with 9 to 18 chromosomes (Romagosa, 1980). As also reported in earlier studies (Levan, 1942; Mochizuki, 1953), male transmission of gametes with an abnormal chromosome constitution is feasible and zygotes with 18 to 27 chromosomes were found in the progeny of triploid x triploid and triploid x diploid crosses. However, due to discrimination against unbalanced gametes and zygotes, most of the offspring (up to 70%) contained 18 or 19 chromosomes and most viable seeds were produced on plants with 18, 19, 20, 26 or 27 chromosomes. Restitution gametes occurred at a frequency that depended on genotype and environment and produced dyads and triads, resulting in giant pollen grains (Romagosa, 1980). It was later suggested that chromosome 5 was involved in this meiotic abnormality (Romagosa, 1983).

1.4.2.4 Tetraploids

Due to the presence of four homologous chromosomes, pairing configurations other than bivalents commonly occur at metaphase of the first meiotic division in tetraploid sugar beet. As reported by Bosemark (1966), the average number of bivalents, quadrivalents, trivalents, and univalents was 9.1, 4.4, 0.1 and 0.2 per meiocyte respectively. Most of the quadrivalents were rings with no zigzag orientation while some were open chains and a few assumed a cruciate configuration. Predominant unequal chromosome distribution at the first anaphase, due mainly to nondisjunctional quadrivalents, resulted in the production of unbalanced gametes. This is why aneuploidy is a common occurrence in tetraploid sugar beet germplasm (Levan, 1942; Mochizuki, 1953; Nagao and Takahashi, 1953; Bosemark, 1966). However, due to severe certation of unbalanced gametes and aneuploid zygotes, the aneuploid frequency in this material is much lower than expected, based on the proportion of unbalanced gametes. Even when only eutetraploids were included in the original population, the aneuploid frequency reached an equilibrium characteristic of the specific population within a few generations, (Mitsuishi, 1965; Rommel, 1966a, b; Bosemark, 1966). Aneuploid frequencies in germinated seed samples ranged from 20 to 45%, with a majority of 37-chromosome plants and a much lower frequency of 35-chromosome plants, particularly under field conditions (Bosemark, 1966). Rommel (1966a, b) also reported rare genomic deviations (diploids, polyhaploids) in the offspring of tetraploid plants.

Although gametophytic selection for euploidy is highly effective when tetraploid pollinators are used in triploid hybrid production, some aneuploids, in the range of 8 to 15%, are always present in commercial triploid hybrid varieties. Theoretically, these aneuploids would cause a yield loss of around 5% but the loss in the field is approximately halved due to competition favoring eutriploid plants (Bosemark, 1966). Apart from aneuploidy, inferior production and quality of pollen on tetraploid plants is mainly due to postmeiotic nuclear disturbances, resulting in nonviable supernumerary microspores. Both meiotic and postmeiotic stability appear to be regulated by a common genetic mechanism (Bosemark, 1967). When tetraploids are used as female parents for the production of "reverse" triploid hybrids, which have a slight yield advantage (Hecker et al., 1970; Bosemark, 1977), the consequences of aneuploidy are magnified since discrimination against chromosomally unbalanced gametes is less effective in megaspores than in microspores.

1.4.2.5 Meiosis in related Beta species

Several studies of chromosome pairing during meiosis of *Beta* species and of interspecific hybrids have been conducted, mainly to reveal relationships between and within sections. A completely diploidized meiosis was observed in tetraploid *Beta macrocarpa,* suggesting that allopolyploidy was involved. When crossed with diploid sugar beet, the triploid hybrids were almost always sterile, but in crosses with tetraploid sugar beet some fertile tetraploid offspring were produced. The predominance of multivalents at meiosis in these hybrids suggested a suppression of the diploidized meiosis (Lange and De Bock, 1989). In a study of meiosis in the *Procumbentes* section (Sobek et al., 1994), *B. procumbens, B. webbiana,* and their hybrids had completely normal bivalent pairing, reflecting their close relationship. There were a few quadrivalents in tetraploid *B. patellaris* and frequencies of trivalents were almost identical in its triploid hybrids with the other two species of the section. Based on these results, it was suggested that *B. patellaris* could be an autotetraploid with a genetic mechanism suppressing multivalent formation and that it evolved separately and earlier than the other two diploid species. Meiosis in species of the *Procumbentes* and *Corollinae* sections and in interspecific hybrids within the sections and with sugar beet has been summarized by Szota (1995). The close similarity between *B. procumbens* and *B. webbiana* was again demonstrated whereas cytological analysis of the interspecific hybrids of the section with sugar beet provided further evidence for their remote relationships. Diploid amphimictic *B. lomatogona* and *B. macrorhiza* species, despite a regular meiotic course, occasionally produced unreduced gametes resulting in higher ploidy level hybrids. Chromosome pairing in tetraploid *B. corolliflora* suggested its origin was either through autopolyploidy or amphidiploidy. In crosses of *B. corolliflora* with sugar beet, sporadically observed homeologous chromosome pairing resulted in hybrids resistant to mosaic virus. Meiosis in fertile hybrids of crosses between *B. macrorhiza* and sugar beet was normal.

1.4.3 Primary Trisomics

Trisomic analysis is a basic cytogenetic tool for assigning genes to chromosomes and establishing linkage relationships (Khush, 1973; Tsuchiya, 1983; Lange et al., 1993). Development of a primary trisomic series, in a broad genetic base sugar beet germplasm, was difficult initially and not practically exploited (Levan, 1942; Butterfass, 1964; Kaltsikes and Evans, 1967; Bormotov et al., 1973). In addition, only the trisomic series of Butterfass (1964) was maintained, but the masking effect of heterogeneity on phenotypic differences among the trisomics and the existence of structural aberrations limited its efficiency in linkage studies. Subsequently, all nine primary sugar beet trisomics, isolated from homozygous inbred germplasm, were identified, based on morphological and somatic karyotype analysis (Romagosa et al., 1986, 1987). Plant vigor of the trisomic types did not correlate with the length of the extra chromosome involved and despite the different genetic background, a degree of similarity (at least for 4 types) was apparent between this second trisomic series and that of Butterfass. Meiosis in seven trisomics was regular and there were no significant differences among these types in chromosome configurations at the first metaphase. Although 42.4% of the microspores carried the extra chromosome, transmission to progeny of crosses was very low (0-5%), while the highest respective figure for transmission through female gametes was 32.6% (Romagosa, 1983). Maintenance of this series has been facilitated through *in vitro* multiplication (Casas et al., 1986).

As indicated in Section 4.2, several linkage maps based on different kinds of molecular markers, are available for sugar beet. The primary trisomics series of Butterfass (1964) was used for the assignment of previously established linkage groups to the nine sugar beet chromosomes (Schondelmaier and Jung, 1997).

1.4.4 Identification of Alien Chromosomal Additions

Due to the relatively narrow genetic base of sugar beet, interspecific hybridization to its wild relatives is of utmost importance, specifically for the introduction of genes for resistance to biotic and abiotic stresses. Except for crosses within the section *Beta*, desirable gene transfer is severely hampered due to chromosome homology barriers. Monosomic addition lines, harboring the chromosome from a wild relative with the desired trait, are generally recognized as an efficient means for gene introduction into the sugar beet genome. Successful use of monosomic addition lines in applied breeding has been the transfer of beet cyst nematode resistance from the *Procumbentes* section. A very important consideration in establishing full sets of addition lines is the ability to distinguish among the different types. Morphological traits used in the past to identify alien addition lines proved insufficient because affected by both genotypic and environmental factors. Also, as mentioned above, classical cytological methods, including various banding techniques, cannot be successfully employed in sugar beet and related species due to the small size and similarity of the chromosomes. Recently developed molecular techniques provide the best available alternative for alien chromatin detection and characterization.

A full set of *B. procumbens* monosomic addition lines in sugar beet, initially described morphologically (Lange et al., 1988), was subsequently characterized by means of isozyme markers (van Geyt et al., 1988). Isozyme markers have also been used, along with morphological markers, to identify nine different monosomic addition lines from *B. webbiana* (Reamon-Ramos and Wricke, 1992). DNA fingerprinting with species-specific DNA probes was successful in identifying and characterizing monosomic addition lines from *B. procumbens* and *B. patellaris* (Mesbah et al., 1997; Kleine et al., 1998). The alien chromosome in three *Procumbentes* monosomic addition lines was detected by fluorescence *in-situ* hybridization (FISH) with a genome-specific satellite probe and YAC DNA (Schmidt et al., 1997), later FISHing with single-copy genes and YACs were applied to nematode-resistant translocation sugar beet lines to localize the resistance gene and flanking sequences (Schmidt et al., 1998).

More recently, eight unique monosomic addition lines from *B. corolliflora* have been characterized by DNA fingerprinting with alien species-specific repetitive sequences and by fluorescence *in situ* hybridization with species-specific and ribosomal-DNA probes (Gao et al., 2000, 2001). Lastly, genomic *in situ* hybridization (GISH), a very powerful tool in precisely identifying the inserted region from an alien species, was used to identify parental chromosomes in interspecific hybrids of sugar beet with *B. procumbens* and *B. corolliflora*, at mitosis and meiosis (Desel et al., 2002).

1.5 Sources of Genetic Variation, Genus *Beta* (B. Ford-Lloyd)

The old world genus *Beta* L., supposedly named because the shape of the seed ball when the seed matures takes the form of the letter of the Greek alphabet (β) is mainly, but not entirely confined to Europe and Western Asia. It is divided into four sections; *Beta*, *Procumbentes*, *Corollinae* and *Nanae* (**Table. 1.1**). Section *Beta* is the most widely distributed and most important because it includes all cultivated forms. While its greatest naturally occurring diversity is centered in Europe and Western Asia, wild forms also extend as far east as China, and as far west as California, USA.

Table 1.1 Sections of the genus *Beta* L.

Beta L. section *Beta* (syn. *Vulgares* Ulbrich)
Beta L. section *Corollinae* Ulbrich
Beta L. section *Nanae* Ulbrich
Beta L. section *Procumbentes* Ulbrich (syn. *Patellares* Transhel)

The question as to how many species exist within the genus lacks a straightforward answer. The number has varied enormously depending on the taxonomic treatment and various accounts are given below. This taxonomic uncertainty is largely due to problems within section *Beta*, cultivated material and associated wild forms. Recent taxonomic assessments have successfully aimed at simplification (Lange et al., 1999) and use of information at the DNA level has substantially improved not only our ability to differentiate species accurately (**Fig. 1.7**), but also to understand species relationships.

1.5.1 Early Descriptions of Beet

It is possible to trace knowledge of beets etymologically. Beet apparently had no name in Sanskrit, but the Arabic term *selg*, the Nabatean *silg,* and the Babylonian *silga* owed their origin to the Greek word *sicula* used by

Theophrastus (ca. 300 B.C.) to mean beet from Sicily. It seems, however, that the earliest Greek words for beet are *teutlon* and *teutlorizon* (syn. *seutlon*). The Greek word for squid or octopus is *teutlis*, so it is tempting to think that the words for beet originated because of the "sprangled" root or twisted flowering stalks of wild beets. Thus the earliest known descriptions of beet originated in ancient Greece around 350 B.C. and were seemingly of wild beets used for culinary or medicinal purposes. While leaf beets have reportedly been cultivated in China for 2,500 years (Shun et al., 2000), no records are available, so the earliest descriptions of beet originated from Aristotle ca. 350 B.C. (Scaliger, 1566) and Theophrastus (Theophrastus, ca. 300 B.C.). Not until the sixteenth century in Europe is there any record of beets which must have been cultivated for their swollen roots (Ruellius, 1536; Matthiolus, 1554, 1586; Ford-Lloyd et al., 1975). From then on, an increasing number of leaf and root forms were described by many herbalists and early horticulturists: beet with white and black (dark) leaves, swollen midribs, a variety of root shapes and colors, and some which were very sweet (Ford-Lloyd et al., 1975). The culmination of such development and selection was the sugar beet in the late eighteenth and early nineteenth centuries. It is such genetic diversity that fuels current-day crop improvement but concomitantly gives those involved in the classification of cultivated beets severe headaches!

1.5.2 Classification History

Among sections *Nanae*, *Corollinae* and *Procumbentes* (**Table. 1.2**) difficulties arise in determining the number of species that can be reliably recognised. The section *Nanae* has only one species, namely *B. nana* which is diploid and has a very restricted distribution on a few mountain tops in Greece. Section *Corollinae* is found across areas of the Balkan Peninsula, Turkey, Transcaucasia, and Iran, a distribution that falls within that of section *Beta* but is generally at higher altitudes. Clarification of the species in this section has been hampered by the occurrence of polyploidy and apomixis. Taxonomic treatments from Transhel onwards (Transhel, 1927) have not achieved unanimity, but clarification of some of the problems has been provided by Jassem (Jassem, 1992). There is no doubt that the cause of apomixis in the section and the underlying reason for taxonomic difficulty is the extent of hybridisation and allopolyploidy. Species of section *Corollinae* described as 4x and 6x *B. trigyna*, multigerm *B. lomatogona*, *B. intermedia*, *B. penta* etc are likely to be allopolyploids with different combinations of three genomes: L = *lomatogona*, M = macrorhiza, and C = *corolliflora* (Jassem, 1992).

There is somewhat less confusion regarding species within *Procumbentes,* in which three species endemic to the Canary Islands are recognized (**Table. 1.2**). However, the two diploid species that are self-

Table 1.2 Species of genus *Beta*, including the revised taxonomy of section *Beta*. (After Letschert, 1993).

Section *Beta*	
B. vulgaris L.	"Basic" diploid, triploid, tetraploid
B. macrocarpa Guss.	Diploid
B. patula Ait	Diploid
Section *Corollinae*	
B. lomatogona Fisch. et May.	"Basic" diploid, autotetrapoid
B. macrorhiza Stev.	"Basic" diploid
B. corolliflora Zos. ex Buttler	Autotetraploid
Agamic complex:	
B. intermedia Bunge	Allotetraploid
B. trigyna Waldst. et Kit.	Allotetraploid, allohexaploid
Section *Nanae*	
B. nana Boiss. et Heldr.	Diploid
Section *Procumbentes*	
B. procumbens Sm.	Diploid
B. webbiana Moq.	Diploid
B. patellaris Moq.	Autotetraploid

sterile, *B. procumbens* and *B. webbiana,* clearly form a continuum of morphological types in the field. Ease of hybridization between the two forms experimentally, and the occurrence of spontaneous hybridization between extremes in experimental plots (Szota, 1964) strongly suggest that these are not distinct species (Buttler, 1977; Curtis, 1968; Wagner et al., 1989) although this situation has not been formally recognized as yet. The third species, *B. patellaris* is tetraploid and self-fertile and there is evidence to suggest that it is an autotetraploid derived from the *B. procumbens* morphological form. Artificially produced autotetraploids of *B. procumbens* have been found to be fully fertile and resemble *B. patellaris* morphologically (Szota and Jassem, 1967).

Section *Beta*, while being the most interesting grouping to plant breeders, has perhaps posed the greatest taxonomic problems. Linnaeus provided the first valid binomial for beet, namely *Beta vulgaris* (Linnaeus, 1753), and included 8 varieties. In 1762 he divided the species into wild (*Beta maritima*) and cultivated (*Beta vulgaris*) and in 1767 further subdivided the cultivated into *Beta vulgaris* and *Beta cicla* (Linnaeus, 1762, 1767). The term *maritima* adopted by Linnaeus differed from the words formerly used for wild beet (*sylvestris*, *communis*, *viridis* etc.). The name was probably taken from Girolamo (1735), who used the Italian adjective *marittima* (marine) for the description of wild beets growing around the harbor of Malamocco, an island near Venice. Since the formulation of the four sections by Transhel (1927) and Ulbrich (1934), the composition of section *Beta* has become variable and complex, bringing together all cultivated and some of the wild material of the genus (Ford-Lloyd, 1986).

The taxonomic confusion within section *Beta* has arisen for a number of reasons. One is the considerable morphological phenotypic variation that exists among wild forms with different levels of inbreeding, resulting in microspeciation. Also, there have been attempts to classify cultivated beets, with uses of different taxonomic ranks by plant breeders and taxonomists (**Table. 1.3**). For example, Helm (1957) used subspecies, convariety, provariety and forma, whereas Mansfeld (1959) used only subspecies, convarieties, varieties, and forma in different quantities. The difficulty in coming to any agreement as to what taxonomic rank should be assigned to different forms is exemplified by Coons (1954) who argued that because *B. maritima* has sprangled roots, distinctive growth habit etc. "no useful purpose seems to be served by classing *B. maritima* as a variety or subspecies of *B. vulgaris* unless all the other members of this section are similarly classed as subspecies." Fortunately, the recent Cultivated Plant Code (Trehane et al., 1995) has allowed for the development of a much more workable system for handling cultivated material within section *Beta* (see below) and research involving genomics and molecular markers is providing a clearer picture of the genetic relationships of the wild forms in the section (see below).

Table 1.3 Taxonomic treatments of section *Beta*.

Author	*sp.*	*ssp.*	*convar.*	*provar.*	*var.*	*forma*
(Komarov, 1936)	2	4			10	
(Helm, 1957)	1	5	2	6		16
(Mansfeld, 1959)	1	4	5		9	10
(Davis, 1967)	4			4		
(Davis, 1967; Ford-Lloyd and Williams, 1975)	1	8			8	

Ultimately the problem has been that beet within section *Beta* are predominantly outcrossing because of self-incompatibility, resulting in much continuous variation in characters often modified by the environment; few if any useful characters exist that will actually discriminate between what people have tried to describe as different taxa. This applies equally to wild taxa and cultivated beets alike (Ford-Lloyd et al., 1975; Frese, 1991; Letschert, 1993; Letschert and Frese, 1993; Tan, 1993). Making matters worse taxonomically has been the occasional breakdown of self-incompatibility in some wild populations giving rise to significant levels of population differentiation. Just as an example,

Northern European wild maritime beets are highly variable morphologically when observed in the field, some of the variation being due to specific environmental conditions such as salt, temperature and daylength. Interaction of life history genes affecting the annual, biennial or perennial nature of populations with environment can result in substantial morphological variation that can mislead taxonomists. On the other hand, Mediterranean forms within the section such as *macrocarpa*, *adanensis* and *trojana* can be considered genetically distinct as a result of inbreeding and the resultant increase in population differentiation, but may not show sufficient morphogenetic differentiation to be classified as separate species.

1.5.3 New Classification for Section *Beta*

In 1999, proposals (Lange et al., 1999; Letschert and Frese, 1993) for the reclassification of *Beta vulgaris* taking into account the new system for the nomenclature of cultivated plants defined in the 1995 International Code of Nomenclature for Cultivated Plants were published. They not only defined the culta to be used in accordance with the Cultivated Plant Code, but also reassessed the taxonomic rankings of wild taxa that may or may not fall within the species *Beta vulgaris* (**Table. 1.4**) following the earlier revision by Letschert (1993).

Table 1.4 Revised taxonomy of *Beta* section *Beta*. (After Letschert, 1993).

Beta	*vulgaris* L.
	ssp. *vulgaris* (all cultivated forms)
	ssp. *maritima* (L.) Arcang.
	ssp. *adanensis* (Pamuk.) Ford-Lloyd & Williams
Beta	*macrocarpa* Guss.
Beta	*patula* Ait.

B. macrocarpa and *B. patula* are given specific ranks separate from *B. vulgaris* for a number of reasons. *B. patula* is confined in its distribution to a small island off Madeira and has a distinctive morphology. *B. macrocarpa* is considered to be even more distinct in its morphology as well as allozyme pattern (Letschert, 1993) that is maintained even when growing sympatrically with *maritima* (Lange et al., 1999; Letschert, 1993). Because subspecies *maritima* contains a large complex of morphological types (see above) further subdivision is regarded as inappropriate and while ssp. *adanensis* can mostly be recognized as annual with some distinguishing features, it warrants only subspecific ranking.

For dealing with the variation within subspecies *vulgaris*, determination of culta as defined in the Cultivated Plant Code is appropriate. The objective in this process would be to define large and unambiguous cultivar groups that can be internationally understood and accepted. This means that ploidy level (diploid, triploid etc), seed characters (monogerm, multigerm), morphological characters (shape of root, color of root or leaves) would be generally unsuitable. Most appropriate to meet the objectives is the actual use of the beet in horticulture, agriculture and industry. Four cultivar groups or culta have therefore been proposed (**Table. 1.5**).

Table 1.5 Cultivar groups (culta) of *Beta vulgaris* L. ssp. *vulgaris* (After Lange et al., 1999).

Cultivar group (culton): delimitation

Leaf beet group:

- leaves and petioles used as vegetable (includes spinach beet, chard and Swiss chard)
- *Garden beet group:*

  swollen hypocotyls used mostly as a vegetable (includes beetroots, and those which may have a pink, orange or red swollen root)
- *Fodder beet group:*

  large swollen hypocotyl and root used as cattle fodder (includes mangolds and forage beets)
- *Sugar beet group:*

  swollen root used for the production of sucrose (those types newly developed for the production of ethanol or fructan are included for the time being)

This cultivar taxonomy has been adopted by the World *Beta* Network and has been fully utilized by the International Database for *Beta* (Germeier and Frese, 2000). However, the UPOV denomination classes for cultivated beets do not map on to this classification directly as there are five classes in contrast to the four culta. Some adjustment will be needed for harmonization by the National Lists of varieties. Examples of different ways that a beet variety could be classified under the new culton system of the Cultivated Plant Code is given in **Table 1.6**.

Table 1.6 Acceptable ways of referring to a cultivar of sugar beet.

- *Beta* L. 'Humber' (Sugar beet group), or *Beta* L. (Sugar beet group) 'Humber'
- *Beta vulgaris* L. 'Humber' (Sugar beet group), or *Beta vulgaris* L. (Sugar beet group) 'Humber'
- *Beta vulgaris* L. ssp. *vulgaris* 'Humber' (Sugar beet group), or *Beta vulgaris* L. ssp. *vulgaris* (Sugar beet group) 'Humber'
- Sugar beet 'Humber'

Weed beet classification is a complex issue as there is strong evidence to indicate that so-called weed beets have more than one origin. Most predominant in Europe are weed beets that arose from crosses between sugar beet and wild ssp. *maritima* (Bartsch et al., 1999; Desplanque et al., 2002). In the USA however, two types can be identified. First there are those that arose in a manner similar to those in Europe, as a result of sugar beet and ssp. *maritima* hybridization. Second, it appears that *B. macrocarpa* exists as a pure species similar to that found in Europe, but in rare cases it may also have hybridized with sugar beet (Bartsch and Ellstrand, 1999). As weed beet populations, whatever their origin, have arisen from hybridization with cultivated material, and also maintain their identity as separate, distinct populations, they can be classified as spontaneous members of *B. vulgaris* ssp. *vulgaris* (Bartsch and Ellstrand, 1999; Lange et al., 1999), further demonstrating the value of the classification proposed by Letschert (1993).

1.5.4 The Future of *Beta* Classification using DNA

If traditional taxonomy needs to reinvent itself in order to survive in the genomic and postgenomic eras, then taxonomic assessment of the genus *Beta* could be an example of the way forward. Many molecular genetic studies have been carried out on different taxa of beet for a variety of purposes, many of them not primarily concerned with classification or taxonomy. These studies however, have invariably revealed findings important in a taxonomic sense and generally support and clarify the taxonomic relationships already identified. Molecular genetic information has therefore been supportive of traditional taxonomic views but has also provided greater clarity than could previously be achieved.

Genus *Beta*, along with other members of the *Chenopodiaceae* has, in common with all other *Caryophyllales* studied, lost the *nad4-i2* intron of the mitochondrial genome (Itchoda et al., 2002; Letschert, 1993). At the sectional level there is little problem in distinguishing between members of different sections. Shen et al., (1998) were able to develop ITS1 (internal transcribed spacer region 1 of rDNA) SNP-type markers that would allow for discrimination of any species in each of the sections (**Fig. 1.7**). There were indications also that *B. nana* is more closely allied to section *Corollinae* than to any other section, a conclusion also supported by Gao et al., (2000) using repetitive DNA. Further evidence which distinguishes between sections, and indicates that section *Procumbentes* is more distant to both sections *Corollinae* and *Beta* (**Fig. 1.8**) has come from characterization of the nuclear rDNA units (Santoni and Berville, 1992), ITS1 sequencing, and RAPD markers (Shen et al., 1996) and RFLPs (Mita et al., 1991). This more distant relationship of the *Procumbentes* species is strongly highlighted by *in-situ* hybridization of three repetitive DNA families

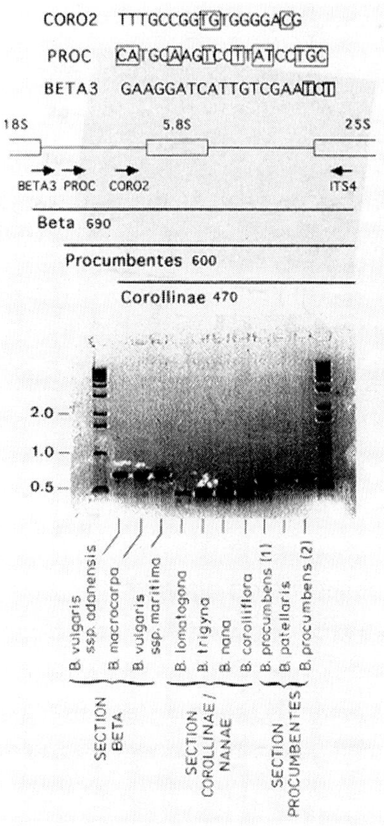

Fig. 1.7 Section differentiation by PCR using primer 'ITS4' and a mixture of three section-specific primers: BETA3, PROC and CORO2 (After Shen et al., 1998).

that are specific to the *Procumbentes* (Schmidt and Heslop-Harrison, 1996), and the distance in terms of age of section *Procumbentes* from *Nanae*, *Corollinae,* and *Beta* is shown using a highly abundant restriction satellite family (Kubis et al., 1997).

What of the specific status of *B. patula* and *B. macrocarpa*? *B. patula* has not been studied often using molecular markers, but using AFLPs, Hansen et al., (1999) showed clear differences between *B. patula* and *B. vulgaris* ssp. *maritima*. Various molecular marker and molecular genetic studies demonstrate the distinctiveness of *B. macrocarpa* from other members of section *Beta*. Shen et al. (1996) used RAPD markers to study variation among different annual taxa of section *Beta* in order to determine how reliable field identification of these taxa might be, given the difficulties that most germplasm collectors experience. The greatest distinction was

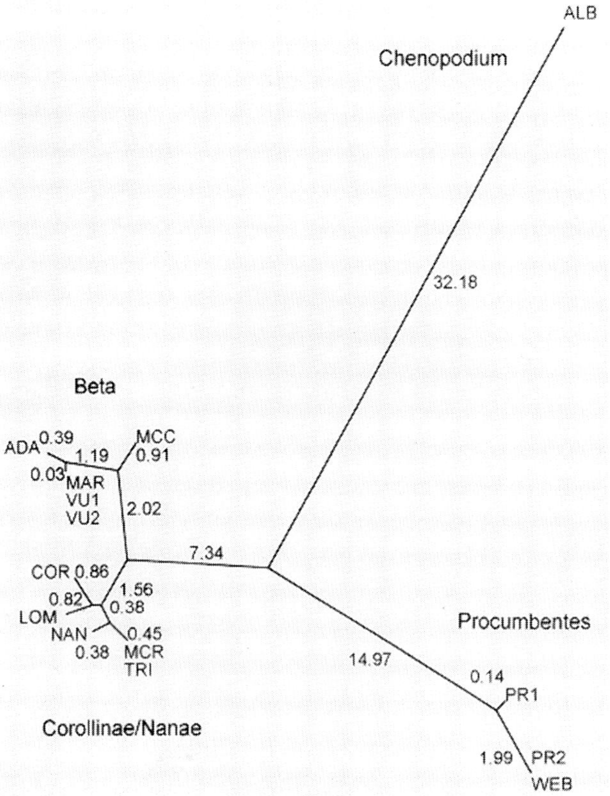

Fig. 1.8 Phylogenetic tree generated from ITS1 sequence data using the neighbor-joining method with the Kimura two-parameter distance displayed (After Shen et al., 1997).

achieved for *B. macrocarpa*, although separation of ssp. *vulgaris*, ssp. *maritima* and ssp. *adanensis* was also possible. Bartsch and Ellstrand (1999) showed a very clear separation of *B. macrocarpa* from ssp. *maritima* as well as the possibility of gene exchange in some material between the two using allozymes.

To conclude, we now have available for genus *Beta* a robust and practical taxonomic framework where phenetic data traditionally used by taxonomists is not in conflict with data generated using molecular methods. Reliable identification of taxa is possible using this framework and molecular genetic approaches together with simple molecular marker systems can taxonomically fingerprint material more accurately within the formal framework. Full adoption of the culton system of classification for cultivated beets together with the new classification system for section *Beta* is recommended.

1.6 Plant Introduction and Genetic Diversity (L. Panella and R.T. Lewellen)

Beet was originally domesticated as a leafy pot herb and was mentioned in early Greek and Roman texts. It is not until the Middle Ages, however, that we find reference to the root being used both as a vegetable and medicinal herb, and in 1822, William Morgan provided a historical account of the development of fodder beet (reviewed by De Bock, 1986; Ford-Lloyd et al., 1975). Sugar beet is not only a relatively new crop, but a relatively new beet crop, developed once a method for measuring sucrose allowed for mass selection based on this criterion. The main source of germplasm for early sugar beet varieties was, most likely, fodder beet germplasm of the type grown in Silesia, the 'White Silesian' (Fischer, 1989). For this reason it has been suggested that the genetic base of sugar beet is probably narrower than that of most open pollinated crops (Bosemark, 1979, 1989;). However, there was speculation that Knauer's *Beta Imperialis*, developed in 1858, originated from unintentional crosses between the Silesian beet and wild sea beet (*Beta vulgaris* ssp. *maritima*) from the North Atlantic Coast (Bosemark, 1979, 1989). There was also speculation that sugar beet originated from spontaneous crosses between fodder beet and chard (Zossimovich, 1940; Fischer, 1989, cited by Coons et al., 1955)

Early sugar beet development and production took place in the temperate climate of northern Europe, which was relatively disease free and therefore posited little pressure to find or maintain high levels of host plant resistance (Lewellen, 1992). As sugar beet production moved east into Russia, south into the Mediterranean region, and west into England and North America, a host of endemic diseases that limited its production was encountered (Lewellen, 1992) and plant breeders were confronted with diseases for which there were few known sources of resistance. It is in these areas that the first attempts to screen exotic and wild beet germplasm, primarily for disease resistance, were initiated at the beginning of the twentieth Century.

By the late 1800s, there was speculation that *B. v.* ssp. *maritima* might provide a reservoir of resistance genes that could be utilized in sugar beet breeding. In the Po Valley of Italy, high humidity and warm night temperatures provide an almost perfect environment for leaf spot of sugar beet (caused by *Cercospora beticola*). The first documented instance of collecting wild beet germplasm (*B.v.* ssp. *maritima*) to use in a sugar beet breeding effort is that of Ottavio Munerati who looked to the wild sea beet growing in the Po estuary as a source of host plant resistance to leaf spot (Munerati, 1908; Munerati et al., 1913). The germplasm that emerged from this program, 'Cesena', 'Mezzano' and the Rovigo series (R148, etc.), has been used worldwide and forms the basis of much of the Cercospora-resistant germplasm in use today (Munerati, 1932; Biancardi and De Biaggi, 1979).

In other areas of Europe, researchers studied *B.v.* ssp. *maritima* and crossed it to sugar beet (Rasmussen, 1932, 1933; Tjebbes, 1933). There were efforts to develop cercospora leaf spot resistant varieties as Munerati had done (Schlösser, 1957; Stehlik, 1949; Zossimovich, 1939), and varieties with resistance to other diseases (Margara and Touvin, 1955; reviewed by Asher et al., 2001), but it is difficult to estimate the extent *B.v.* ssp. *maritima* germplasm was used in commercial breeding programs, especially because of undesirable traits from the wild beet, e.g., fangy roots, annualism, high fiber, elongated crowns, red pigment (in root, leaf or petiole) and lower sucrose production (reviewed in Coons, 1975) that were introduced into these varieties.

In the United States, most sugar beet seed came from Europe until transport from Germany was interrupted by World War I, and a domestic seed production industry was begun. After the war, seed shipment from Europe was resumed but by the end of the 1930s domestic seed production accounted for about one-third of the needs of the USA (Coons, 1936). George H. Coons of the USDA was familiar with Ottavio Munerati's work and made collection trips in 1925 and 1935 to look for sources of leaf spot resistance and curly top resistance in wild beet accessions (Coons et al., 1931) as well as in the other species in genus *Beta* (Coons, 1975). In the US, there was some effort to evaluate this material, and material collected by Dewey Stewart in 1969 (all of which was deposited with Gerald Coe in Beltsville, Maryland), for resistance to cercospora leaf spot (Bilgen et al., 1969), rhizoctonia root rot (caused by *Rhizoctonia solani*) and black root (caused by *Aphanomyces cochlioides*) (Schneider and Gaskill, 1962). This germplasm found its way into USDA-ARS breeding programs throughout the USA; however, storage conditions at Beltsville were poor and by the time the collection was sent to Salinas, CA for regeneration by John McFarlane, much of it was not viable. Nonetheless, this collection was extensively evaluated later and found to possess many useful traits (Whitney, 1989a, b; Lewellen and Whitney, 1993; Yu et al., 1999; Lewellen and Schrandt, 2001)

In the first half of the twentieth century, breeders at Great Western Sugar Co. and other US sugar beet seed companies were also testing hybrids of Milpitas wild beet and sugar beet. The Milpitas wild beet is the weedy type that grows wild around San Jose, CA and is thought to have been introduced by the early Spanish settlers and escaped. It was evaluated by Dahlberg and Brewbaker (1948). This early work by Dahlberg and Brewbaker at Great Western Sugar Co. also led to the open pollinated germplasm GW359 (Dahlberg, 1938, 1940). The early development of germplasm in the US has been reviewed by Lewellen (1992). Although Brewbaker and coworkers discussed the testing of hybrids with *B.v.* ssp. *maritima* parentage (Brewbaker et al., 1950), it is likely that most of the *B.v.* ssp. *maritima* germplasm that found its way into U.S. commercial varieties was from either Cesena or Mezzano stations in Italy out of Munerati's program in Italy (Lewellen, 1992). Some of the germplasm developed by the USDA-ARS programs during these years had *B.v.* ssp. *maritima* parentage but it is difficult to say how much has been used in commercial hybrid varietal development. Additionally, of course, there was continual contact and germplasm exchange among sugar beet breeding programs worldwide, but, most of the exchanges were informal and impossible to document (Lewellen, 1992).

A number of changes in sugar beet breeding came together in the 1960s. The cytoplasmic male sterility (CMS) and genetic fertility restoration system developed out of V.F. Owen's research (Owen, 1954) was introduced into CMS and O-type maintainer lines to produce commercial monogerm, CMS hybrid varieties (Savitsky, 1952). Use of a few sources for monogerm and a single source of CMS led to a genetic bottleneck in this time period, which exacerbated growing disease pressure due to an increase in cultivated area and shortening of rotation between sugar beet crops. Until the 1980s, there seemed to be a reluctance to use wild beet germplasm, perhaps because of earlier experiences that resulted in the introgression of many undesirable traits from the exotic germplasm (Frese et al., 2001). The need for increased resistance to disease and insect pests and a greater productivity rekindled interest in *B.v.* ssp. *maritima* and other exotic sources of germplasm (Lewellen, 1992).

In 1983, the Sugar Beet Crop Advisory Committee (now Crop Germplasm Committee-CGC) was formed to represent the sugar beet germplasm user community in the US. Although it was instituted through the American Society of Sugar Beet Technologists (ASSBT), it was an integral part of the USDA-ARS National Plant Germplasm System (NPGS) (reviewed by Janick, 1989). The initial committee consisted of scientists from the sugar beet seed industry, the university research system, and USDA-ARS. Since its establishment, the Sugar Beet CGC has aggressively managed the *Beta* germplasm evaluation of the USDA-ARS NPGS.

Meanwhile in Europe, Bosemark (1989) developed a theoretical framework for effectively introgressing new germplasm into elite breeding programs, and Frese expanded his ideas into a strategy to broaden the germplasm base of the sugar beet gene pool (Frese, 1990). This was implemented through the World *Beta* Network, founded in 1989 with the goal of improving international collaboration among users and curators of germplasm collections throughout the world. Members supported the development of a central data base of all *Beta* accessions throughout the World International Database for *Beta* (IDBB) maintained at the Federal Centre for Breeding Research on Cultivated Plants (BAZ) Gene Bank (in Braunschweig, Germany) under the direction of Lothar Frese.

Around the world there was an increasing realization that wild germplasm was being lost (Doney et al., 1995; Pignone, 1989) and that these wild relatives could be of great value in the improvement of the sugar beet crop (De Bock, 1986; Doney and Whitney, 1990; Doney, 1993; Lewellen and Skoyen, 1991; Van Geyt et al., 1990). Using the data from the Sugar Beet CGC evaluations, the USDA-ARS plant breeders began introgressing wild germplasm into the sugar beet gene pool and releasing this germplasm to US sugar beet seed companies and international seed companies (Doney, 1998, Lewellen, 1991, 1997, 2000a, b; Lewellen and Whitney, 1993; Panella, 1998, Yu, 2002). Currently, the Genetic Resources Information Network (GRIN) Data base of NPGS *Beta* collection includes everything from wild relatives to heritage open-pollinated varieties (McGrath et al., 1999); and germplasm registered in *Crop Science* (Doney, 1995). Among the GRIN system's 2,447 *Beta* accessions, some of the most interesting are in the taxon *Beta vulgaris* ssp. *maritima* (Lange et al., 1999), a rich source of disease resistance genes that is being vigorously evaluated. As of 2003, about 25,000 evaluation records (descriptors times database accessions evaluated) are in the database. These and other data in the GRIN database can be accessed at: www.ars-grin.gov/npgs (Panella et al., 2003; Panella and Frese, 2003).

Concomitantly, in Europe, public and private plant breeders, working collaboratively through the IIRB Genetics and Breeding Working Group, started developing "Doggett" buffer populations (Doggett and Eberhardt, 1968) to introgress different sources of disease resistance into the sugar beet gene pool (described in Frese et al., 2001). Additionally, Frese (2000) used an international core collection comprising 805 accessions of the IDBB held by various gene banks in Europe and around the world to facilitate germplasm evaluation funded through the European Union project GENRES CT95 42. Between 300 and 700 accessions of the core collection were used by project partners in Germany, Greece, Sweden, the United Kingdom, Italy, and the Netherlands to screen for resistance to seedling diseases (caused by *Aphanomyces cochlioides, Phoma betae*), leaf diseases (caused by *Cercospora beticola, Erysiphe betae*, Beet yellows virus, Beet mild yellowing virus), the root diseases rhizomania (caused by *Beet necrotic yellow vein virus*) and rhizoctonia root and crown rot (caused by *Rhizoctonia solani*), as well as drought resistance (Panella and Frese,

2003). Additionally, the "Promotion of the use of East European *Beta vulgaris* germplasm collections" project has screened horticultural traits, nitrogen accumulation and characterized beet collections held by the Vavilov-Institute in St. Petersburg, Russia, and the Czech genebank in Prague. Data from both can be accessed at: http://ice.zadi.de/idbbonline/beta.php and users can query passport, characterization and evaluation data (including statistical parameters) and download the results (Panella and Frese, 2003). Private and public plant breeders in Europe and throughout the world have taken the results of these evaluations and are beginning to introgress these newly discovered sources of disease resistance into sugar beet (Asher et al., 2001; Biancardi et al., 2002; Luterbacher et al., 2000).

1.7 History of Sugar Beet Breeding (E. Biancardi)

The first beet that Achard worked with was probably a fodder beet known locally as *Rünkelrübe*; it was heterogeneous for root form, shape and color (Knapp, 1958). We cannot discount the possibility that the first sugar beet might have come from a cross between fodder and wild beet (Frickenhaus, 1864) or from a casual hybridization between fodder beet and Swiss chard (Fischer, 1989).

Achard (1803) asserted that conical white roots which barely protruded from the soil produced higher sugar yields. Through mass selection he obtained the variety "White Silesian" whose sugar content ranged from 5-7% of root weight (Coons, 1936). Knauer subjected the variety to additional cycles of mass selection and in 1850 released the "*Imperial Rübe*" with a sugar content that was sometimes as high as 14% (Ware, 1880; Knapp, 1958; Winner, 1993).

Using "White Silesian" as a base population, Philippe André de Vilmorin obtained further improvements with mass selection based on root morphology (Coons, 1936). The experiments were continued by his son Louis, who not only introduced progeny testing into breeding practice, but also improved the densimetric analysis of the sugar content of the roots (de Vilmorin, 1923). Before that time, roots had always been selected on the basis of form. To obtain an estimate of their relative sugar content, roots were immersed in a saline solution. Floating beets had less sugar and were discarded. With the same purpose in mind, gravimetric scales were used to take measurements of the pressed juice (McFarlane, 1971). As stated before, de Vilmorin improved the use of the saccharimeter. This instrument had been utilized for some time in sugar factories and allowed faster and more precise measurements than previous methods described; after 1860 it became widely used for selection purposes (Coons *et al.,* 1955). De

Vilmorin was the first to attempt identification of useful traits in sea beet and to transfer these traits to cultivated varieties (Coons, 1936).

With de Vilmorin's improved method, selected plants were reproduced in pairs (Oltmann et al., 1984), and seed from every plant planted in field trials. A population was formed using the seed of plants whose progenies gave the highest yield (Winner, 1993). Progeny testing not only allowed faster improvements than mass selection (Coons, 1936), but also identification of productive capability. In fact, varieties with high root weight and low sugar content or low root weight and high sugar were identified. Those with low sugar content were designated E (*Ertrag*), those with high sugar content Z (*Zucker*), and the intermediates N (*Normal*). Each type was suitable for specific growing situations (Oltmann et al., 1984).

In the first years of the twentieth century, thanks to Mendel's discoveries the practice of inbreeding began with selection of lines used to produce synthetic varieties after combining ability tests (Oltmann *et al.*, 1984). Agronomic evaluation of the genotypes, undoubtedly the most delicate phase in any selective system, improved rapidly with the application of Fischer's discoveries in the field of statistical analysis. Introduction of variance analysis (ANOVA) of yield data and reduction of experimental error increased the certainty of recognizing desirable genotypes.

Again in the first years of the twentieth century, after ascertaining that resistance to disease could be hereditary (Ellerton and Arnold, 1982), curly top resistance breeding began in the USA (McFarlane, 1971). In the same period in Italy, selection of varieties with cercospora leaf spot resistance was begun (Skaracis and Biancardi, 2000). Around 1950, virus yellows resistance breeding began in England (McFarlane, 1971). Genetic variability for response to rhizomania was discovered in Italy in 1958 by Bongiovanni and Lanzoni (1964). However, varieties which produced well in infected soil were not widely used until 20 years later (De Biaggi, 1987; Biancardi *et al.*, 2002). In recent decades, genetic resistance to pathogenic agents such as nematodes, *Rhizoctonia,* etc. has been sought but with less success.

Apart from productivity increases and improved resistance traits obtained through breeding, the genetic base of cultivated varieties remained that of "White Silesian" for a long time. A radical change began in the 1940s (Bosemark, 1993). It is not by chance that progress in breeding work on sugar beet was possible in the USA rather than Europe, since collaboration between and assignment of specific objectives to different USDA stations and private companies permitted studies impossible in Europe (Coons *et al.*, 1955). In fact, the most important discoveries were made in public research institutes where solutions to basic problems were the result of adequate funding and collaboration. The discovery of monogermity would have been delayed were it not for the combined efforts of USDA breeders and the sugar industry. Understanding genetic cytoplasmic male sterility (CMS) would not have been possible without the genetic and statistical analyses developed in the USA for corn breeding (Desprez, 2000). Desprez speaks of an "American school" which, in recent times, has made significant contributions, especially in enhancing genetic resistance to major diseases (Lewellen, 1992).

References

Achard, F.C. 1803. Anleitung zum Anbau der zur Zuckerfabrication anwendbaren Runkelrüben und zur vortheilhaften Gewinnung des Zuckers aus denselben. Reprinted in: Ostwald's Klassiker der exacten Wissenschaft. 1907. Engelmann, Liepzig, Germany.

Anonymous. 1998. Russland feiert 200 Jahre Rübenzuckererzengung. Zuckerindustrie 123:69.

Artschwager, E. 1926. Anatomy of the vegetative organs of the sugar beet. J. Agric. Res. 33:143-176.

Artschwager, E. 1927. Development of flowers and seed in the sugar beet. J. Agric. Res. 34:1-25.

Artschwager, E. 1930. A study of the structure of sugar beets in relation to sugar content and type. J. Agric. Res. 40:867-915.

Artschwager, E. and R.C. Starrett. 1933. The time factor in fertilization and embryo development in the sugar beet. J. Agric. Res. 47:823-843.

Asher, M.J.C., M.C. Luterbacher, and L. Frese. 2001. Wild *Beta* species as a source of resistance to sugar-beet pests and diseases. Proc. IIRB 64:141-152.

Bartsch, D. and M. Schmidt. 1997. Influence of sugar beet breeding on populations of *Beta vulgaris* ssp. *maritima* in Italy. J. Veget. Sci. 8:81-84.

Bartsch, D. and N.C. Ellstrand. 1999. Genetic evidence for the origin of Californian wild beets (genus *Beta*). Theor. Appl. Genet. 99:1120-1130.

Bartsch, D., M. Lehnen, J. Clegg, M. Pohl-Orf, I. Schuphan, and N.C. Ellstrand. 1999. Impact of gene flow from cultivated beet on genetic diversity of wild sea beet populations. Molec. Ecol. 8:1733-1741.

Battle, J.P. and W.J. Whittington. 1969. The relation between inhibitory substances and variability in time to germination of sugar beet clusters. J. Agric. Sci. Camb. 73:337-346.

Biancardi, E. 1984. La barbabietola da zucchero. Sci. Amer. (Italian ed.) 184:120-130.

Biancardi, E. 1999. Miglioramento genetico. *In:* B. Casarini, E. Biancardi, and P. Ranalli (eds.) La Barbabietola da Zucchero in Ambiente Mediterraneo. Edagricole, Bologna, Italy, pp. 57-45.

Biancardi, E. and M. De Biaggi. 1979. *Beta maritima* L. in the Po delta. *In:* Proc. Convegno Tecnico Internazionale in Commemorazione di Ottavio Munerati. Rovigo, 5-6 Oct. 1979. ISCI, Bologna, Italy, pp. 183-185.

Biancardi, E., R.T. Lewellen, M. De Biaggi, A.W. Erichsen, and P. Stevanato. 2002. The origin of rhizomania resistance in sugar beet. Euphytica 127:383-397.

Bilgen, T., J.O. Gaskill, R.J. Hecker, and D.R. Wood. 1969. Transferring Cercospora leaf spot resistance from *Beta maritima* to sugarbeet by backcrossing. J. ASSBT 15:444-449.

Bongiovanni, G.C. and L. Lanzoni. 1964. La rizomania della bietola. Progresso Agricolo 2:209-220.

Bormotov, V.E., A.M Scherbakova, and M. Luneva. 1973. Cytological characteristics in trisomic types of sugar beet. Nauka Tekh. 1973:27-34.

Bosemark, N.O. 1966. On the origin and consequences of aneuploidy in the triploid and tetraploid sugar beet. J. IIRB 2:9-34.

Bosemark, N.O. 1967. Post meiotic nuclear fragmentation and pollen sterility in autotetraploid sugar beet. Hereditas 57:217-238.

Bosemark, N.O. 1977. Use of tetraploid monogerm male-steriles in triploid-seed production. Proc. IIRB 40:271-287.

Bosemark, N.O. 1979. Genetic poverty of the sugar beet in Europe. p. 29-35. *In:* A.C. Zeven (ed.) Proc. Conf. Broadening Genet. Base of Crops. Pudoc, Wageningen, Netherlends.

Bosemark, N.O. 1989. Prospects for beet breeding and use of genetic resources. p. 89-97. *In:* Report of an International Workshop on *Beta* Genetic Resources. International Crop Network Series 3. International Board for Plant Genetic Resources, Rome, Italy.

Bosemark, N.O. 1993. Genetics and breeding. *In:* D.A. Cooke and R.K. Scott, (eds.) The Sugar Beet Crop. Chapman & Hall, London, UK, pp. 67-119.

Bosemark, N.O. and V.E. Bormotov. 1971. Chromosome morphology in a homozygous line of sugar beet. Hereditas 69:205-212.

Brewbaker, J.L. 1967. The distribution significance of binucleate and trinucleate pollen grains in the angiosperms. Amer. J. Bot. 54:1069-1083.

Brewbaker, H.E., H.L. Bush, and R.R. Wood. 1950. A quarter century of progress in sugar beet improvement by the Great Western Sugar Company. Proc. ASSBT 6: 202-207.

Briem, H. 1895. Der praktische Rübenbau. Hofbuchhandlung Wilhelm Frick, Vienna.

Butterfass, T. 1964. Die chloroplastenzahlen in verschiedenartigen Zellen trisomer Zuckerruben *(Beta vulgaris* L.). Z. Bot. 52:46-77.

Buttler, K.P. 1977. Revision von *Beta* Sektion *Corollinae* (*Chenopodiaceae*) I. Selbsterile Basisarten. Mitt. Bot. München 13:255-336.

Campbell, L.G. and J.W. Enz. 1991. Temperature effects on sugar beet seedling emergence. J. Sugar Beet Res. 28:129-140.

Capron, I., F. Corbineau, F. Dacher, C. Job, D. Côme, and D. Job. 2000. Sugar beet seed priming: effects of priming conditions on germination, solubilization of 11-S globulin and accumulation of LEA proteins. Seed Sci. Res. 10:243-254.

Casas, A., J. M. Sanz, and J. M. Lasa. 1986. Multiplicacion *in vitro* de trisomicos primarios de remolacha azucarera, *Beta vulgaris* L. XXII Jornadas de Genetica Luso-Espanolas. Oviedo, Spain, Abstr., p. 41.

Chrobozek, E. 1934. A study of some ecological factors influencing seed-stalk development in beets (*Beta vulgaris* L.). Mem. Cornell Univ. Agric. Exper. Stat. 154:1-84.

Cistue, L., I. Romagosa, T. Tsuchiya, and J.M. Lasa. 1985. Karyotype analysis in haploid sugar beet. Bot. Gaz. 146:259-263.

Coons, G.H. 1936. Improvement of the sugar beet. In: Yearbook of Agriculture, USDA, Washington, DC pp. 625-656.

Coons, G.H. 1954. The wild species of *Beta*. Proc. ASSBT 8:142-147.

Coons, G.H. 1975. Interspecific hybrids between *Beta vulgaris* L. and the wild species of *Beta*. J. ASSBT 18: 281-306.

Coons, G.H., D. Stewart, and H.A. Elcock. 1931. Sugar beet strains resistant to leaf spot and curly top. In: Yearbook of Agriculture, USDA, Washington, DC pp. 493-496.

Coons, G.H., F.V. Owen, and D. Stewart. 1955. Improvement of the sugar beet in the United States. Adv. Agron. 7:89-139.

Copeland, L.O. and M.B. McDonald. 2001. Seed Science and Technology. Kluwer Academic Press, Boston, MA (4th ed.)

Crosthwaite, S.K. and G.I. Jenkins. 1993. The role of leaves in the perception of vernalizing temperatures in sugar beet. J. Exper. Bot. 44:801-806.

Curth, P. 1962. Spezielle Untersuchungen zu Vernalisation und Photoperiodismus der Zuckerrübe. Z. Pflanzenzucht. 47:254-276.

Curtis, G.J. 1968. Observations of fruit shape and other characters in the species of section *Patellares*, genus *Beta*. Euphytica 17:485-491.

Dahlberg, H.W. 1938. Some observations on the wild beet (*Beta maritima*). Proc. ASSBT 1:76-79.

Dahlberg, H.W. 1940. A study of sugar beet hybrids. Proc. ASSBT 2:143-144.

Dahlberg, H.W. and H.E. Brewbaker. 1948. A promising sugar beet hybrid of the Milpitas wild type x commercial. Proc. ASSBT 5:175-178.

Davis, P. 1937. Flora of Turkey. Edinburgh Univ. Press. Edinburgh, UK, vol. 2.

De Biaggi, M. 1987. Methodes de selection—Un cas concret. Proc. IIRB 50:157-161.

De Bock, T.S.M. 1986. The genus *Beta*: Domestication, taxonomy and interspecific hybridization for plant breeding. Acta Horti. 182: 335-343.

De Jong, J.H. and T.S.M. De Bock. 1978. Use of haploids of *Beta vulgaris* L. for the study of orcein and Giemsa stained chromosomes. Euphytica 27:41-47.

De Jong, J.H., and P. Stam. 1985. The association of centromeres of nonhomologous chromosomes at meiotic prophase in *Beta vulgaris* L. Can. J. Genet. Cytol. 27:165-171.

De Vilmorin, J.L. 1923. L'hérédité chez la betterave cultivée. Gauthiers-Villars et Cie, Paris, France.

Desel, D., R. Jansen, G. Dedong, and T. Schmidt. 2002. Painting of parental chromatin in *Beta* hybrids by multi-colour fluorescent in situ hybridization. Ann. Bot. 89:171-181.

Desplanque, B., N. Hautekeete, and H. Van Dijk. 2002. Transgenic weed beets: possible, probable, avoidable? J. Appl. Ecol. 39:561-571.

Desprez M. 2000. Evolution de la génétique et de l'amelioration de la betterave sucriére au cours des vingt derniéres anneés. Proc. IIRB 63:15-31.

Doggett, H. and S.A. Eberhardt. 1968. Recurrent selection in sorghum. Crop Sci. 8:199-121.

Doney, D.L. 1979. Seedling physiology and sugar beet yield. J. ASSBT 20:399-418.

Doney, D.L. 1993. Broadening the genetic base of sugarbeet. J. Sugar Beet Res. 30:209-220.

Doney, D.L. 1995. USDA-ARS sugarbeet releases. J. Sugar Beet Res. 32:229-257.

Doney, D.L. 1998. *Beta* evaluation and sugar beet enhancement from wild sources. *In:* L. Frese, L. Panella, H.M. Srivastava, and W. Lange (eds.). Report 4th Int. *Beta* Genetic Resources Workshop and World *Beta* Network Conf. Izmir, Turkey, 28 Feb.-3 Mar. 1996. IPGRI, Rome, Italy, pp. 62-72.

Doney, D.L., and E.D. Whitney. 1990. Genetic enhancement in *Beta* for disease resistance using wild relatives: A strong case for the value of genetic conservation. Economic Botany 44:445-451.

Doney, D.L., B.V. Ford-Lloyd, L. Frese, and A. Tan. 1995. Scientists worldwide rally to rescue the native beets of the Mediterranean. Diversity 11:124-125.

Dureau, G. 1886. Traitè de la Culture de la Betterave. Bureaux du J. des Fabricantes de Sucre, Paris, France (2nd ed.).

Durr, C. and J. Boiffin. 1995. Sugar beet seedling growth from germination to first leaf stage. J. Agric. Sci. Camb. 124:427-435.

Elamrani, A., P. Raymond, and P. Saglio. 1992. Nature and utilization of seed reserves during germination and heterotrophic growth of young sugar beet seedlings. Seed Sci. Res. 2:1-8.

Ellerton, S. and M.H. Arnold. 1982. Selection de la betterave a sucre. *In :* Cinquante ans d'historie de l'IIRB. IIRB, Brussels, Belgium, pp. 19-34.

Elliott, M.C. and G.D. Weston. 1993. Biology and physiology of the sugar-beet plant. *In:* D.A. Cooke and R.K. Scott (eds.). The Sugar Beet Crop. Chapman & Hall, London, UK, pp. 34-66.

Esau, K. 1977. Anatomy of Seed Plants. John Wiley & Sons, New York, NY (2nd ed.).

Fischer, H.E. 1989. Origin of the "Weisse schlesische Rübe" (White Silesian beet) and resyntnesis of sugar beet. Euphytica 41:75-80.

Ford-Lloyd, B.V. 1986. Infraspecific variation in wild and cultivated beets and its effect upon infraspecific classification. *In:* B.T. Styles (ed.) Infraspecific Classification of Wild and Cultivated Plants. Systematic Association, Clarendon Press, Oxford, UK, vol. 29, pp. 331-334.

Ford-Lloyd, B.V., A.L.S. Williams, and J.T. Williams. 1975. A revision of *Beta* section *Vulgares* (*Chenopodiaceae*), with new light on the origin of cultivated beets. Bot. J. Linn. Soc. 71:89-102.

Frese, L. 1990. The world *Beta* network cooperation. Proc. IIRB 53:161-171.

Frese, L. 1991. Variation patterns in a leaf beet (*Beta vulgaris, Chenopodiaceae*) germplasm collection. Pl. Syst. Evol. 176:1-10.

Frese, L. 2000. The synthetic *Beta* core collection — state of the art. J. Sugar Beet Res. 37:1-10.

Frese, L., B. Desprez, and D. Ziegler. 2001. Potential of genetic resources and breeding strategies for base-broadening. *In:* H.D. Cooper, C. Spillane, and T. Hodgkin (eds.). Broadening the Genetic Base of Crop Production. FAO, Rome, Italy, pp. 295-309.

Frickenhaus H. 1864. Über Rübenzucht. Zeitschrift Vereines Rübenzucker-Industrie 14:387-393.

Gao, D., T. Schmidt, and C. Jung. 2000. Molecular characterization and chromosomal distribution of species-specific repetitive DNA sequences from *Beta corolliflora*, a wild relative of sugar beet. Genome 43:1073-1080.

Gao, D., D. Guo, and C. Jung. 2001. Monosomic addition lines of *Beta corolliflora* Zoss in sugar beet: cytological and molecular-marker analysis. Theor. Appl. Genet. 103: 240-247.

Germeier, C.U. and L. Frese. 2000. The international database for *Beta* — state of the art. *In:* L. Maggioni, L. Frese, C. Germeier, and E. Lipman (eds.). 1st Report of Working Group on *Beta*. Broom's Barn, UK, 9-10 Sept. 1999. IPGRI, Rome, Italy.

Geschwind, L. and E. Sellier. 1902. La Betterave. Association des Chemistes de Sucrerie, Paris, France.

Girolamo, G. 1735. Istoria delle Piante che nascono ne'Lidi intorno a Venezia. Presso Antonio Bortoli, Venice, Italy.

Gummerson, R.J. 1986. The effect of constant temperatures and osmotic potentials on the germination of sugar beet. J. Expor. Bot. 37:729-741.

Hallden C, D. Ahrén, A. Hjerdin, T. Säll, and N.O. Nilsson. 1998. No conserved homoeologous regions found in the sugar beet genome. J. Sugar Beet Res. 35:1-13.

Hansen, M., T. Kraft, M. Christiansson, and N.O. Nilsson. 1999. Evaluation of AFLP in *Beta*. Theor. Appl. Genet. 98:845-852.

Hecker, R.J., R.E. Stafford, R.H. Helmerik, and C.W. Maag. 1970. Comparison of the same sugarbeet F_1 hybrids as diploids, triploids, and tetraploids. J. ASSBT 16:106-116.

Heinisch, O. 1960. Die Zuckerrübe. Akademie Verlag, Berlin, Germany.

Helm, J. 1957. Versuch einer morphologisch-systematischen Gliederung der Art *Beta vulgaris* L. Zuchter 27:203-22.

Hjerdin, A., T. Säll, N.O. Nilsson, C.H. Bornmann, and C. Hallden. 1994. Genetic variation among wild and cultivated beet of the Section *Beta* as revealed by RFLP analysis. J. Sugar Beet Res. 1:59-67.

Inoue, K. and R. Yamamoto. 1974. The growth inhibitors in sugar beet seed balls. I. Isolation of monosodium oxalate as a root growth inhibitor. Proc. Crop Sci. Soc. Jpn. 43:439-444.

Inoue, K. and R. Yamamoto. 1975. The growth inhibitors in sugar beet seed balls. II. Isolation of potassium nitrate as the germination inhibitor and hypocotyl stimulating substance. Proc. Crop Sci. Soc. Japan 44:465-470.

Itchoda, N., S. Nishizawa, H. Nagano, T. Kubo, and T. Mikami. 2002. The sugar beet mitochondrial *nad4* gene: an intron loss and its phylogenetic implication in the Caryophyllales. Theor. Appl. Genet. 104:209-213.

Janick, J. 1989. Plant Breeding Reviews, the National Plant Germplasm System of the United States. Timber Press, Portland, OR.

Jassem, B. 1992. Species relationship in the genus *Beta* as revealed by crossing experiments. *In:* 2nd Intl. *Beta* Genetic Resources Workshop. Braunschweig, Germany, 24-28 June 1991, pp. 55-61.

Jung, C., K. Pillen, L. Frese, and A.E. Melchinger. 1993. Philogenetic relationship between cultivated and wild species of the genus *Beta* revealed by DNA fingerprinting. Theor. Appl. Genet. 86:449-457.

Kaltsikes, P.J. and L.E. Evans. 1967. Production and identification of trisomic types in *Beta vulgaris*. Can. J. Genet. Cytol. 9:691-699.

Kao, T.H. 1992. Molecular genetics of gametophytic self-incompatibility in Solanaceae. *In:* E. Ottaviano (ed.) Angiosperm Pollen and Ovules. Springer-Verlag, New York, NY, pp. 66-74.

Khush, G.S. 1973. Cytogenetics of Aneuploids. Academic Press, New York, NY.

Kleine, M., H. Voss, D. Cai, and C. Jung. 1998. Evaluation of nematode-resistant sugar beet (*Beta vulgaris* L.) lines by molecular analysis. Theor. Appl. Genet. 97:896-904.

Klotz, K.L. and F.L. Finger. 2002. Contribution of invertase and sucrose synthase isoforms to sucrose catabolism in developing sugar beet roots. J. Sugar Beet Res. 39:1-24.

Knapp, E. 1958. *Beta* Rüben. *In:* T. Roemer and W. Rudorf (eds.) Handbuch der Pflanzenzüchtung. Paul Parey, Berlin, Germany, vol. 3.

Komarow, V.L. 1936. Flora USSR. Science Acad. USSR, Moskow, USSR, vol. 6

Krasochkin, V.G. 1959. Review of the species of the genus *Beta*. Trudy Po Prikladnoi Botanike, Genetike i Selektsii 32:3-35.

Kubis, S., J.S. Heslop-Harrison, and T. Schmidt. 1997. A family of differentially amplified repetitive DNA sequences in the genus *Beta* reveals genetic variation in *Beta vulgaris* subspecies and cultivars. J. Molec. Evol. 44:310-320.

Lane, H.C., H.M. Cathey, and L.T. Evans. 1965. The dependence of flowering in several long-day plants on the spectral composition of light extending the photoperiod. Amer. J. Bot. 52:1006-1014.

Lange, W. and T.S.M. De Bock. 1989. The diploidised meiosis of tetraploid *Beta macrocarpa* and its possible application in breeding sugar beet. Plant Breed. 103:196-206.

Lange, W., W.A. Brandenburg, and T.S.M. De Bock. 1999. Taxonomy and cultonomy of beet (*Beta vulgaris* L.). Bot. J. Linn. Soc. 130: 81-96.

Lange, W., T.S.M. De Bock, J.P.C. Van Geyt, and M. Oleo. 1988. Monosomic additions in beet (*Beta vulgaris*) carrying extra chromosomes of *B. procumbens*. II. Effects of the alien chromosomes on *in vivo* and *in vitro* plant development. Theor. Appl. Genet. 76:656-664.

Lange, W., T.S.M De Bock, G.J. Speckmann, and J.H. DeJong. 1993. Disomic and ditelosomic alien chromosome additions in beet (*Beta vulgaris*), carrying in extra chromosome of *Beta procumbens* or telosome of *Beta patellaris*. Genome 36:261-267.

Larsen, K. 1977. Self-incompatibility in *Beta vulgaris* L. I. Four gametophytic, complementary S-loci in sugar beet. Hereditas 85:227-248.

Lawrence, D.M., P. Halmer, and D.J. Bowles. 1990. Mobilization of storage reserves during germination and early seedling growth of sugar beet. Physiol. Plant. 78:421-429.

Leach, L.D. and Bainer R. 1942. Seed treatment of segmented seed. Proc. ASSBT 3:213-220.

Letschert, J.P.W. 1993. *Beta* section *Beta*: biogeographical patterns of variation and taxonomy. PhD. diss., Wageningen Agric. Univ. Wageningen, Netherlands.

Letschert, J.P.W. and L. Frese. 1993. Analysis of morphological variation in wild beet (*Beta vulgaris* L.) from Sicily. Genet. Resources Crop Evol. 40:15-24.

Levan, A. 1942. The effect of chromosomal variation in sugar beets. Hereditas 28:345-399.

Lewellen, R.T. 1991. Registration of rhizomania-resistant germplasm of *Beta vulgaris*. Crop Sci. 31:244-245.

Lewellen, R.T. 1992. Use of plant introductions to improve populations and hybrids of sugarbeet. *In:* Use of Plant introductions in Cultivar Development, Part 2. Crop Science Soc. of America, Madison, WI, pp. 117-135.

Lewellen, R.T. 1997. Registration of 11 sugarbeet germplasm C79 lines with resistance to rhizomania. Crop Sci. 37:1026.

Lewellen, R.T. 2000a. Registration of powdery mildew resistant sugarbeet germplasm CP01 and CP02. Crop Sci. 40:1515.

Lewellen, R.T. 2000b. Registration of rhizomania resistant sugarbeet x *Beta vulgaris* subsp. *maritima* germplasms C26, C27, and C51. Crop Sci. 40:1513.

Lewellen, R.T. and I.O. Skoyen. 1991. Improvement and performance of populations of sugarbeet x *Beta maritima*. J. Sugar Beet Res. 28:79.

Lewellen, R.T. and E.D. Whitney. 1993. Registration of germplasm lines developed from composite crosses of sugar beet x *Beta maritima*. Crop Sci. 33:882-882

Lewellen, R.T., and J.K. Schrandt. 2001. Inheritance of powdery mildew resistance in sugar beet derived from *Beta vulgaris* subsp. *maritima*. Plant Dis. 85:627-631.

Lexander, K. 1978. Seed composition in connection with germination and bolting of *Beta vulgaris* L. (sugar beet). Proc. Easter School Agric. Sci., Univ. Nottingham, England, UK pp. 271-291.

Lexander, K. 1980. Present knowledge of sugar beet bolting mechanisms. Proc. IIRB 43:245-258.

Lexander, K. 1985. *Beta vulgaris* L. *In:* A.H. Halevy (ed.) Handbook of Flowering, vol. 2. CRC Press, Boca Raton, FL, pp. 24-32.

Lexander, K. 1993. Present understanding of the physiology of sugar-beet seed germination. Proc. IIRB 56:387-394.

Linnaeus, C. 1753. Species Plantarum. Stockholm, Sweden (1st ed.).

Linnaeus, C. 1762. Species Plantarum. Stockholm, Sweden (2nd ed.).

Linnaeus, C. 1767. Species Plantarum. Stockholm, Sweden (3rd ed.).

Loptien, H. 1985. Breeding nematode-resistant beets. III. C-banding patterns on mitotic chromosomes in *Beta vulgaris* L. and wild species of the section *Patellares* Tr. Z. Pflanzenzucht. 94:41-49.

Lundquist, A., U. Østerbye, K. Larsen, and I. Linde-Laursen. 1973. Complex self-incompatibility systems in *Ranunculus acris* L. and *Beta vulgaris* L. Hereditas 74:161-168.

Luterbacher, M.C., J.M. Smith, M.J.C. Asher, and L. Frese. 2000. Disease resistance in collections of *Beta* species. J. Sugar Beet Res. 37:39-47.

Maletsky, S.I. and N.J. Weisman. 1978. A population genetic analysis of self- and cross-incompatibility in sugar beet (*Beta vulgaris* L.). Theor. Appl. Genet. 52:21-28.

Mansfeld, R. 1959. Vorlaufiges Verzeichnis landwirtschaftlich oder gartnerisch kultivierte Pflanzenarten. Kulturpflanze 2:38-45.

Margara, J. and H. Touvin. 1955. Sur le possibilité d'obtention de types de betteraves tolerants au virus de la jaunisse. C.R. Acad. Sci. France 41:650-655.

Marggraf, A.S. 1767. Chemische Versuche einen wahren Zucker aus verschiedenen Pflanzen die in unseren Ländern wachsen zu ziehen. Reprinted in: Ostwald's Klassiker der exacten Wissenschaft. 1907. Engelmann, Liepzig, Germany.

Martin, S.S. 1983. Sugar beet. *In:* I.D. Teare and M.M. Peet (ed.) Crop-Water Relations. Wiley, New York, NY, pp. 423-443.

Matthiolus, P.A. 1554. Commentarii in sex Libros Pedacii Dioscoridis. Venice. Italy.

Matthiolus, P.A. 1586. De Plantis Epitome utilissima. Frankfurt, Germany.

McFarlane, J.S. 1971. Variety development. p. 402-435. *In:* R.T. Johnson, J.T. Alexander, G.E. Bush, and G.R. Hawkes (eds.) Advances in Sugar Beet Production. Iowa State Univ. Press, Ames, IA.

McFarlane, J.S. 1975. Factors affecting sugar beet seed germination in North America. Proc. IIRB 7:1-9.

McGinnis, R.A. 1982. Sugar-beet Technology. Beet Sugar Development Foundation, Fort Collins, CO.

McGrath, J.M., C.A. Derrico, and Y. Yu. 1999. Genetic diversity in selected, historical US sugarbeet germplasm and *Beta vulgaris* ssp. *maritima*. Theor. Appl. Genet. 98:968-976.

Mesbah, M., T.S.M. De Bock, J.M. Sandbrink, R.M. Klein-Lankhorst, and W. Lange. 1997. Molecular and morphological characterization of monosomic additions in *Beta vulgaris*, carrying extra chromosomes of *B. procumbens* or *B. patellaris*. Molec. Breed. 3:147-157.

Milford, G.F.J. 1973. The growth and development of the storage root of sugar beet. Ann. Appl. Biol. 75:427-438.

Milford, G.F.J. and B.J. Houghton. 1999. An analysis of the variation in crown size in sugar-beet (*Beta vulgaris*) grown in England. Ann. Appl. Biol. 134:225-232.

Milford, G.F.J., T.O. Pocock, and J. Riley. 1985a. An analysis of leaf growth in sugar beet. II: Leaf appearance in field crops. Ann. Appl. Biol. 106:173-185.

Milford, G.F.J., T.O. Pocock, J. Riley, and A.J. Messem. 1985b. An analysis of leaf growth in sugar beet. III: Leaf expansion in field crops. Ann. Appl. Biol. 106:187-203.

Mita, G., M. Dani, P. Casciari, et. al. 1991. Assessment of the degree of genetic variation in beet based on RFLP analysis and the taxonomy of *Beta*. Euphytica 55:1-6.

Mitchell, E.D. and N.E. Tolbert. 1968. Isolation from sugar beet fruit and characterization of cis-4-cyclohexene-1,2-dicarboximide as a germination inhibitor. Biochemistry 7:1019-1025.

Mitsuishi, S. 1965. Equilibrium system of eu and aneuploid in the tetraploid lines of sugar beets. Jpn. J. Breed. 15:33-39.

Miyamoto, T. 1957. The germination inhibitor in sugar beet seed balls. Quart. Bull. Mich. St. Univ. 39:518-523.

Mochizuki, A. 1953. Genetic and cytological studies on the genus *Beta*. *In:* S. Matsumura (ed.). Improvement of Sugar Beets by Means of Triploidy. Science-Sha, Tokyo, Japan, pp. 14-39.

Munerati, O. 1908. La Coltivazione della Bietola Zuccherifera. Tipografia Sociale Editrice, Rovigo, Italy.

Munerati, O. 1932. Sull'incrocio della barbabietola coltivata con la beta selvaggia della costa adriatica. L'Industria Saccarifera Italiana 25:303-304.

Munerati, O., G. Mezzadroli, and T. V Zapparoli. 1913. Osservazioni sulla *Beta maritima* L. nel triennio 1910-1912. Staz. Sper. Agric. Ital. 46:415-445.

Nagao, S. and M. Takahashi. 1953. Studies in polyploid varieties of sugar beets. *In:* S. Matsumura (ed.) Improvement of Sugar Beets by Means of Triploidy. Science-Sha, Tokyo, pp. 40-75.

Nakamura, C. and T. Tsuchiya. 1982a. Cytogenetics of alien addition trisomics in sugar beets. I. Meiotic chromosome behavior in nematode-resistant trisomics. Biol. Zbl. 101:227-240.

Nakamura, C. and T. Tsuchiya. 1982b. Pachytene chromosome morphology in diploid sugar beet, *Beta vulgaris*. Z. Pflanzenzucht. 89:229-244.

Nakamura, C., G.N. Skaracis, and I. Romagosa. 1991. Cytogenetics and breeding in sugar beet. *In:* T. Tsuchiya and P.K. Gupta (eds.). Chromosome Engineering in Plants: Genetics, Breeding, Evolution. Part B. Elsevier, Amsterdam, Netherlands, pp. 295-314.

Olbrich H. 1989. Zucker Museum. Druckhaus Hentrich, Berlin, Germany.

Oltmann, W., M.Burba, and G. Boltz. 1984. Die Qualität der Zucherrübe: Bedeutung, Beurteilungskriterien und züchterische Massnahmenzh ihrer Verbesserung. Fortschritte der Pflanzenzüchtung, Paul Parey, Berlin, Germany, vol. 12.

Owen, F.W. 1942. Inheritance of cross- and self-sterility and self-fertility in *Beta vulgaris* L. J. Agric. Res. 64:679-698.

Owen, F.V. 1954. Hybrid sugar beets made by utilizing both cytoplasmic and mendelian male sterility. Proc. ASSBT 8:64.

Owen, F.V., E. Carsner, and M. Stout. 1940. Photothermal induction of flowering in sugar beets. J. Agric. Res. 61:101-124.

Panella, L 1998. Screening and utilizing *Beta* genetic resources with resistance to rhizoctonia root rot and cercospora leaf spot in a sugar beet breeding program. *In:* L. Frese, L. Panella, H.M. Srivastava, and W. Lange (eds.). Report of 4th *Beta* Genetic Resources Workshop and World *Beta* Network Conf. Izmir, Turkey, 28 Feb.-3 Mar. 1996. IPGRI, Rome, Italy.

Panella, L. and L. Frese. 2003. *Beta* germplasm evaluation data in two databases GRIN & IDBB. Proc. IIRB/ASSBT 1:233-241.

Panella, L., R.M. Hannan, and A. Hodgdon. 2003. *Beta* genetic resources: North American activities. *In:* L. Frese, L. Maggioni, C. Germeier, and E. Lipman (edS.) Report of 2nd Working Group on *Beta* and World *Beta* Network, Bologna, 23-26 October, 2002. IPGRI, Rome, Italy.

Perry, D.A. and J.G. Harrison. 1974. Studies on the sensitivity of monogerm sugar beet germination to water. Ann. Appl. Biol. 77:51-60.

Pignone, D. 1989. Wild *Beta* germplasm under threat in Italy. FAO/IPGRI Pl. Gene. Res. Newsl. 77:40.

Pliny (Gaius Plinius Secundus). 77. Naturalis Historia, Lib. XIX. Rome, Italy.

Rasmussen, J. 1932. Näjra undersokiningen over *Beta maritima*. Bot. Not. 13:33-36.

Rasmussen, J. 1933. Some observations on *Beta maritima*. Botaniska Notiser 14:316-324.

Reamon-Ramos, S.M. and G. Wricke. 1992. A full set of monosomic addition lines in *Beta vulgaris* from *Beta webbiana*: morphology and isozyme markers. Theor. Appl. Genet. 84:411-418.

Robertson-Scott, J.W. 1911. Sugar Beet: Some Facts and Some Illusions. Horace Cox, London, U.K.

Roemer, T. 1927. Handbuch des Zuckerrübenbaues. Velagbuchhandlung Paul Parey, Berlin, Germany.

Romagosa, I. 1980. Meiosis in triploid sugarbeet. MS thesis, Colorado State Univ. Fort Collins, CO.

Romagosa, I. 1983. Primary trisomics in sugar beet. PhD diss., Colorado State Univ., Fort Collins, CO.

Romagosa, I., L. Cistue, T. Tsuchiya, J.M. Lasa, and R.J. Hecker. 1987. Primary trisomics in sugarbeet. II. Cytological identification. Crop Sci. 27:435-439.

Romagosa, I., R.J. Hecker, T. Tsuchiya, and J.M. Lasa. 1986. Primary trisomics in sugarbeet. I. Isolation and morphological characterization. Crop Sci. 26:243-249.

Rommel, M. 1966a. Cytogenetics of autotetraploid sugar beet *(Beta vulgaris* L.). Part II: The type of numerical chromosome reproduction. Züchter 36:121-126.

Rommel, M. 1966b. A self-fertile polyhaploid plant in an autotetraploid variety of sugar beet. Nature 111:648-649.

Ruellius, J. 1536. De Natura Stirpium. Paris, France.

Santoni, S. and A. Berville. 1992. Characterization of the nuclear ribosomal DNA units and phylogeny of *Beta* L. wild forms and cultivated beets. Theor. Appl. Genet. 83:533-542.

Santos, D.S.B. and M.F.A. Pereira. 1989. Restrictions of the tegument to the germination of *Beta vulgaris* L. seeds. Seed Sci. Tech. 17:601-611.

Savitsky, V.F. 1952. Methods and results of breeding work with monogerm beets. Proc. ASSBT 7:344-350.

Scaliger, J.C. 1566. In Libros de Plantis Aristoteli Inscriptos Commentarii. Geneva, Switzerland.

Schacht, H. 1859. Über einige Feinde der Rübenfelder. Zeitschrift des Vereins für die Zuckerindustrie 9:175-179.

Schlösser, L.A. 1957. Cercoploy-ein Fortschritt in der Cercospora-Resitenzzüchtung. Zucker 10:489-492.

Schmidt, T. and J.S. Heslop-Harrison. 1996. High-resolution mapping of repetitive DNA by in situ hybridization: Molecular and chromosomal features of prominent dispersed and discretely localized DNA families from the wild beet species *Beta procumbens*. Plant Molec. Biol. 30:1099-1113.

Schmidt, T., C. Jung, J.S. Heslop-Harrison, and M. Kleine. 1997. Detection of alien chromatin conferring resistance to the beet cyst nematode (*Heterodera schachtii* Schm.) in cultivated beet (*Beta vulgaris* L.) using *in situ* hybridization. Chromosome Res. 5:186-193.

Schmidt, T., C. Desel, D. Cai, and C. Jung. 1998. FISHing nematode-resistant sugar beets. VI Conference "Plant and Animal Genome". W156. (www.intl-pag.org/6/abstracts/261.html).

Schneider, C.L. and J.O. Gaskill. 1962. Tests of foreign introductions of *Beta vulgaris* L. for resistance to *Aphanomyces cochliodes* Drechs. and *Rhizoctonia solani* Kühn. J. ASSBT 11:656-660

Schondelmaier, J. and C. Jung. 1997. Chromosomal assignment of the nine linkage groups of sugar beet (*Beta vulgaris* L.) using primary trisomics. Theor. Appl. Genet. 95:590-596.

Schondelmaier, J., G. Steinrücken, and C. Jung. 1996. Integration of AFLP markers into a linkage map of sugar beet (*Beta vulgaris* L.). Plant Breed. 115:231-237.

Shen, Y., H.J. Newbury, and B.V. Ford-Lloyd. 1996. The taxonomic characterisation of annual *Beta* germplasm in a genetic resources collection using RAPD markers. Euphytica 91:205-212.

Shen, Y.L., H.J. Newbury, and B.V. Ford-Lloyd. 1998. Identification of taxa in the genus *Beta* using ITS1 sequence information. Pl. Molec. Biol. Reporter 16:147-155.

Shun, Z.F., S.Y. Chu, and L. Frese. 2000. Study on the relationship between Chinese and East Mediterranean *Beta vulgaris* L. subsp. *vulgaris* (leaf beet group) accessions. p. 65-69. Ist Intl. *Beta* Genetic Resources Workshop. Broom's Barn, UK, 9-10 September 1999. IPGRI, Rome, Itely.

Simmonds, N.W. 1976. Evolution of Crop Plants. Longman, London, UK.

Skaracis, G.N. 1980. Meiosis in diploid sugarbeet. MS thesis, Colorado State Univ., Fort Collins, CO.

Skaracis, G.N. and E. Biancardi. 2000. Breeding for cercospora resistance in sugar beet. *Cercospora beticola* Sacc. Biology, agronomic influence and control measures in sugar beet. *In:* Advances in Sugar Beet Res. vol. 2. IIRB, Brussels, Belgium, pp. 177-195.

Sliwinska, E., H.C. Jing, C. Job, 1999. Effect of harvest time and soaking treatment on cell cycle activity in sugar beet seeds. Seed Sci. Res. 9:91-99.

Smith, G.A. 1980. Sugar beet. *In:* W.R. Fehr and H.H. Hadley (eds.). Hybridization of Crop Plants. ASA, CSSA, and SSSA, Madison, WI, pp. 60-616.

Smith, G.A. 1987. Sugar beet. p. 577-625. *In:* W.R. Fehr (ed.) Principles of Cultivar Development, vol. 2: Crop Species. Macmillan Publ., New York, NY, pp. 577-625.

Sobek, R., A Brandes and G. Wricke. 1994. Meiotic chromosome pairing of species and hybrids in section *Procumbentes* genus *Beta. In:* L. Frese and D.L. Doney (eds.) Report 3rd Int. Beta Genetic Resourses Workshop and World Beta Network Conference, IPGRI, Rome, p.16.

Srivastava, S. and H.M. Srivastava. 1999. Meiotic chromosome mechanism of diploid *Beta* species. Sugar Tech. 1:146-147.

Srivastava, S. and H.M. Srivastava. 2000. Cytological and karyotypic studies in four *Beta* species. J. Sugar Beet Res. 37:135-142.

Stehlik, V. 1949. Essais de croisment de betterave sucrière avec *Beta* trigyna. IIRB Publ. Tech. 15:103-108.

Stout, M. and B. Tolman. 1941. Factors affecting the germination of sugar beet and other seeds, with special reference to the toxic effects of ammonia. J. Agric. Res. 63:687-713.

Surface, G.T. 1918. The Story of Sugar. Appleton, New York, NY.

Szota, Z. 1964. The investigation into hybrids between sugar beet and wild beets, *B. procumbens* and *B. webbiana*. Biul IHAR 1-2:17-20.

Szota, M. 1995. Meiosis in the species of the *Procumbentes* and *Corollinae* sections and in interspecific hybrids within these sections and with sugar beet (*Beta vulgaris* L.). Hodowla Roslin Aklimatyzacja i Nasiennictwo pp. 3-59.

Szota, Z. and B. Jassem. 1967. Artificially obtained tetraploids of wild *Beta webbiana* and *Beta procumbens* species, and their use in resistance breeding. Biul IHAR 3/4.

Szota, M. and Goska, M. 1990. Meiosis in haploid sugar beets (*Beta vulgaris* L.) obtained in *in vitro* cultures. Genet. Polon. 31:175-178.

Tan, A. 1993. A numerical taxonomic study of wild beets (*Beta vulgaris* L. *s.l.*) in the Mediterranean. Anadolu 3:1-12.

Terry, N. 1968. Developmental physiology of sugar beet, I. The influence of light and temperature on growth. J. Exper. Bot. 19:795-811.

Theophrastus. ca. 300 B.C. Enquiry into Plants. Transl. by Sir Arthur Hort, 1916, London, UK.

Tjebbes, K. 1933. The wild beets of the North Sea region. Botaniska Notiser 14: 305-315.

Tolman, B. and M. Stout. 1940. Toxic effects on germinating sugar beet of water soluble substances in the seed ball. J. Agric. Res. 61:817-830.

Transhel, V.A. 1927. Obzor vida roda *Beta*. Trudy prikl. Bot. Genet. Selek. 17:203-223.

Trehane, P., C.D. Brickell, B.R. Baum, 1995. The International Code of Nomenclature for Cultivated Plants. Regnum Vegetabile No. 133. Quarterjack Publ. Wimborne, UK.

Tsuchiya, T. 1983. Aneuploidy and chromosome mapping in barley. p. 251-281. *In:* M.S. Swaminathan, P.K. Gupta, and U. Sinha (eds.). Cytogenetics of Crop Plants. MacMillan India, New Delhi.

Ulbrich, E. 1934. *Chenopodiaceae. In:* A. Engler and H. Harms (eds.). Die natürlichen Pflanzenfamilien, vol. 16c. Verlag von Wilhelm Engelmann, Leipzig, Germany, pp. 379-584.

Van Geyt, J.P.C., M. Oleo, W. Lange, and T.S.M. De Bock. 1988. Monosomic additions in beet (*Beta vulgaris*) carrying extra chromosome of *Beta procumbens,* I. Identification of the alien chromosomes with the help of isozyme markers. Theor. Appl. Genet. 76:577-586.

Van Geyt, J.P.C., W. Lange, M. Oleo, and T.S.M. De Bock. 1990. Natural variation within the genus *Beta* and its possible use for breeding sugar beet: A review. Euphytica 49:57-76.

Von Lippmann, E.O. 1925. Geschichte der Rübe (*Beta*) als Kulturpflanze. Verlag Julius Springer, Berlin, Germany (2nd ed.).

Von Lippmann, E.O. 1929. Geschichte des Zuckers. Verlag Julius Springer, Berlin (2nd ed).

Von Proskowitz, E. 1893. Über die Kulturversuche mit *Beta maritima*. Oester Ung. Zeitschr. f. Zuckerind u. Landw. 88:493-528.

Wagner, H., E.M. Gimbel, and G. Wricker. 1989. Are *Beta procumbens* Chr. Sm. and *Beta webbiana* Moq. different species? Pl. Breed. 102:17-21.

Ware, L.S. 1880. The Sugar Beet, Including a History of the Sugar Beet Industry in Europe. Henry Carey Baird, Philadelphia, PA.

Watson, L. and M.J. Dallwitz. 2000. The families of flowering plants: descriptions, illustrations, identification, and information retrieval [on-line]. Available at http://biodiversity.uno.edu/delta/ (posted 14 Dec. 2000; verified 24 Oct. 2002).

Whitney, E.D. 1989a. Identification, distribution and testing for resistance to rhizomania in *Beta maritima*. Plant Dis. 73:287-290.

Whitney, E.D. 1989b. *Beta maritima* as a source of powdery mildew resistance in sugar beet. Plant Dis. 73:487-489.

Winner, C. 1993. History of the crop. *In:* D.A. Cooke, and R.K. Scott (eds.). The Sugar Beet Crop. Chapmann & Hall, London, pp. 1-35.

Wyse, R.E. 1982. The sugar beet and chemistry: The sugar beet and sucrose formation. *In:* R.A. McGinnis (ed.). Beet Sugar Technology. Beet Sugar Development Foundation, Fort Collins, CO, pp. 17-24 (3rd ed.).

Yu, M.H. 1977. Preliminary study of pachytene morphology in a homozygous line of sugarbeet. Crop Sci. 17:833-836.

Yu, M.H. 1980. Meiotic chromosome behavior in monoploid sugarbeet. Can. J.Genet. Cytol. 22:375-380.

Yu, M.H. 2002. Registration of M1-2 beet germplasm resistant to root-knot nematode. Crop Sci. 42:317-318.

Yu, M.H., W. Heijbroek, and L.M. Pakish. 1999. The sea beet source of resistance to multiple species of root-knot nematode. Euphytica 108:151-155.

Zossimovich, V.P. 1939. New hybrids between wild and sugar beets that are resistant to *Cercospora.* Selektsiya Semenovodstvo USSR 1:1-16.

Zossimovich, V.P. 1940. Wild species and origin of cultivated beets. Sveklovodstro 1:17-44.

2

Objectives of Sugar Beet Breeding

2.1 Introduction (E. Biancardi)

The principle objective of all sugar beet breeding programs is development of varieties with the maximum yield potential at the lowest economic and environmental costs possible (Knapp, 1958). From this general objective, several secondary aims that vary according to the environment arise. It is necessary to produce varieties adapted to the unique local agronomic conditions and pathological threats which occur in each production area (Barocka, 1985).

Heritable traits controlled by one or multiple pairs of alleles are designated qualitative (Mendelian), or quantitative (additive) respectively. More than 40 recognized qualitative traits are listed in **Table 2.1** (Smith, 1987); among them are annuality, monogermity, Mendelian male sterility, self-fertility and some forms of resistance to rhizomania (Lasa and Romacrosa, 1992). Sugar yield, processing quality, germination potential, bolting resistance, and most disease resistances are qualitative traits. Improving quantitative traits requires complex breeding schemes, compared to those used for qualitative traits.

Breeders must also be concerned with traits related to seed multiplication and production of commercial hybrids (**Fig. 2.1**). It is important that the parental components not only produce high seed yields, but also uniform seed with high germination potential and seedling emergence. Seed size must be commercially acceptable, with a low frequency of empty seeds and false monogerms. To optimize seed yield and quality, CMS maternal lines and the pollinator lines they are to be crossed which should flower simultaneously. For high yields, the branches of seed-bearing plants should produce numerous flowers close to each other.

Table 2.1 Monogenic traits and gene symbols in sugar beet (After Smith, 1980, modified).

Character	*Genetic symbols*	*Source*
Yellow pigment	Y, Y^r, y	Kajanus (1917) Keller (1936)
Hypocotyl color	R, R^t, R^p, R^h, r	Kajanus (1917) Keller (1936)
Colored leaf	*CL, cl*	Owen and Ryser (1942)
Trout or spotted leaf	*Tr, tr*	Owen and Ryser (1942)
Colored vein	*Cv, cv*	Deming (cited by Abegg, 1940) Owen and Ryser (1942)
Annual habit	*B, b*	Munerati (1931) Abegg (1936) Owen et al., (1940)
Variegated foliage	v_1, V_1	Owen and Ryser (1942)
Partially dominant curly top resistance	*C, c*	Abegg and Owen (1936)
Crinkled foliage	*cr, Cr*	Steward (cited by Abegg, 1940)
Nonproduction of color	*p, P*	Linde-Laursen (1972)
Monogerm seed	m, M, M^{Br}, M^1, M^2	Savitsky (1950, 1952)
Late bolting	*lb*	Savitsky (1952)
Partial curly top resistance	C_2	Savitsky and Murphy (1954)
Genetic male sterility	a_1, A_1	Owen (1952)
Lutescens green cotyledons	lu_2, Lu_2	Theurer (1968a)
Russet root	*ru, Ru*	Owen and Ryser (1942)
Chlorina cotyledons	ch_2, Ch_2	Savitsky (1957)
Chlorina cotyledons	ch_1, Ch_1	Savitsky (1940)
Delay in chlorophyll production	vi_4, Vi_4	Stander and Theurer (1970)
Restores male fertility in sterile cytoplasm	*X, x*	Owen (1945)
Produces partial fertility in sterile cytoplasm	*Z, z*	Owen (1942, 1945)
Pollen restorer	Rf_1	Theurer and Ryser (1969)
Annual pollen restorer	Rf_2	Theurer (1971)
Weak pollen restorer	Rf_3	Roundy and Theurer (1974)
Enhances pollen production	*Sh, sh*	Hogaboam (1957)
Yellow leaf	*yl, Yl*	Roundy and Theurer (1974)
Black root	*bl, Bl*	Munerati and Costa (1930)
Dwarf plant	*d, D*	Theurer (1968b)
Feather shape, no petiole	*fl. Fl*	Theurer (Cited by Smith, 1980)
Plantain leaf	*pl, Pl*	Abegg (1936)
Plantain leaf, semiparallel venation	Pl_2	Theurer (Cited by Smith 1980)
Self-fertility	S^S, S^F	Owen (1942)
Self-sterility	S^a, S^b	Owen (1942)
Self-incompatibility	S_a, S_b, S_c, S_d	Larsen (1977)
Albino lethal	*w, W*	Steward (cited by Abegg, 1940)
Variegated cotyledons	v_2, V_2	Abegg (1940)

(Table 2.1 Contd.)

(Table 2.1 Contd.)

Variegated foliage and root flesh	v_3, V_3	Abegg (1940)
Flaccid leaf	*f, F*	Steward (cited by Abegg, 1940)
Nana plants	*n, N*	Steward (cited by Abegg, 1940)
Miniature, lethal, conflicts with monogerm	*m, M*	Abegg (1940)
New leaves reduced in size	*re, Re*	Savitsky (cited by Abegg, 1940)
Albino seedlings	w_1, W_2	Savitsky (1957)
Golden-yellow cotyledons	*Au, au*	Savitsky (1957)
Lutescens leaves	*lu, Lu*	Savitsky (1957)
Virescens, golden-yellow seedlings	vi_1, vi_2, vi_3	Savitsky (1957)
Rhizomania resistance (Rizor and Holly types)	Rz_1, rz_1	Lewellen (1988)
Rhizomania resistance (WB42 type)	Rz_2, rz_2	Scholten et al., (1999)

The germination ability of a seed lot influences field emergence, uniformity of plant populations, and has a significant impact on sugar yield (Desprez, 1993). Uniformity of spacing between seedlings is directly related to field emergence (Kästner, 1990). Germination potential is also associated with embryo and operculum weight and is limited, especially in triploids, by anomalies in early endosperm and embryo development (Jassem, 1976).

Seedling establishment depends on the seeds ability to germinate quickly and the size of embryo. It is also influenced by processing/pelleting methods and chemical protectants that reduce seedling diseases (Lexander, 1993). As a consequence of these improvements, field emergence is continuously increasing and in several countries has surpassed 80% (Prince and Durrant, 1990; Desprez, 1993).

The seedlings need to develop quickly so that leaves cover the space between rows as soon as possible. From this moment, accumulation of sugar begins in the roots. Early development of the leaves, especially important in varieties with a prostrate growth habit, helps reduce development and competition of weeds (Paolini et al., 1999). According to these authors, varieties with rapidly developing canopies could reduce the need for herbicides. On the other hand, excessive canopy development can result in shading and reduce the efficiency of sucrose accumulation (Röver, 1994).

It is also imperative that plants not initiate flowering during commercial production. In fact, bolting, among other things, lowers sugar yields and reduces processing quality. Planting of bolting resistant varieties is important with early spring sowing and especially so in areas where autumn

POLLINATOR | **CMS SEED PARENT**

SEED PRODUCTION

POLLINATOR:

+ Combining ability
I
+ Pollen fertility
I
+ Pollen production
I
+ Concurrent flowering with CMS

CMS SEED PARENT:

+ Combining ability
I
+ % Monogermity
I
+ % Male sterility
I
+ Concurrent flowering with pollinator
I
- Stalk fasciation

→ **SEED**

I
I
+ Seed yield
I
+ % Seed of commercial size
I
- % Empty or false monogerm seed
I
- % Twin embryos

SOWING

I
I
+ % Germination
I
+ % Field emergence

↓ **PLANTS**

+ Seeding vigor
I
+ Rapid canopy development
I
- Bolting
I
- Fitness to environment
I
+ Biotic and abiotic stress resistances
I

(Fig. 2.1 Contd.)

(Fig. 2.1 Contd.)

+ Root yield
I
+ Sucrose yield
I
+ % Extractable sugar
↓
ROOTS
I
I
+ Harvest Losses
HARVEST I
- Soil and crown tare
I
I
+ Storage rot resistances
I
STORAGE + Sucrose concentration
I
- Respiration rate
I
I
+ Sucrose concentration
I
- % Woody root
PROCESSING I
- Soil and crown tare
I
- Impurity concentration

Fig. 2.1 Breeding targets (+ = more; – = less) in the different stages of hybrid sugar beet production (After Biancardi, 1999, modified).

sowing is practiced. In recent years noticeable improvement in bolting resistance has taken place even though its expression is complex due to genotype × environment interactions.

In many areas, growers need varieties with resistance to both rhizomania and cercospora leaf spot (l.s.), and in the USA additional genetic resistances to curly top virus and other diseases are required. These traits not only eliminate or reduce yield losses, but also reduce the use of pesticides and their impact on the environment, as in the case with cercospora l.s. Seed companies have developed lines and varieties resistant to broad-spectrum herbicides, which could reduce the cost, toxicity, and environmental damage associated with herbicides (Brants et al., 1995).

As stated above, the primary goal of sugar beet breeders is to improve sugar yield. This may be achieved by increasing photosynthesis efficiency, selecting for traits that favor sucrose accumulation in the roots, finding resistance to diseases, and so on.

Moreover, it is necessary to improve traits related to: i) mechanical harvesting; ii) reduction in postharvest storage losses; and iii) efficiency of factory operations. Breeders and growers desire roots with small crowns and rounded form to facilitate mechanical harvesting. The ideal root should have a smooth surface and superficial or absent root grooves to reduce soil tares. It should be possible to modify the physical and chemical characteristics of roots to improve extraction efficiency. Improving processing quality through selection is difficult because of complex interactions with cultural practices, the environment, and extraction procedures. Nevertheless, in all countries processing quality has increased slowly but continuously (Delasalle and Dhellemmes, 1996; Noe and Mantovani, 1996).

2.2 Polyploidy (G.N. Skaracis and E. Biancardi)

Efforts to modify the number of chromosomes in sugar beet cells date back to Dahlberg in 1933 (Coons, 1936). The difficulties encountered were overcome when the mutagenic properties of alkaloid colchicine were discovered (Schwanitz, 1938; Rasmusson and Levan, 1939). The first tetraploid sugar beet families, having twice ($2n = 4x = 36$) the normal ($2n = 2x = 18$) chromosomes, were obtained in this manner. These families were easy to reproduce and differed in several characteristics from the initial diploid material. Plants from the two ploidy levels could be crossed, producing triploid ($2n = 3x = 27$) hybrids (Bosemark, 1993).

Early results from field evaluations of tetraploid lines were encouraging (Peto and Boyes, 1940). Tetraploid families were characterized as having better root shape and fewer but larger leaves with shorter and stronger petioles (Lasa and Romagosa, 1992). Flowers, seed balls and pollen grain diameter were also larger. Seed germination and root development were, on the average, slower, providing a probable explanation for inconsistent yield results, compared to their diploid counterparts. Bolting resistance however, was almost always improved at the tetraploid level.

Triploid sugar beet hybrids, manifesting morphological characteristics intermediate to the parental ploidy levels, were of more interest commercially because they were almost completely sterile due to reproductive abnormalities caused by the odd number of chromosomes. As far as advantages for production characteristics are concerned, triploid hybrids display a stronger disease resistance when resistance is quantitatively controlled (Nakamura et al., 1992), as has been demonstrated with *Rhizoctonia* (Hecker and Ruppel, 1976) and *Cercospora* resistance (Skaracis and Smith, 1987).

Meiosis in tetraploids is characterized by abnormal pairing and unequal chromosome distribution, which leads to the formation of gametes with a lower or higher than normal ($2n = 2x = 18$) chromosome number. As a consequence, 30-35% and 10-15% aneuploid plants are normally present in tetraploid populations and triploid hybrids respectively (Bosemark, 1993). Embryos produced by aneuploid gametes germinate with considerable difficulty, thus lowering the quality of commercial seed. In addition, aneuploid plants are not as competitive as euploid sugar beet plants. When interspaced among euploid plants of the same ploidy level, aneutriploids and aneutetraploids yielded 50% and 30% less than the euploid plants respectively. According to Bosemark (1967) however, because competition effects are contained in such measurements, yield reduction is greatly overestimated and the real differences are probably around half of that estimated from mixed plantings. Furthermore due to overcompensation by euploid plants, total elimination of aneuploid plants in a triploid hybrid would improve sugar yield by less than 2-3%. Despite several attempts to capitalize on such relatively small, but economically important improvement, elimination of aneuploids has not been possible.

Abnormalities due to autotetraploidy necessitate rigorous and frequent chromosome observation to eliminate aneuploid individuals during the basic seed production of tetraploid pollinators. Chromosome counts, almost totally replaced by flow cytometric analyses, are also indispensable for the elimination of diploid plants within tetraploid families since diploids, due to the higher vitality of haploid pollen, would cause a rapid return to the diploid level (Lasa and Romagosa, 1992).

Triploid hybrids display a productive superiority, sometimes yielding 10% more than the parental average yield. This advantage was used initially in the production of anisoploid, often called polyploid varieties. In multiplication fields, seed of these varieties were obtained by unrestricted intercrossing between diploid and tetraploid plants transplanted in a 1:3 ratio. The higher proportion of tetraploid plants compensated for the considerably lower competitiveness of tetraploid pollen. The resultant seed had a proportion of triploid hybrid plants as high as 50%, thus ensuring a superior sugar yield potential. The remaining seed was composed of various proportions of diploid and tetraploid seeds. Anisoploid varieties were used exclusively in the years following the Second World War, up until cytoplasmic male sterility was employed for hybrid variety production.

2.3 Monogerm Seed (E. Biancardi and G.N. Skaracis)

Prior to 1940, all sugar beet seed planted was multigerm. Each seed cluster contained 3-4 embryos, necessitating manual thinning to achieve desired plant populations. Hand-thinning required about 100 man-hours per hectare. Consequently, particularly in areas with expensive hand labor, it became necessary to reduce or even eliminate such operations (Knapp, 1958).

Efforts initiated in the USA in 1903 to select plants producing monogerm seed were unsuccessful (Brewbaker et al., 1946). Mechanical separation of multigerm fruits into single-germ pieces was achieved in 1940 and the resultant technically monogerm seed sown at predetermined spacings (Knapp, 1958). Distribution of emergent plants was adequate and hand-thinning was not necessary. The usefulness of this technique was obvious but seed produced in this manner, called "precision seed", had poor germination and a high proportion of remnant bigerm seeds. In addition, seed losses during processing were high. Despite these disadvantages, technically monogerm seed enjoyed some success, even in Europe.

A significant milestone was achieved in the USA when Victor Savitsky was provided the opportunity to continue studies on genetic monogermity initiated in Russia. In 1948, in a field of "Michigan Hybrid 18", he identified and selected five plants with predominantly monogerm flowers and fruits. The character was expressed best in plants originating from two very similar self-fertile lines with obvious signs of inbreeding depression, SLC101 and SLC107 (Savitsky V.F., 1950).

The monogerm plants could be distinguished by the presence of either a lateral branch or a single flower, but never both as normally seen in multigerm plants that originated in the leaf axil. It was later established that the monogerm character depends on a pair of alleles designated *Mm* and is expressed in the homozygous recessive state. However, the alternate multigerm allele is not completely dominant. In fact, F_1 hybrids between multigerm and monogerm plants produce fewer embryos per seed than the multigerm parent. As early as 1945, Savitsky realized the inadequacy of the theoretical model and hypothesized the presence of modifying genes (Savitsky, 1954; Knapp, 1967), a conclusion also accepted by Tatlioglu (1974). Contrarily, Maletsky et al., (1991) proposed that expression of the trait also depends on a second dominant gene not belonging to the *M*-allelic series.

SLC101 has been available to American and European breeders since 1951 and eventually the monogerm character was transferred to most cultivated varieties. Initial problems with poor germination, productivity, and absence of disease resistance were overcome or diminished by appropriate backcrossing programs. Commercial dissemination of monogerm varieties started in the early 1960s. At present, monogerm seed is the only type of seed planted in countries with an advanced agriculture.

Commencing in 1934, Savitsky and his assistant M.G. Bordonos studied a second source of monogerm seed, labeled "Russian". Interesting lines with a high proportion of monogerm plants matured up to two months later than usual and the seed was never completely monogerm (Knapp, 1967). This type of monogermity, repeatedly studied in an effort to find an alternative to the American source, proved difficult to handle due to its polygenic inheritance (Brewbaker et al., 1946); hence, its practical applications have been extremely limited.

In Russia, Savitsky and Bordonos also identified a monogenic recessive type of monogermity (Bordonos, 1939), similar to the one discovered a few years later in the USA. It has not been determined whether these two monogerm sources are related (McFarlane, 1993).

2.4 Male Sterility (G.N. Skaracis and E. Biancardi)

Synthesis of commercial hybrids became a reality after the discovery of useful types of male sterility that are infrequent in *Beta*. Male sterile plants produce atrophic, empty whitish anthers that dry rapidly without pollen emission. Intermediate types of sterility exist as well; in some the anthers turn yellow, similar to fertile anthers, but do not dehisce and pollen either is not dispersed or, if so, is not viable. These intermediate types of sterility are so influenced by environmental conditions that predictions regarding anther fertility are impossible (Owen, 1945; Bosemark, 1993). Two types of male sterility are in use in sugar beet: one results from the combined action of genetic and cytoplasmic factors while the other depends only on genetic (or nuclear) factors. The first type of male sterility provides complete control of pollination and thus has played a pivotal role in the production of commercial hybrid varieties. The second type is used by breeders to facilitate cross-pollination in some breeding populations.

The mechanism of genetic-cytoplasmic male sterility (CMS) was discovered by Owen (1945). As reported in his fundamental publication, male sterility was maternally transmitted, independent of characteristics of the male parent. In fact, in contrast to genetic male sterility, seed harvested on the two parents gave varying progenies. It was not possible, even with existing models of cytoplasmic male sterility, to precisely explain the transmission of the character. After several years of experimentation, an S (sterile) type cytoplasm was found, as opposed to the normal or N type that results in male sterility only when two genetic factors, designated *X* and *Z*, are in a homozygous recessive state. In other words, CMS plants have an (S)*xxzz* genotype whereas all other combinations, even with one dominant *X* or *Z* allele, produce fertile plants or plants with partial fertility. The N type cytoplasm always produces fertile progenies. For this reason, reproduction of cytoplasmic male sterile lines requires employment of a maintainer that has the N type cytoplasm and the nuclear factors *X* and *Z* in the homozygous recessive state. In sugar beet, these maintainers are called O-types in memory of Owen, though he modestly denoted them as Zero-types. As described later, these lines have a fundamental role in the production of genetically monogerm sugar beet hybrid varieties.

Owen's hypothesis is generally valid and widely accepted. However, some exceptions are well known to experienced sugar beet breeders. In fact, some CMS lines produce a low percentage (2-3%) of fertile individuals that are extremely difficult to eliminate, even when single plants of the corresponding O-types are repeatedly selfed and crossed to CMS testers, in order to eliminate the individuals producing fertile test-cross progenies. According to Owen's model, selfing should result in a rapid reduction of the dominant alleles. The recognition that fertile plants are also monogerm excludes the possibility of pollen contamination.

Genetic nuclear male sterility (NMS), on the other hand, depends on a pair of alleles *Aa*, and is expressed in the homozygous recessive condition (Owen, 1952). In contrast to cytoplasmic male sterility, it does not provide for the production of a high percentage of hybrid seed and consequently its employment is limited to certain selection schemes.

2.5 Annual and Biennial Growth Habit, Bolting Resistance (G. Steinrücken)

All cultivated beets are biennial, requiring a vernalization period to enter the reproductive phase. Environmental pressure forced the ssp. *vulgaris* of *Beta vulgaris to* evolve drought and salt resistance as well as a reproductive cycle that would ensure survival in a harsh, quickly changing environment (Van Dijk et al., 1997; Debenham, 1999). However, wild beet populations, *Beta vulgaris* ssp. *maritima*, depending on latitude, exhibit a great variation in life cycle behavior ranging from perennial/biennial populations in the northern area, i.e. the UK, the Netherlands, and Belgium, to annual populations in the Mediterranean region (Letschert, 1993). True annual populations are difficult to find, however.

2.5.1 Annual Growth Habit

Annuality is considered by some researchers to be of perennial origin and the biennial trait to be an intermediate form (Battey 2000). A dominant gene, the *B* gene (Munerati, 1931; Abegg, 1936; Owen, 1954), causes annuality. Bolting (stem elongation) and flowering in plants homozygous for the *B* gene is under long day control while heterozygous plants from annual × biennial crosses exhibit a more complex response to environmental factors (Lichter and Vieweg, 1969; Shimamoto et al., 1990; Boudry et al., 1994; Abe et al., 1997). The product of the *B* gene is still not known but recent high-resolution mapping of annual × biennial crosses has produced very tightly linked markers to the *B* gene (El-Mezawy *et al.*, 2002).

2.5.2 Biennial Growth Habit

Some northern wild beet populations as well as commercially grown sugar beet are biennial plants. They require a combination of environmental stimuli to bolt and flower, i.e. long-day conditions preceded by a period of low temperature (vernalization). Short day conditions or high temperatures after vernalization will revert the plant back to the vegetative state (devernalization). Bolting may also occur without being followed by flowering (Chroboczek, 1934; Curth, 1955; Heide, 1973; Debenham, 1999).

Flowering is indeterminate, continuously giving rise to new flowers, and starts in spring to early summer, dependent on latitude and vernalization requirement, and continues through the rest of the growing period. Thus mature fruits shedding on the soil germinate over a long period of time. A large part of the seed will germinate during autumn or the following spring but germination can be further delayed through the occurrence of long-lived seed banks in which the seed may remain viable for more than five years.

The time between shedding and germination is further enhanced by difference in size, mechanical strength, and level of germination inhibitors of the fruit wall of the individual seed (Battle and Whittington, 1969; Lexander, pers. comm.). Germination shows a strong relationship with thermal time (base temperature 3°C) if water and oxygen availability is not limiting (Gummerson, 1986). After the seedling has emerged and greened the plant enters a phase of leaf initiation. In sugar beet very little of the assimilates are directed to the root until the plant reaches the 8-10 leaf stage, but from this stage on the root will start to grow, increasing the root/shoot ratio until light and temperature become limiting for growth (Scott et al., 1974).

2.5.3 Bolting and Flowering Regulation

The switch from vegetative growth to reproductive development is the major developmental transition in plants. To maximize reproductive success timing is essential and regulated by favorable environmental conditions. Thus, bolting and flowering is a genetically regulated multifactorial process (Bernier et al., 1993) and many physiological factors, e.g. temperature, light, sugars, calcium, polyamines, sterols (membrane fluidity), and hormones are believed to be involved in the transition process. Regulation and physiology of biennial sugar beet flowering have been comprehensively described by several authors (Curth, 1960; Margara, 1960; Lexander, 1975, 1985, 1987) and, more recently discussed in relation to the emerging genomic data (van Roggen et al., 1998).

The complexity revealed by physiological studies favors a multifactorial hypothesis as do the mounting data evolving from characterization of plant genomes using molecular techniques. During the last decades *Arabidopsis thaliana* mouse-ear cress has been used as a model plant in many studies related to the flowering process. Several pathways related to flowering control have been defined in *Arabidopsis,* and there is little reason to think that regulation of flowering is less complex in beet.

The step between *Arabidopsis* and sugar beet may not be as long as it first seems. Both are rosette plants in which flowering, or time to flowering, is affected by photothermal conditions. Some *Arabidopsis* ecotypes are winter annuals, i.e. flowering is delayed if long-day and vernalization conditions are not supplied. They will, unlike biennial sugar beet eventually flower, even without environmental stimuli.

2.5.4 Major Flowering Pathways

The pool of data regarding molecular control of floral transition is rapidly increasing, revealing a complex genetic interaction. Significant consensus has been reached regarding the main flowering pathways in *Arabidopsis* (Mouradov et al., 2002; Simpson and Dean, 2002 and references therein). Pathways for autonomous or constitutive promotion, photoperiodic promotion, and vernalization promotion, as well as currently known repressors have all been described.

In sugar beet the vernalization response was identified early as one of the most important factors for flower induction (Chroboczek, 1934 and references therein). Curth (1963) reported an optimum at +8°C in light and 2 to 4°C in darkness. Beets can be partly vernalized already as seeds or even as unripe seeds (Lexander, 1969 1975) and apparently no strictly juvenile stage exists. The required duration of cold treatment is genotype dependent (Heide, 1973) and selection for bolting-resistant material has increased the demand for longer vernalization periods. Currently a 14- to 18- week cold requirement is not unusual.

In other species it is known that low temperatures increase the content of unsaturated fatty acids of the plasma membrane (Bulder et al., 1989) and make it more fluid (Nozaws et al., 1974). Also, the fatty acid composition present before vernalization affects the bolting frequency (Sadeghian and Lexander, 1993).

Temperature changes cause a range of altered enzyme activities, not necessarily related to flower induction but to adaptation to the new environmental conditions. Increased enzymatic activity of invertase, pyruvate kinase, isocitrate-dehydrogenase, malate-dehydrogenase, glutamate-pyruvate-transaminase, glutamate-oxaloacetate-transaminase was monitored after four weeks of cold treatment (Schmid, 1974 a, b). Glucose-6-phosphate-dehydrogenase, aldolase, and peroxidase displayed only minor changes in activity. Reproductive development enhanced mobilization of accumulated sucrose and enzyme activities in the tricarboxilic cycle were simultaneously altered. Between vegetative and reproductive plants differences were most pronounced in the epicotyls and roots while the hypocotyls displayed relatively small changes (Schmid, 1974a).

Low-temperature induction of gibberellin (GA) genes is likely in sugar beet since the concentration of endogenous gibberellins increases upon vernalization (Radley, 1975). Endogenous production of GA stimulates elongation of the stem (Lexander 1985, 1987) and the gibberellin pathway appears to be in close interaction with the vernalization pathway (van Roggen et al., 1998).

Photoperiodic promotion is another major pathway in *Arabidopsis* and *Beta* (Simpson and Dean 2002; Mouradov et al., 2002; Smit, 1983; Lexander, 1985). Sugar beet is a long-day plant (LDP) and as such is characterized by three general features. LDPs have an action spectrum at 710-720nm, increased flowering when far-red (FR) light is added simultaneously with an increased length of day, and exhibit a periodic sensitivity for red (R) and FR light during each daylength cycle (Thomas, 1991). Thus, in *Arabidopsis* R inhibits and FR promotes flowering (Martinez-Zapater et al., 1994); Lane et al., (1965) reported the importance of addition of R or FR incandescent light, preferably in the morning prior to 8 hours' daylight to promote flowering in annual sugar beet. The best treatments had an R/R+FR-ratio between 0.2 - 0.3. Curth (1960) monitored the interaction of light conditions and temperature applied; as temperature increased the demand for higher light intensities increased.

The autonomous pathway described in *Arabidopsis* (Simpson and Dean 2002; Mouradov et al., 2002) may not play a major role in biennial beets. Sugar beet has no strict juvenile stage (Lexander 1969 1975) but younger plants need a longer cold period (Curth 1955, 1960; Boudry et al.,

2002), suggesting that the required length of cold treatment correlates negatively with age of the plant. Also, a decline in LD requirement with increasing leaf age has been observed (Crosthwaite and Jenkins, 1993). These findings support the existence of an autonomous pathway, even if of limited importance for regular sugar beet breeding and farming.

It has been suggested that the various flowering pathways may converge to a common floral integrator; moreover the *Arabidopsis* model is continuously updated and revised (Levy and Dean, 1998 and references therein; Mouradov et al., 2002; van Roggen et al., 1998).

Temperature, daylength, duration of cold period, plant, age and postvernalization temperatures affect sugar beet flower induction. The apex of the stem senses the low temperature (Chroboczek, 1934) and subsequently membrane conformations (Lexander, 1975; Sadeghian and Lexander, 1993; Crespi et al., 1993), increased gibberellin metabolism (Radley, 1975), and signaling take place. Light quality (quantity and periodicity) is perceived by the leaves and the resultant products are transported to the apex where they affect a change in cell divisions (Vince-Prue, 1975) and the floral transition phase commences.

Conditions for bolting and flower induction in sugar beet are genotypic dependent with complex genetics (Sadeghian, 1993). Bolting susceptibility is affected by plant characteristics present already before the onset of flower induction (Lexander, 1987; Sadeghian and Lexander, 1993). This implies that some of the genes responsible for increasing bolting susceptibility are active before bolting-inducing conditions occur (Lexander, 1987).

2.5.5 Flowering Traits in Breeding

Within the breeding process there are different needs for regulating the flowering process, such as timing of pollinators and CMS-types in crosses and the need for early and even maturation. In practice, timing may be achieved by planting the vernalized plants at different times enabling simultaneous flowering, cutting of the elongating stem, or when possible reducing or promoting growth by altering the temperature, etc.

2.5.6 Breeding for Bolting Resistance

Given the complexity of flower regulation in biennials, surprisingly straightforward selection has given very good effects on the improvement of bolting resistance in sugar beet. Lysgaard (1978) was able to reduce bolting susceptibility by up to 98% in three cycles of selection. Bolting susceptibility was shown to be highly heritable, when full-sib progenies of diploid sugar beet developed using a factorial mating design were tested; narrow-sense heritability was estimated to be 0.93 to 0.96 (Sadeghian and

Johansson, 1993). However, biennial beets differed in response to vernalization treatment and the genetics of bolting resistance is still not clear. Early studies by Marcum (1948) indicated that bolting resistance is under the control of several genes with different degrees of dominance. That same year, McFarlane and colleagues (1948) published data showing that mainly recessive genes control bolting resistance. It was later shown that an additive–dominance model was able to explain the bolting response in offspring of a cross between a susceptible and a resistant inbred line (Le Cochec, 1989). Other studies showed that genes with additive effects were of major importance but also that dominance and interaction effects were important in some crosses (Sadeghian and Johansson, 1993; Sadeghian et al., 1993). In a few cases bolting resistance was found to be dominant to bolting susceptibility (Le Cochec, 1989) and in others the opposite (Sadeghian et al., 1993). Consequently, in areas with low spring temperatures, breeding has evolved toward more bolting resistant varieties. This has enabled a change in farming practice toward earlier drilling, resulting in a prolonged growth period and improved yield (Westerdijk and Tick, 1991).

Acknowledgments: The author thanks his colleagues Mats Levall and Kenneth Fredlund for their collaboration in editing this material.

2.6 Self-fertility and Self-incompatibility (M. De Biaggi)

Sugar beet is normally characterized as having a high degree of self-incompatibility, with self-pollination rare (Bosemark, 1993). In some populations, it is possible to find completely self-sterile plants, completely self-fertile plants, and everything in between. The higher the self-fertility, the less frequent its occurrence (Schwanitz, 1940). In commercial germplasm, self-sterility is much more common than self-fertility (Owen, 1942).

Self-sterility was used routinely to enhance and maintain heterosis in multigerm varieties prior to the discovery of male sterility (Owen, 1945). Release of the monogerm SLC lines, characterized as having a high degree of self-fertility (Savitsky H., 1950), led to the introduction of self-fertility into commercial germplasm and it became common for breeders to develop two complementary programs—one using self-incompatible germplasm and the other using self-fertile monogerm O-type germplasm (Bosemark, 1993; Hecker and Helmerick, 1985; Smith, 1987).

2.6.1 Self-incompatibility

Allogamy maintains heterozygosity in sugar beet populations. The breeding of multigerm self-incompatible (self-sterile) families has provided adequate levels of hybridization without the employment of male sterility.

Development of diploid and anisoploid synthetic varieties using half-sib and full-sib families is explained in chapter 3.

Self-sterility generally does not prevent germination of incompatible pollen grains on the surface of the stigma, but growth of the pollen tubes slows down or ceases inside the pistils (Savitsky H., 1950). The author recognized two other physiological mechanisms capable of explaining the failure of self-fertilization: i) death of eventually formed zygotes; ii) growth of normal embryos stops and degeneration takes place.

Owen (1942) studied the heritability of self-sterility in the curly top resistant variety US1, a variety which has a preponderance of self-sterile plants. He obtained clones of three plants and made crosses to validate the theories of Prell (1921), East and Mangelsdorf (1925), and Lawrence (1930). These authors speculated that the self-sterility mechanism could be explained by identical or duplicate multiple alleles S^1-S^n and Z^1-Z^n acting gametophytically and carried on different chromosomes. The hypothesis assumed that a single *S* or *Z* factor carried by the pollen and not present in the tissue of the stigma results in fertility. Owen (1942) accepted this hypothesis: as an example a $S^1S^1Z^1Z^2$ female plant (producing S^1Z^1 and S^1Z^2 gametes) would be successfully fertilized with S^1Z^3 or S^3Z^2 pollen, because at least one of the *S*- or *Z*- alleles would not encounter its corresponding allele in the style. Owen concluded that the results were consistent with the hypothesis of duplicate factors involved in cross-sterility. Lack of complete agreement with the theoretical results suggested interactions with modifying genes.

According to Larsen (1977), the self-sterility system present in sugar beet is conditioned by four gametophytic *S*-loci with complementary interaction, i.e. four *S* genes in the pollen have to match, in allelic form, genes in the pistil to cause sterility. The four loci were designated S^a, S^b, S^c, and S^d. Larsen also emphasized that additional *S* loci may exist. He (1978a,b) confirmed the multigenic inheritance of self-incompatibility in crosses of sugar beet with fodder beet and sea beet. Bosemark (1993) further confirmed that self-sterility is controlled by four *S* loci in sugar beet. Maletskii et al., (1991) reported that the self-incompatibility locus *S* and the *M* locus that controls the formation of clustered or separate flowers were linked (0.25-0.26 recombination frequency).

In many cases, self-sterility is partial or incomplete and the plant produces some seed after selfing. This behavior, referred to as pseudocompatibility, is due to a breakdown of the sterility mechanism under unique climatic conditions or during late-season flowering (Owen, 1942; Bosemark, 1993). Confirming Schwanitz (1940), Larsen (1977) asserted that some self-sterile plants may set a few seeds in isolation. Under low temperatures, pseudocompatibility becomes more frequent, permitting the production of sufficient seed for many breeding purposes (Vaisman, 1981). Pseudocompatibility increases at low (around 15°C) and high (around 35°C) temperatures. In these situations, incomplete expression of self-sterility results in the production of some seed in multigerm diploid families when selfed in isolation. Although not yet fully understood, pseudocompatibility is likely under polygenic control (Larsen, 1977).

2.6.2 Self-fertility

The existence of self-fertile sugar beet plants was noted by Schwanitz(1940) and the heritability of self-fertility studied by Owen (1942). Owen demonstrated that self-fertility was controlled by a single Mendelian factor, S^F. The genetic composition of homozygous self-fertile plants is S^FS^F, and S^FS^a (or S^FS^b or S^FS^x) for heterozygous self-fertile plants, while S^aS^b represents the self-sterile condition. The designations S^a, S^b, S^x are necessary because diverse self-sterile sugar beet sources may carry many different *S* allelomorphs. In the hybrids described by Owen, the factor S^F came either from the curly top resistant strain 1167 or from selections derived from strain 2769. The progeny of crosses between self-sterile S^aS^b seed parents and heterozygous self-fertile S^FS^x pollen parents segregated in a 1 self-fertile:1 self-sterile ratio, consistent with the hypothesis of a single factor. Owen (1942) also pointed out that with selfing heterozygous S^FS^x individuals it was possible to obtain only self-fertile plants because the pollen with allelomorphs S^a, S^b (or S^x) was inhibited, resulting in self-fertilization. Lewellen (1989) described an intense self-fertility gene expression in a curly top resistant type-O line, C600, that made hybridization with other lines difficult.

Self-fertility has been used extensively in breeding programs for 30 years, often in combination with monogenic male sterility (NMS). The S^F trait is employed to develop new recurrent populations in both monogerm and multigerm germplasm (Bosemark, 1971) and also in a reciprocal recurrent selection scheme outlined by Doney and Theurer (1978). Plants carrying the S^F gene, either in heterozygous or in homozygous forms, usually produce 90 to 95% selfed seed even under conditions of open pollination. Hence the S^F gene expedites development of inbred lines (Bosemark, 1993).

2.7 Seed Quality (E. Biancardi and M. De Biaggi)

Seed germination is a complex character that has a significant effect on crop productity. Rapid and regular field emergence increases the uniformity of stands and mature roots (Fauchere, 2001; Stibbe and Märländer, 2002). In fact, an adequate number of uniformly distributed plants optimizes light interception by the canopy and reduces both the impact of weeds and root losses during mechanical harvesting (Snyder, 1963). The switch from multigerm to genetic monogerm varieties about 40 years ago, has enhanced the importance of seed germination traits since there is no way to increase the stand once sowing has taken place (Biancardi, 1999).

Normal seedling development does not depend solely on environment, pathological factors and cultural practices after sowing, but also on conditions during seed production, processing and storage (Decoux and Weenen, 1962). Germination is also affected by genetic factors and some genetic variability has been observed (Sedlmayr, 1960; Longden, 1990). Inheritance of traits related to germination rate has been summarized by Battle and Wittington (1971). Sadeghian and Khodaii (1998) confirmed the importance of additive dominance in germination, vigor, and 1000 seed weight. Variance of the first two characters was substantially influenced by environmental factors.

It is theoretically possible therefore, to improve germination and vigor and reduce the discrepancy between percentage of seeds that germinate in laboratory and seedling emergence in the field (Longden, 1990). Mass selection is often used to improve such characteristics. It involves elimination of individuals and lines below a standard level and propagating individuals that carry the desired traits.

Many attempts have been made to improve germination through selection, although the results have not been particularly encouraging. Some authors have associated high germination with a low concentration of inhibiting substances in the seed cluster (Doxtator and Finkner, 1958; De Kock *et al.*, 1953). Others relate the differences in germination to morphological differences in the operculum: as its thickness increases, seedling emergence becomes more difficult (Sedlmayr, 1960; Hogaboam, 1961; Peto, 1964). Sedlmayr also claimed that all genetic variability is caused by tissues of maternal origin, such as the pericarp; a theory backed by Snyder (1963). In modern F_1 hybrid varieties, the influence of the female parent clearly outweighs that of the pollinator (Bychkova and Krivoklyakin, 1993).

Smith (1952), evaluating low-temperature (6°C) germination percentage and vigor of four genotypes, found notable differences from values obtained at 24°C. By propagating plants that germinated quickly, he increased germination in the progenies, particularly in genotypes initially deficient in this character. Doxtator and Finkner (1958) reached the same conclusions after selecting plants that germinated early in a water-inositol solution.

Uniformity of field stands depends, in part, on the percentage of empty seeds planted. Empty seeds are normal shaped but contain no embryo, or the embryo is not fully developed or unable to germinate (Tekrony and Hardin, 1969). A high frequency of empty seeds was observed in triploid monogerm hybrids. In this case, gametes with different ploidy levels caused disorders during development of the endosperm and embryo. These anomalies increased the frequency of empty or nonviable seeds (Jassem, 1976) and the author made several recommendations for reducing this phenomenon.

Another undesirable trait is the occurrence of two or more embryos in a single monogerm seed (false monogermity). After sowing, two or more seedlings emerge, as with bigerm or multigerm seed. This character is inherited and is particularly widespread among tetraploid genotypes and hence in triploid hybrids (Fischer, 1956). There are also diploid CMS monogerm lines which produce a relatively high percentage of twin embryos. This trait is especially troublesome if these lines are used as seed parents of commercial hybrid varieties.

Nemazi and Nielson (1967) obtained lines with a reduced frequency of false mongerm seeds. In general, the character has a positive correlation with seed weight and therefore can be reduced, but not completely eliminated, by using a gravity table. False monogerms often appear in lines that develop flowers with an excess of pistil lobes (5-6 instead of the normal 2-3). This anomaly often correlates with the presence of two or more ovules in the same ovary (Shavrukov et al., 2000), each producing an embryo. According to these authors, it is possible to reduce the occurrence of false monogerms by removing plants with an abnormal number of lobes.

Improvements in seed characteristics and emergence of commercial varieties in the field have been slow but steady (Longden, 1990). Much of this progress is due to improved techniques for growing and processing seed and protection of young plants with pesticides incorporated in the pellet.

2.8 Root, Hypocotyl, and Leaf Color (E. Biancardi)

The color of cultivated beet roots may be: i) varying intensities of red in garden beet; ii) shades of red, green, orange, yellow, or white in fodder beet; and iii) ivory-white in sugar beet **(Plate 1.1)**. The ivory-white color was selected to interfere as little as possible with extraction of sugar (Goodshall et al., 1991), which should be as white as possible. Also, it is desirable that root color not interfere with determination of sugar concentration, the accuracy of which is compromised if color in the pulp cannot be eliminated.

Therefore, in sugar beet color needs to be uniform and any colored roots are considered contaminated (Biancardi, 1999).

Root and hypocotyl color depend on at least two genes: *Y* with the allelic series *Y*, Y^r, *y* and *R* with the allelic series *R*, R^t, R^P, R^h, and *r*. The *R* allele induces red hypocotyls, petioles, and reddish leaves, while R^t produces red stripes on the petioles. The *rr* plants have green petioles and leaves. In beet breeding, hypocotyl color at the cotyledon stage (color differences tend to disappear in successive growth stages) is frequently used to identify hybrid progeny (Bosemark, 1993). F_1 progeny of green hypocotyl plants pollinated with pollen from a line with colored hypocotyls will have colored hypocotyls, whereas selfs or sibs will have green hypocotyls and can be discarded.

The *Y* genes control yellow coloration and are modified by genes at the *R* locus. If the genes at both the *Y* and *R* loci are recessive (*rryy*), the roots are white and the hypocotyls green. If a dominant *R* gene is present (*RRyy or Rryy*), the roots are white and the hypocotyls red. Other combinations with *Y* alleles are associated with other root colors and may modify hypocotyl color (Pedersen, 1944; Theurer 1968a, Smith, 1987).

2.9 Morphological and Physiological Traits (E. Biancardi)

During and after the mechanical harvest of the roots, a great deal of sugar is lost due to root breakage, imperfect topping, and brushing. Several breeders have attempted to modify the form of the root to make it more suitable for harvest operations and concomitantly diminish the damage and resultant loss of sugar yield. Leaf material that remains attached to the roots after topping and excessive soil adherent to the taproot can cause significant problems in sugar factories.

Harvest losses depend, in part, on the form of the taproot. If the root is too long, losses may arise due to cracks around the tail. Overdeveloped crowns make topping difficult. Excessive fanging of the roots, due to genetic traits or the soil structure, causes difficulties during harvest and retention of soil. Deep root grooves make cleaning in the field and factory difficult, especially in the presence of heavy soils (Mesken, 1987a,b).

In 1913, Kajanus proposed the first hypothesis regarding genetic factors that determine the morphology of the sugar beet root. He suggested four pairs of genes were involved, two for length and two for shape. By crossing sugar beet and fodder beet, Stehlik (1933) and Olak (1950) established that the sugar beet shape was dominant. Other authors realized that root form is a quantitative character. Mesken (1987a.b) with coworkers (1988) selected lines with tare losses (crown + soil) that were 40% to 60% below the check. He began by crossing red beet and sugar beet and selected for a shortened round form. The hybrids were almost completely free of root grooves and were subsequently backcrossed with sugar beet.

The root yield of improved "globe shaped" genotypes is comparable to that of traditional hybrids but sugar content and processing quality relatively low (Theurer, 1993). Rounded roots do not anchor well in the soil and may be susceptible to uprooting during topping, particularly when the soil is damp or the roots larger and protruding from the soil more than average. Cutting the crown is difficult or irregular under these conditions and the whole root often breaks up and is lost (Biancardi, 1999). With mass selection, Mesken and Dieleman (1988) reduced crown height, i.e. the part protruding from the soil consisting of the hypocotyl and the epicotyl. In 2n genotypes, selection reduced crown height about 18% per cycle; heritability of the trait was 0.4. For tetraploid genotypes reduction was 12.5% per cycle and heritability 0.3 (Mesken, 1987a,b).

It is possible to improve or maintain a suitable root form with all breeding procedures by eliminating mother beets that do not conform to the ideotype. It is relatively easy to reduce the depth of the grooves even though the genetic basis of the character is not known. It is more difficult to improve taproot shape, due to the influence of soil variability, climate, and field stand (Biancardi, 1999). However, reasonable progress has been made; some varieties used in France in recent years have soil tares 10-20% less than the average of normal varieties (ITB, 1995). Differences of the same order have also been observed in Germany (Koch and Jung, 1995).

High fiber content, often referred to as woodiness, is an undesirable characteristic of the root. It occurs when the layers of woody fibers surrounding the vascular vessels are more developed than usual. Woody roots dull the slicing machines that produce the cossettes from which sugar is extracted. Fiber content depends mainly on environmental factors. Schlösser (1961) found a higher percentage of woody or fibrous roots in tetraploid genotypes, especially in type Z varieties, i.e. those with a relatively high sugar content. The author stated that this negative character could be reduced with a few cycles of mass selection. Woody roots can also be found in beet seeded in the autumn, especially in bolting plants (Rosnowsky et al., 1961). These authors concluded that two genetic factors, B and B^1, and climatic factors controlled woodiness.

Stem (stalk) fasciation is an undesirable trait in seed-bearing plants because of its detrimental effects on seed yield and germination. Fasciation is frequently associated with the monogerm seed trait (Hasegawa *et al.*, 1977).

2.10 Resistance to Parasites

2.10.1 Introduction (E. Biancardi)

Host-plant resistance has allowed crops to thrive in the presence of otherwise devastating diseases and pests which cannot be controlled by other means. Notable examples in sugar beet include resistance to the viral diseases curly top and rhizomania. The disease control obtained with genetic resistance is often incomplete, as seen in cercospora l.s. (Skaracis and Biancardi, 2000). In this case, however, use of resistant varieties complements other control methods and reduces fungicide usage. For some diseases, useful levels of resistance have not been found and in other cases reduced yield and processing quality have accompanied the introduction of resistance, as in the case of resistance to nematodes (*Heterodera schachtii*).

Different terms are used to characterize the response of a plant or a variety to a disease. Tolerance refers to a plant's ability to maintain productivity while supporting a parasite or pest population that would severely affect development or reduce yields of other plants or varieties. Resistance, on the other hand, indicates the ability of a plant to preclude multiplication and/or spread of a pathogenic agent. In order to characterize resistance, it is necessary to determine the quantity or concentration of the pathogen in the diseased tissues, while data regarding growth and productivity are required for demonstrating tolerance. In general, the measurements necessary to document resistance are less subject to error (Agrios, 1988). With many sugar beet diseases, it is appropriate to speak of genetic resistance since a lower concentration of the pathogen is detectable in resistant plants; examples include rhizomania and cercospora l.s. (Paul, 1993). Genetic resistance may be complete or partial, depending on the intensity of its protective mechanisms. The highest level of resistance is immunity; lower resistance levels are often incorrectly referred to as tolerance (Agrios, 1988).

2.10.2 Viruses

2.10.2.1 Curly Top (L. Panella)

Beet curly top virus (BCTV) is a gemini virus transmitted by the beet leafhopper, *Circulifer tenellus,* that attacks sugar beet throughout the semi arid areas in the western United States, Southwestern Canada, Mexico, the Mediterranean basin, Turkey, and Iran (Bennett and Tanrisever, 1958; Bennett, 1971; Duffus, 1986; Duffus and Ruppel, 1993; Mukhopadhyay, 1987).

The sugar beet industry in California was begun in the 1870s and BCTV symptoms were observed shortly thereafter (Bennett, 1971). In

1912 the first beets were grown in southern Idaho; in 1919 yields dropped almost 50% from the previous year due to BCTV (Murphy, 1946). During this time sugar beet yield in the western USA was contingent upon pressure from the beet leafhopper on the crop; when populations were high early in the season, yield was devastated (Carsner, 1926b; Murphy, 1946). By the 1920s it was obvious that if the sugar beet industry were to survive in the western USA, it would be necessary to develop varieties with resistance to the curly top virus. By 1929, the sugar beet industry was in serious trouble west of the Rocky Mountains.

Although host plant resistance to BCTV was observed in the early 1900s (Coons, 1936; Townsend, 1908), breeding programs were not initiated until 15 years later by Carsner in Utah and California (Carsner, 1926a), Esau in California (Esau, 1930), and other sugar companies in the western USA. Mass selection of roots showing resistance in heavily infested fields proved effective (Coons et al., 1931) and the first resistant open-pollinated variety US1 was released in 1933 by Carsner and Pack (Carsner, 1933) and was in extensive use by 1934 (Coons, 1936). US33 and US34 were based on mass selections from heavily infested fields of US1; they were replaced by US12 and US22, which were improved as US22/2 and US22/3 (Coons et al., 1955). Murphy (1942) and McFarlane (1969) demonstrated the increased resistance of the newer varieties under severe curly top pressure.

Although twenty years of mass selection had been successful in producing resistant open-pollinated populations, with the advent of hybrid varieties it was realized that inbreeding and progeny testing were necessary for continued improvement of varieties and conversion of open-pollinated varieties to monogerm O-type parents (McFarlane, 1969). Much of the breeding for BCTV resistance was done through selfed family evaluation using the self-fertility gene (Owen, 1942) and genetic male sterility (Owen, 1952). The breeding effort was helped by the development and refinement of techniques to create an artificial epiphytotic of BCTV in the field (Mumford, 1974; Murphy, 1942).

Early studies by Abegg and Owen (1936) and Abegg (1938) described a partially dominant genetic factor, *C*, linked to the gene for crown color, *R*, by 20 to 30%. Later studies by Murphy and coworkers (1952) indicated a more intermediate (additive) resistance in F_1 hybrids under mild to moderate curly top pressure. But this intermediate resistance broke down in later generations, especially in crosses including SLC101, a highly susceptible parental line (Murphy and Savitsky, 1952). The work by Murphy and Savitsky (1952) confirmed the linkage between BCTV resistance and the *R* gene. Under severe curly top exposure, and when hybrids between more curly top susceptible parents were studied, the genetic nature of BCTV appeared more complex. Savitsky and Murphy (1954) estimated that more than two genes were responsible for the resistance. Although resistance is thought to be polygenic, there are no definitive estimates of heritability or number of loci controlling resistance in the literature.

At the USDA-ARS Station in Salinas CA, McFarlane made a major effort to produce germplasm resistant to BCTV and other important diseases, e.g., virus yellows (McFarlane et al., 1969). This effort has continued under the directorship of R.T. Lewellen (Lewellen et al., 1985a,b; Lewellen, 1994a,b). BCTV is managed in California today by combining vector control, beet free time, and chemical protection of the crop (Wisler and Duffus, 2000). Use of resistant varieties is critical in maintaining yield under severe disease pressure (Kafka et al., 2002).

With the closing of the USDA-ARS station in Logan UT in the late 1970s, the only other public breeding program in the USA for BCTV resistance is the USDA-ARS program at Fort Collins CO. The work begun by Gaskill in combining curly top resistance with cercospora and rhizoctonia resistance (Gaskill et al., 1967, 1970) was continued by Hecker and Smith along with Ruppel and, through inbreeding, resulted in the FC600 series of monogerm O-type, and CMS equivalent releases (Panella, 1998). This work is carried on in the breeding program of Panella and Hanson (Panella and Ruppel, 1997). Because of the continued pressure from BCTV, the commercial seed companies that sell sugar beet seed west of the Rocky Mountains have active BCTV resistance breeding programs. In the USA, resistance is assessed in a cooperative curly top nursery managed by the Beet Sugar Development Foundation and planted in Kimberly ID. BCTV resistant varieties are also grown in Turkey and the Near East.

2.10.2.2 Beet Yellows (M. De Biaggi)

Virus yellows was recognized as an infectious entity of sugar beet in Europe in the 1930s (Roland, 1936; van Schreven, 1936) and was first documented in California in 1951 (Duffus, 1978). The disease is characterized by a yellowing of the leaves, which often thicken and became brittle (Duffus and Ruppel, 1993). According to Russell (1958, 1960) and Duffus (1960, 1964) the symptoms are caused by one or more of the following viruses: the closterovirus beet yellows (BYV), and the luteoviruses beet mild yellows (BMYV) and beet western yellows (BWYV) (Russell, 1958, 1960; Duffus, 1960, 1964). All three viruses can be transmitted by *Myzus persicae* (green peach aphid), while *Aphis fabae* (black aphid) can inoculate BYV but not BMYV or BWYV (Russell, 1958). Virus yellows reduces photosynthetic efficiency and consequently sugar yield. BYV infection reduced root length but did not affect nutrient uptake in glasshouse experiments (Allison and Clover, 1996).

BYV and BMYV are the prevalent yellowing viruses of sugar beet in Europe (Russsell, 1958), while in the USA the disease is generally caused by BWYV (Duffus, 1960). There is evidence that BMYV and BWYV are similar but neither is related to BYV (Duffus and Russell, 1975). One of the most important hosts for virus yellows is the sugar beet seed crop. Therefore, an effective method of reducing both the spread and the damage associated with the disease is to geographically separate the sugar and seed crops (Duffus and Ruppel, 1993).

Bennet and Leach (1971) observed that BYV transmission by *Myzus persicae* and *Aphis fabae* is semipersistent. The minimum feeding period for acquisition of the virus and its inoculation is 5-10 minutes (Duffus and Ruppel, 1993). However, aphids attain the maximum ability to transmit after feeding 4 to 6 hours, up to 24 hours and occasionally 72 hours (Bennet and Leach, 1971). BMYV is recognizable as polyhedric particles vectored by *Myzus persicae*; these aphids, after a feeding time of 24 to 48 hours on infected plants, are able to transmit the virus throughout their life cycle (Casarini, 1999). Bennet and Leach (1971) also concluded that BWYV remains active in *Myzus persicae* during the life of the vector.

Early studies showed that virus yellows reduced root yields from 2 to 47% and sucrose concentration from 0.1 to 3.1 degrees (Duffus, 1978). In glasshouse experiments, Bennet and McFarlane (1967) reported root yield losses of 24 to 41.7% for BYV and 10.8 to 18.2% for BWYV. In general, sugar losses caused by BYV and BWYV depend on the time of infection; early infection can severely decrease yield and quality, late infection only moderately (Duffus and Ruppel, 1993). Success of foliar insecticides in controlling aphids and minimizing virus spread has been erratic. In the last decade, Imidacloprid incorporated in the seed pellet has been successful in protecting young plants from the aphids and consequently the viruses (Wewers and Heijbroek, 1994). The effectiveness of this new class of insecticides has slowed virus yellows resistance breeding efforts (Biancardi, 1999).

The first attempts to select for resistance to virus yellows were carried out independently in the Netherlands (Rietberg, 1959) and the UK (Hull, 1960b) in the late 1940s. These efforts led to the production of the multigerm variety "Maris Vanguard" released in 1965, and the monogerm variety "Vytomo". Russell (1969) identified three forms of inherited resistance in sugar beet: i) virus tolerance, ii) resistance to virus infection, and iii) resistance to aphids. These three tolerance/resistance mechanisms often complement each other to minimize virus damage and spread. Bosemark (1993) recognized these forms of tolerance/resistance and listed a fourth; resistance to virus multiplication. He attributed the reduced emphasis on breeding for virus yellows resistance to and confirmed that the low attention of the breeders for this disease has been caused also because the low incidence of virus yellows across Western Europe during the 1970s.

Since 1954, thousands of breeding lines, commercial sugar beet varieties, *Beta vulgaris* ssp. *maritima* accessions, and other *Beta* species have been screened for reaction to BYV. No immune plants were found but it was clear that a useful degree of virus tolerance/resistance was present in sugar beet varieties (Campbell, and Russell, 1964). These authors describe a breeding technique, based on inoculation with BYV transmitted by *Myzus persicae,* originally developed by Watson and Russell (1956). Seedlings were scored on a 0 to 5 scale based on visual symptoms, and plants with low scores selected for seed production. Russell (1964) used five viruliferous and apterous *Myzus persicae* per plant in glasshouse trials. Selection based on root weights 10-12 weeks after inoculation was efficient. Plants with larger roots and minor symptoms were selected. A similar screening for tolerance/resistance to BMYV worked better in field nurseries.

Russell (1969) interpollinated inbred lines selected for tolerance/ resistance to BYV or BMYV to produce several varieties for field testing. Tolerance/resistance was apparently dominant in some hybrids and recessive in others. Inheritance of tolerance/resistance to yellowing viruses appeared to be complex, involving several genes. The traits seem to be inherited independently (Russell, 1969). Some plants resisted or had less severe virus yellows infection when exposed to inoculation. Under a severe attack in the field, only 51.9% of plants of the multigerm variety VT40SG exhibited BYV or BMYV symptoms, whereas 79.7% of the control plants were infected. Studies of the mechanisms involved in this type of tolerance/ resistance were inconclusive but it was noted that sugar content of the leaves appeared to be associated with increased aphid feeding (Russell, 1969).

According to McFarlane et al., (1969), tolerance/resistance to BYV and BWYV based on the absence of symptoms and root size, is a relatively effective selection strategy when the parental material includes virus yellows-resistant germplasm. The authors obtained equal yield improvement with self-sterile and self-fertile breeding populations. They reported that hybrids between monogerm male sterile females and selected yellows-resistant pollinators performed well under both severe and moderate infections. Two virus yellows resistant hybrids, USH9A and USH9B, were released as commercial varieties. McFarlane et al., (1969) further observed similar sugar yields in tolerant/resistant hybrids and susceptible hybrids in the absence of infection. In other trials with no virus yellows, a negative relationship between tolerance/resistance and sugar concentration and juice purity was detected (Campbell, G.K.G. and Russell, 1964).

Under diseased conditions, germplasm selected for tolerance/ resistance to BYV and BMYV had higher sugar yields and sugar concentrations and a small increase in sodium and potassium (Campbell, G.K.G. and Russell, 1964). Russell (1964) suggested a possible relationship between virus tolerance/resistance and potassium content. Most of the virus-tolerant/resistant germplasm was also resistant to *Alternaria* infection that caused loss of older leaves in the susceptible control varieties, when infected with the virus. Resistance to *Alternaria* should be considered when breeding for tolerance/resistance to yellowing viruses (Campbell, G.K.G. and Russell, 1964).

2.10.2.3 Beet mosaic (R.T. Lewellen and E. Biancardi)

Beet mosaic virus (BMV), transmitted nonpersistently by the green peach aphid *Myzus persicae*, is widespread especially in temperate regions where sugar beet is grown as a winter crop or overwintered for seed production (Shepherd et al., 1964). BMV is often associated with beet yellows virus (BYV). The pathogenic effects of the two viruses are additive (Shepherd et al., 1964) but the damage caused by BMV is small compared to that caused by BYV (Mukhopadhyay, 1987). The infection appears as small yellow spots on the younger leaves, that develop later into pale green-yellow mottling. A 20% root yield reduction was observed in a very severe attack of BMV (Watson and Watson, 1953).

Feltz and Marx (1960) recognized the existence of BMV-resistant Russian and Hungarian varieties released since 1935. A variety from Hungary "Sedlmayr Beta 242", and its hybrid with the German variety, "KWS CR", were artificially inoculated with BMV using infected green peach aphids. Mass selection in these populations produced the resistant families 231 and 235. No information regarding the inheritance of BMV resistance was provided.

Lewellen (1973) analyzed the BMV resistance of a self-fertile annual line 8500, derived from Owen's annual male sterile tester (Owen, 1948). The annual trait, *B,* was from Munerati germplasm. The CMS equivalent of the line was crossed with a susceptible line C5600. Backcross (parents x F_1), F_2 and F_3 populations were also obtained. Seedlings of these populations were mechanically inoculated with BMV at the two-four true leaf stage. Systemic symptoms appeared 7-10 days after inoculation, thereafter the plants were scored every one or two days. Observations indicated that the BMV resistance carried by 8500 was inherited as a single dominant or incompletely dominant gene, designated *Bm* (*Bm* = resistant; *bm* = susceptible). *Bm* was linked to the locus for genetic male sterility (A = fertile; aa = sterile) (Lewellen, unpubl. data).

The allele *Bm* appears to reduce the inoculation or multiplication rate of BMV and therefore its concentration inside the plant tissues (Lewellen, 1973). This action, which reduces the negative effect of the virus on yield, is additive. As a consequence, in homozygous dominant genotypes BMV concentration is reduced more than in heterozygous individuals. Variations in susceptibility were observed in segregating populations, indicating the presence of minor genes, which could modify the expression of resistance in the absence of the major gene (Lewellen, 1973).

2.10.2.4 Rhizomania (M. De Biaggi)

Beet necrotic yellow vein virus (BNYVV), recently classified a *Benyvirus* by the International Committee on Taxonomy of Viruses, causes rhizomania. BNYVV is carried and transferred to sugar beet roots by *Polymyxa betae*, a soilborne plasmodiophoromycete. The disease has spread into northern European production areas, including England, Denmark, and Sweden, areas considered rhizomania-free a few years ago (Asher, 1993; Richard-Molard, 2002). The virus can cause losses of up to 80% of the potential sugar yield (Casarini et al., 1999). Resistant varieties that currently provide protection against the disease (ITB, 1997) are the product of 20 years of breeding efforts.

Koenig et al., (1986, 1995) and Koenig and Burgermeister (1989) identified three BNYVV pathotypes based on RNA analysis (Koenig et al., 1995). Type A is distributed throughout Europe and type B is confined to Germany and France, often in combination with type A. The third form, type P, was found near Phitiviers, France (Koenig et al., 1995), and more recently in the United Kingdom (Harju et al, 2002). Types A and B of BNYVV are characterized by the presence of 4 RNA particles while P type has an additional RNA particle (Tamada et al., 1989; Koenig et al., 1997). Lemaire et al., (2003) speculated that BNYVV-RNA-5 has also been identified in Kazakhstan, Japan and China by Koenig and Lennefors (2000). As far as we know, RNA-2 codes for the virus coat protein (Lemaire et al., 2003), RNA-5 is putatively involved in disease severity (Tamada et al., 1990), and RNA-3 influences disease symptom expression (Koenig and Burgermeister, 1989). The RNA-3 and RNA-5 sequences appear to be stable over time, suggesting a slow evolution of the virus (Lemaire et al., 2003). Interactions between varieties and locations have been assumed to be caused by corresponding interactions between resistance genes and BNYVV pathotypes (Van Euwijk and Keizer, 1997; Richard-Molard, 2002).

The first source of resistance to rhizomania was discovered in cercospora l.s. resistant Italian germplasm derived from the multigerm variety "Alba P" (Biancardi et al., 2002). The superior performance of "Alba P" was initially observed in trials grown in 1957, before the discovery of the agent causing rhizomania (Bongiovanni and Lanzoni, 1964). Based on segregating F_2 populations, this resistance was classified as quantitative (Lewellen and Biancardi, 1990).

A more resistant variety, "Rizor", was released in 1985 by De Biaggi, breeder at SES, Italy (De Biaggi, 1987). The genotypes used in the initial selections in 1980 were cercospora l.s. resistant and therefore surely originated from families selected by Munerati, or their descendants (Skaracis and Biancardi, 2000). In 1983, a collaborative study had been set up with the Istituto di Patologia Vegetale of Bologna University (Biancardi et al., 2002). The aim was to better understand the action of resistance. It was confirmed that *Polymyxa betae* infected the rootlets carrying the BNYVV with it. This behavior was observed in susceptible genotypes, in the pollinator of "Rizor", and also in "Rizor" itself (Giunchedi et al., 1985). The virus moves rapidly through xylematic tissue of susceptible genotypes; movement in the bundles of "Rizor" appears to be hindered (Giunchedi et al., 1987).

Prior to 1987, the "Rizor" breeding program used mass selection supplemented with artificial infection and ELISA tests on single plants (De Biaggi, 1987; Biancardi et al., 2002). Early in the selection program, "Rizor type" resistance was recognized as monogenic and dominant. In fact, the variety was produced using susceptible CMS seed parent. Subsequently, it became evident that this characterization was not completely true due to the presence of a low but persistent frequency of recessive (susceptible) individuals that proved very difficult to reduce or eliminate in the offspring (Biancardi et al., 2002).

In the summer of 1983, Erichsen, a breeder working for Holly Hybrids, Tracy CA, observed very low sugar yields in a trial, except for three experimental hybrids, which produced five times more than the susceptible control USH11 (Lewellen et al., 1987). An ELISA test on the roots confirmed the presence of BNYVV. The hybrids had different pollinators but the same monogerm CMS seed parent, obviously carrying some unknown resistance (Lewellen et al., 1987; Biancardi et al., 2002). During 1985 and 1986 the three hybrids were evaluated at Salinas CA, under rhizomania attack. Compared with resistant varieties of European and Japanese origin, the trials confirmed the high level of resistance derived from the Holly female line. Just as in the first trial, the three hybrids segregated in a pattern typical for a single dominant gene, subsequently called *Rz* (Lewellen et al., 1987; Lewellen, 1988). The Holly resistance gene reduced the BNYVV replication rate in the root (Lewellen et al., 1987). Wisler et al., (1999) observed that the resistant (*RzRz*) genotypes had significantly lower virus titer than susceptible (*rzrz*) genotypes.

Lines carrying both sources of resistance (Rizor or Holly) frequently exhibit different levels of expression, probably due to the presence of minor genes which interact with the major allele in *Rzrz* individuals (Scholten et al., 1996; De Biaggi et al., 2003). The *Rz* allele is used extensively in commercial hybrids throughout the world. It is designated the "Holly type" resistance to distinguish it from the "Rizor type", which is currently used in a number of European commercial varieties (Biancardi et al., 2002).

Attempts to explain the genealogy of the monogenic sources of resistance have not been successful, but many observations and data obtained through molecular analysis suggest that "Rizor and Holly types" originated from sea beet (Biancardi et al., 2002). Barzen et al., (1997) stated that "Rizor" and "Holly" sources have the same resistance allele.

Soon after the discovery of rhizomania in California, the wild beet (WB) collection of about 60 accessions stored at Salinas was screened for resistance to the disease (Lewellen and Whitney, 1993). Observations confirmed that WB42, an accession collected in Denmark, was resistant in field trials (Lewellen and Whitney, 1993; Lewellen, 1995a). Greenhouse tests employing ELISA confirmed the dominant behavior of the trait. Individual resistant plants were selected and crossed with the susceptible multigerm line C37. After a few cycles of selection, a new line, C48, was released (Lewellen and Whitney, 1993). Continued backcrossing with C37 as the recurrent parent and resistance coming from WB42, led to the release of an improved line, C79-3 (Lewellen, 1995a,b,1997).

Based on germplasm provided by Lewellen, Scholten et al., (1999) reported that WB42 resistance was conditioned by a dominant gene closely linked to the *Rz* gene. The allele was named *Rz2* but in a previous paper it had been suggested that two complementary genes controlled resistance (Scholten, 1997). In greenhouse tests, the WB42 source conditioned lower virus concentration in roots than *Rz* (Scholten et al., 1996). It was reported that variety "Angelina" carries both the Holly (*Rz*) and the WB42 (C48) gene for resistance (Harju and Richard-Molard, 2002). In the severe conditions of Pithiviers, France, in the presence of type P BNYVV, the combined effects of the resistance genes appears to allow "Angelina" to produce higher sugar yields than varieties with *Rz* alone.

From the aforementioned California sea beet collection, it was possible to isolate other apparently different resistance traits (Lewellen and Whitney, 1993). The R22 line, developed from composite crosses of the lines C37 and 60 sea beet accessions, led to the release of C50 (Lewellen and Whitney, 1993). This synthetic version of R22 has improved protection against the combined damaging effects of rhizomania and root rots associated with high temperatures in the Imperial Valley of California (Lewellen, pers. Comm.).

The known types of resistance, both quantitative (Alba) and qualitative (Rizor, Holly, WB42, etc.), appear to be derived from sea beet (Biancardi et al., 2002). The qualitative traits have been mapped in the same chromosomal fragment (Scholten et al., 1999; Biancardi et al., 2002). Therefore, the differences among the so-called monogenic resistances remains an open question complicated by the presence of minor or modifying genes, which interfere with the segregation patterns (De Biaggi et al., 2003). The same authors confirm that combining resistance sources would improve control of the disease, even in the presence of different BNYVV pathotypes or combinations of pathotypes.

Initiating a BNYVV-resistance breeding program with new genotypes or populations with low frequencies of the *Rz* genes would require several years of trials in areas with rhizomania and in rhizomania-free fields. Trials in disease-free areas would be needed to eliminate lines without agronomic interest because the rhizomania-resistant varieties must have competitive yields under healthy conditions. The yield potential of the current resistant varieties is satisfactory. But in some production areas the level of resistance appears to be decreasing. These problems are arising near Ciudad Real, Spain, near Pithiviers, France, and in the Imperial Valley, CA. Some authors attribute these reductions in resistance to interactions between varieties and BNYVV pathotypes (Harju and Richard-Molard, 2002).

Methods based on different systems of plant infection, followed by precise BNYVV detection on single plants, have facilitated development of resistant germplasm (De Biaggi, 1987; Lewellen and Biancardi, 1990). Analyses of the virus can be carried out in different developmental phases and parts of the plant (Casarini-Camangi and Canova, 1987; Burky and Büttner, 1989; Grassi et al., 1988, 1989). Molecular markers provide another method of detecting *Rz* alleles in segregating populations (Barzen et al., 1997).

For improving new genotypes, it is necessary to select mother beets in fields regularly infected with *Polymyxa betae* and BNYVV and avoid interferences with nematodes, *Rhizoctonia* etc. Inoculation of the virus begins when soil temperatures are suitable; hence it may be appropriate to delay planting to increase infection rate. Obviously, the mother beets should be protected from foliar diseases, which might interfere with their reaction to BNYVV.

Just before harvest, it is necessary to inspect the canopy, discarding plants with yellow and palegreen leaves and harvesting plants with no or minor rhizomania symptoms. Then the roots can be carefully evaluated and normal-shaped roots without (or with minor) development of beard selected. Chemical analysis of brei samples from single roots provides valuable information. Roots with relatively high sugar content and minimum deviation from typical concentrations of nonsugars are probably the most resistant.

This type of mass selection is important at the beginning of a program. But because of the dominance of qualitative resistance traits, it is not possible to distinguish between homozygote (*RzRz*) and heterozygote (*RZrz*) plants. Therefore, each mother beet selected should be inbred and resistance of the progeny verified. The first "Rizor" pollinator improved with this method (De Biaggi, unpubl. data). Testing progenies in a BNYVV infected field allows expression of the segregation pattern and detection of homogeneous lines with normal green leaf color. Selections whose progenies do not segregate were assumed to be homozygotes for the *Rz* trait.

It is desirable to have the higher resistance expression in all disease conditions but necessary in the highly damaged areas mentioned earlier. It is very probable that effectiveness of the resistance conferred by the *Rz* genes depends on the severity of rhizomania infection. In other words, lines/hybrids with sufficient resistance for most environments may not provide adequate protection in severely infected fields. For example, it is known that BNYVV-P type are aggressive against varieties that perform well in other situations. It is also necessary to fix the major gene, *Rz,* with a set of positive and stable modifier genes. It is possible to isolate individual plants which carry a complex "major gene (*Rz)* plus modifiers" able to express a high level of resistance under severe rhizomania (De Biaggi, unpubl. data). A high correlation between virus concentration of the roots under greenhouse conditions and sugar yield in infected fields was reported by Giunchedi et al. (1987). In the breeding of rhizomania-resistant genotypes, it is necessary to: (i) precisely measure the BNYVV content in the roots; (ii) screen with all known virus pathotypes; and (iii) use inbreeding to fix the traits.

Some seed companies are using molecular markers to detect the *Rz* alleles in sugar beet lines. Barzen et al., (1997) described a method of selecting individuals with the resistance gene, based on SCAR markers. The immunoenzymatic systems (ELISA) for detecting the virus (Clark and Adams, 1977) are frequently utilized by breeders as an early selection protocol, or to confirm plants are resistant to BNYVV. Giunchedi et al., (1985) indicated that it is important to take two samples - one obtained by grouping the rootlets, the other from the taproot. Relative BNYVV concentrations provide information on the level of resistance of each plant. Plants with positive BNYVV in the rootlets and no virus (zero) in the taproots are generally considered resistant. It is also possible to find plants with very low virus concentrations in the rootlets. The ELISA test is applied to seedlings grown in a greenhouse in soils infected with different BNYVV sources (or with the single A, B, P types) at 20°C. In these conditions, plants with negative ELISA value (zero virus) in both the rootlets and taproot are assumed to have the maximum level of resistance.

Current resistant varieties tested under infected greenhouse conditions showed no more than 80% resistant plants. Improvement of this rate should enable overcoming the most severe rhizomania infections. In other words, it should be possible to obtain varieties in which all plants carry the resistant trait in complete heterozygote form. At least for fields infected with BNYVV-P, we cannot exclude the so-called *Rz2* resistance is necessary to improve the expression of the *Rz* trait. In areas with both rhizomania and cercospora, breeding efforts to combine the *Rz* qualitative genes and the quantitative Alba trait in a single variety would provide protection from both diseases and increase sugar yields.

2.10.2.5 Beet soilborne virus (M. De Biaggi)

Beet soilborne virus (BSBV), a furovirus, is frequently detected in conjunction with BNYVV, the causal agent of rhizomania and is transmitted by the same vector, *Polymyxa betae*. BSBV is common in many sugar beet production areas. In the early 1980s, it was detected in the UK, Scandinavia, Germany, and Italy (Rubies-Autonell et al., 1996; Büttner, 2002). Similarity in particle morphology and the occurrence of rhizomania-like symptoms in sugar beet fields where BNYVV is absent led to speculation regarding the contribution of BSBV to rhizomania (Büttner, 2002). Wisler et al., (2003) reported that in the USA, a *benyvirus,* beet soilborne mosaic virus (BSBMV), also transmitted by *P. betae,* can cause rhizomania-like symptoms.

Rubies-Autonell et al., (1996) found that only 5% of soil samples collected in Italy were positive for BSBV alone, 29% were positive for BNYVV alone, and both BNYVV and BSBV were present in 35%. To date no involvement of BSBV in the expression of rhizomania severity has been demonstrated, but the presence of other beet soilborne viruses with BNYVV in seriously diseased areas indicates a need for further investigation. Liu et al., (2003) suggested that a new *beny*-like virus, tentatively called beet oat-leaf virus (BOLV), may affect rhizomania severity in California. BOLV is serologically distinct from BNYVV but its host range appears to be similar.

Büttner (2002) studied the reaction of a group of sugar beet varieties, one of which appeared to be BSBV resistant in field trials in Germany. He concluded that: (i) no significant losses in beet yield and quality were observed on BSBV contaminated fields *vis-a-vis* fields without the virus; (ii) BSBV does not seem to contribute to the damage caused by rhizomania, therefore BNYVV is the sole causal agent of rhizomania; and (iii) due to the large number of BSBV pathotypes, the existence of pathogenic strains detrimental to sugar beet cannot be ruled out.

2.10.3 Bacteria

2.10.3.1 Bacterial vascular necrosis and rot (L.G. Campbell)

Most beet diseases caused by bacteria do not result in extensive losses; however, one exception is erwinia root rot (bacterial vascular necrosis and rot) caused by *Erwinia carotovora* ssp. *betavasculorum* (Thomson et al., 1981a). Erwinia root rot causes significant losses primarily in western USA, but also occasionally in Europe (Duffus and Ruppel, 1993). The disease is characterized by necrotic or discolored vascular bundles in the root. Severely infected roots may become "hollowed" but plants sometimes remain alive. Aboveground symptoms include black lesions along the petioles, froth in the center of the crown, and wilt (Whitney, 1986b). The bacterium persists in soils where the disease occurs (Thomson et al., 1977) and is transmitted by cultivation, insects, or water splash (Whitney and Lewellen, 1985). Plant injury or cracks in the roots provide infection courts for the pathogen.

Recognition of erwinia root rot as an important disease in the USA coincided with the widespread distribution of two virus yellows-resistant hybrids, USH9 and USH10 (Whitney and Lewellen, 1977). During development of the pollinator lines used in these hybrids the erwinia root rot resistance of the parental population, US75, was lost. This was probably the result of too few plants being used to reconstitute the gene pool during selection for virus yellows resistance (Whitney and Lewellen, 1977). Some cultural practices reduce erwinia root rot severity (Thomson et al., 1981b) but resistant varieties provide the most reliable control (Ruppel et al., 1975b).

Susceptibility to *Erwinia* is common in species of *Beta* in the *Vulgares* and *Corollinae* sections (Whitney, 1982). An absence of disease on three *Beta* species of the *Patellares* section suggests it could be a potential source of resistance (Whitney, 1982). Whitney and Lewellen (1978a) observed that breeding populations responded to selection rapidly with much of the improvement in the first cycle, evidence for control by a single locus. Higher levels of resistance obtained with additional selection cycles suggested that additional genetic factors affect the rate of disease development. Further study (Lewellen et al., 1978b) confirmed that resistance was simply inherited, with a large dominance component. A second, primarily additive component determined the amount of rot in susceptible plants. This additive component may confer useful levels of resistance in the absence of the major resistance gene.

Both field and greenhouse trials have been used to evaluate germplasm and study the pathogen. Disease severity has been quantified using a disease index, percent diseased plants, or percent of rot per diseased plant. Correlations among the three were highly significant (Whitney and Mackey, 1989). Stanghellini et al., (1977) identified biotypes of the pathogen, based upon serological and physiological differences. More recently, differences in aggressiveness among strains have been documented with the recommendation that several highly aggressive strains be used when selecting for durable resistance (Whitney and Mackey, 1989). Ezra et al., (2000) characterized a gene from *E. herbicola* that encodes a protein that acts as a virulence factor on *Gypsophila paniculata* and elicits a hypersensitive response on beet. Further understanding of interactions between host and pathogen may eventually lead to alternative disease management strategies. Meanwhile erwinia root rot resistant germplasm is available (Whitney and Lewellen, 1978b), some with multiple disease resistance (Lewellen et al., 1978a; Whitney, 1989b; Lewellen, 1995c).

2.10.3.2 Yellow wilt (R.T. Lewellen and E. Biancardi)

Yellow wilt is a serious destructive sugar beet disease that occurs only in Argentina and Chile. Symptoms include yellowing and wilting of the leaves, resulting in the death of infected plants, especially when accompanied by drought and high temperatures (Whitney and Duffus, 1986). Yellow wilt is caused by a rickettsia-like organism transmitted by a leafhopper, *Paratanus exitiosus,* and two species of dodder (Bennett et al., 1967). Gaskill and Ehrenfeld (1976) observed a range in disease severity among 381 varieties and breeding lines of European and American origin. Efficient screening and breeding programs must have disease nurseries or screening procedures that guarantee high percentages of infected plants. Mother beets are selected in both the yellowing and subsequent wilting phase (Bennett and Munch, 1946).

Success in breeding for resistance was initially hampered by frequency of escapes in the disease nurseries and difficulties in producing seed on selected plants that were resistant, but not immune. Satisfactory amounts of seed were obtained by repeatedly spraying the flowering plants with antibiotics (Gaskill and Ehrenfeld, 1976). Three sea beet introductions displayed a relatively high resistance to yellow wilt, suggesting sea beet is a potential source of resistance genes. As of now, no variety with appreciable resistance to yellow wilt has been identified.

2.10.4 Fungi

2.10.4.1 Cercospora leaf spot (G.N. Skaracis and E. Biancardi)

Cercospora leaf spot, caused by the fungus *Cercospora beticola*, is prominent in humid temperate zones where sugar beet is cultivated. It is particularly damaging in Greece, northern Italy, northern Spain, Yugoslavia, Austria, southern France, Japan, China, Michigan (USA) and elsewhere (Holtschulte, 2000). The disease appears to have spread in recent years into northern zones where it occurred sporadically in the past. The infection agent provokes characteristic necrotic lesions that enlarge and eventually cause a rapid destruction of the foliar apparatus. Several classes of fungicides have proven effective in limiting the disease spread (Meriggi et al., 2000). Chemical treatment, much like genetic resistance, provides partial protection. When fungicides and resistant varieties are combined in an integrated management system, satisfactory, albeit not complete disease control is achieved (Biancardi and Graf, 1984; Miller et al., 1994; Skaracis and Biancardi, 2000).

Only one source of partial, quantitative, genetic resistance is available for agronomic use. A second, qualitative, type of resistance named C2, has been reported for plants infected with pathogen strains from California (Lewellen and Whitney, 1976). However, when isolates from other areas were used, disease progression was unhampered. Due to the specific and very limited conditions required for its expression, this monogenic resistance has been of no practical interest. Species within the section *Procumbentes* exhibit high levels of resistance with unknown genetic characteristics (Biancardi, unpubl. data).

The partial resistance dates back to crosses Munerati initiated in Rovigo, Italy. Simultaneously around 1915, using *Beta vulgaris* ssp. *maritima* that grows spontaneously along the coast in the Po River delta. In the USA, Tracy was attempting to isolate resistance in commercial genotypes (Coons et al., 1955) through mass selection followed by several cycles of inbreeding. The lines obtained were used by Spencer in 1925 and by Skuderna four years later (Coons, 1936). The commercial varieties US217 and US200 × 215 were subsequently produced (Coons et al., 1955).

After about 20 years of selection and backcrossing, Munerati's more resistant lines, in particular R581, were distributed to public organizations and seed firms in Italy, Europe, and the United States and programs for adaptation to local conditions initiated (Coons, 1975). The task was most actively pursued in the USA where the commercial varieties GW304 and

GW359 were produced and from which, following the introduction of monogermity, GW672 and US201 were derived, the latter selected by USDA geneticists (Coons, 1975; Lewellen, 1992). In Italy, the same materials gave rise to varieties "Cesena" and "Mezzano 71" that were also used extensively in cultivation. Apart from resistance to cercospora l.s., these varieties manifested a moderate tendency to bolting as well as a relatively low root yield, both defects of origin that were gradually diminished. The same was true for a particular tendency to root fanging (with secondary roots emerging at a 90° angle from the tap root) typical of *B.v.* ssp. *maritima* (Coons et al., 1955; Biancardi, 1999).

The cercospora l.s. resistance commonly used in sugar beet breeding is controlled by at least 4 or 5 gene pairs with variable effects depending on the severity of infection (Smith, G.A. and Gaskill, 1970). Broad sense heritability ranges from 60 to 71%, while gene action is of a quantitative nature. Narrow sense heritability was found to be around 24% and environmental variation ranged from 44 to 62% (Smith and Ruppel, 1974). Based on QTL analysis, Koch (1997) agreed to some extent with the aforementioned results, attributing a part of the difficulties encountered in selection to recessive genes controlling the trait expression.

A major difficulty for improvement programs is the negative correlation between resistance and sugar yield (Koch, 1970; Christmann, 1983; Smith and Campbell, 1996). A further negative factor was introduced with genetically monogerm hybrids, in which the seed parents almost invariably are more susceptible to cercospora l.s. than the pollinators. Consequently, the hybrids are generally more susceptible than their multigerm parent (Skaracis and Biancardi, 2000). The passage from multigerm to genetically monogerm seed was accompanied by a noticeable deterioration in cercospora l.s. resistance. To cope with the problem, several male sterile (CMS) lines with improved disease resistance were developed. Among these lines, specific reference should be made to those developed at Fort Collins CO (FC502/2, FC504, FC604, FC605, FC607 and FC609).

Some CMS lines with an acceptable level of resistance exhibit a low combining ability when crossed to resistant pollinators. This, most probably, reflects a lack of genetic diversity for the trait; the common resistance background in both parents reduces heterosis (Hecker, 1985). To overcome such difficulties, it becomes necessary to employ new sources of resistance.

More dependable criteria are needed in order to recognize and classify resistant individuals (Coons et al., 1955). The difficulty in properly characterizing disease resistance based solely on foliage observations, apart from genotypic variation, is due to the substantial environmental influence on the trait. Several researchers have attempted to develop a more objective selection index. Bosemark (1993) verified that resistant varieties are distinguished by a higher concentration of dihydroxy phenylalanine as earlier reported by Hecker et al., (1975). On the other hand, disease resistance has been related to two phytoalexin-like fungitoxic isoflavones, namely betagarin and betavulgarin, that are more abundant in leaves of inoculated resistant plants (Martin, 1977, 1990).

Evidence of various responses to disease infection, depending on the resistance level of the host plant, was recently verified (Rossi et al., 1997). In susceptible varieties, for example, production of fungal spores is higher, incubation period shorter, and diameter of necrotic lesions larger than in resistant varieties. Some of these characters, less environmentally influenced and more objectively measured, may facilitate recognition and isolation of true resistant plants and improve selection efficiency.

The importance of selecting under controlled conditions in order to reduce environmental influence, has further been confirmed by *in vitro* techniques in which infection progress was evaluated in sugar beet leaf disks under sterile conditions. Results proved quite reliable despite considerable variation among leaves of the same plant as well as among plants of the same line (Koch, 1997).

During the juvenile stage (up until 80-90 days from emergence), sugar beet plants frequently appear immune to *Cercospora* attack, suggesting an inhibitory mechanism for fungal establishment and spread on the leaves during that period. Several explanations have been postulated, for example reduced stomatal density and size or lack of synchronization between hyphae elongation and stomatal openness (Canova, 1959; Solel and Minz, 1971). Ruppel (1972) however, was not able to confirm a relationship between cercospora resistance and these mechanical hindrances.

2.10.4.2 Powdery mildew (L.G. Campbell)

Powdery mildew occurs worldwide on sugar beet, causing root yield losses, reduction in sucrose concentration (Hills et al., 1975), and increased impurity concentrations (Skoyen et al., 1975). The disease is most damaging in warm arid climates and frequent rains or sprinkler irrigation (Ruppel et al., 1975a) impede disease progress. Susceptibility increases with plant age (Ruppel and Tomasovic, 1977). The causal fungus, *Erysiphe betae* (*E. polygoni*) (Weltzien, 1963), is specific to *Beta* species (Ruppel and Tomasovic, 1977; Francis, 2002). Mildew first appears as dispersed whitish mats of hyphae on older leaves, then spreads rapidly over the upper, and sometimes lower leaf surfaces until all leaves appear dusty white. Infected leaves eventually develop chlorosis and senesce prematurely (Ruppel, 1986; Francis, 2002). Powdery mildew also occurs in greenhouses and enclosures used for seed production. The fungus overwinters on escaped or wild beet and spreads via windborne conidia (Ruppel et al., 1975; Ruppel and Tomasovic, 1977; Ruppel, 1986). Sexual reproduction has a very limited role in the disease cycle but could facilitate development of more virulent or better adapted strains (Byford and Bentley, 1976) and provide an additional overwintering form.

According to Francis (2002), powdery mildew on sugar beet in Europe was first documented in 1903. In England the occurrence of powdery mildew increased from a disease rarely seen in the late 1940s to a significant problem since the early 1970s. Weltzien (1963) documented the recognition and spread of sugar beet powdery mildew in numerous European and Middle East countries. In North America, powdery mildew was discovered on sugar beet in 1934 but not recognized as a serious disease until an outbreak in 1974 (Kontaxis et al., 1974; Ruppel et al., 1975; Ruppel and Tamasovic, 1977).

In response to the increased prevalence of powdery mildew, breeders began searching for sources of resistance. Russell (1965, 1966a) identified lines, including two constituent lines of a commercial hybrid, with some powdery mildew resistance among lines in the virus yellows (Beet virus yellows and Beet mild yellowing virus) breeding program. Powdery mildew incidence and severity in most European varieties was lower than in lines in the US breeding program at Salinas, California, in 1974 (Skoyen, 1975). In general, *Beta* species within the section *Beta* are susceptible to powdery mildew (Van Geyt et al., 1990; Luterbacher et al., 1998; Asher et al., 2001a). Among *B.v.* ssp. *maritima* accessions, frequency distributions are skewed toward susceptibility (Asher et al., 2001b). Whitney (1989) identified seven *B.v.* ssp. *maritima* accessions with resistance. Resistance within fodder beet, garden beet, and leaf beet lines is uncommon (Luterbacher et al., 2000). Species within the *Corollinae* section are very susceptible and some species within the *Procumbentes* section appear to have partial resistance.

Whitney (1989) reported that F_1 hybrids of a mildew susceptible variety and resistant *B.v.* ssp. *maritima* were resistant. Significant genetic variation and broad-sense heritabilities between 54 and 75% (Whitney et al., 1983) indicated that selection of moderately resistant lines was feasible (Lewellen, 1995c). Resistance was characterized as slow mildewing or horizontal. More recently, a dominant gene (*Pm*) that confers a high level of resistance was identified in two *B.v.* ssp. *maritima* accessions, WB97 and WB242 (Lewellen, 2000; Lewellen and Schrandt, 2001). The allelism of the two sources has not been determined. Although lines carrying the *Pm* gene have high levels of resistance, powdery mildew was observed on mature leaves late in the season. Janssen et al., (2003) mapped the resistance gene from WB242 to chromosome II. Francis (2002) identified two molecular markers associated with the quantitatively expressed resistance from PI 504236, a *B.v.* ssp. *maritima* accession. Five markers in four linkage groups that explained 27% of the phenotypic variation in resistance have also been identified (Janssen et al. 2003).

Both greenhouse and field evaluations have been effective in screening for resistance. Whitney et al., (1983) found the highest consistent correlations between field and greenhouse observations when greenhouse plantings were evaluated five weeks after inoculation of 6-week-old plants. However, the difference between the most resistant and most susceptible entries remained fairly constant between 2 and 5 weeks after inoculation, and greenhouse seedling trials could identify highly susceptible lines. Russell (1965) proposed a 0 to 5 rating scale based on incidence of mycelium on old, middle-aged, and young leaves, while Bayford (1978) described a 0 to 5 scale that incorporated percentage of infected plants. Hills et al., (1980) quantified disease intensity on mature leaves using a 0 to 5 scale with large intervals in the intermediate classes and small intervals on both ends. The average rating of multiple observations of individual plants were then converted to a percent mature leaf area diseased. A linear 0 to 4 scale and an expanded 0 to 9 scale (Whitney et al., 1983) corresponding to leaf area covered have gained wide acceptance.

Magyarosy et al., (1976) observed that powdery mildew had a negative effect on photosynthesis similar to Diuron, a substituted urea herbicide. Francis (2002) cited research by Hamacher and Weltzien that includes callous formation under penetration sites, hypersensitive cell death, and fungitoxins as defense mechanisms. A better understanding of the plant systems affected, the mechanisms behind resistance, and gene transfer technology would hasten development of durable resistance and reduce dependence on chemical control.

2.10.4.3 Downy mildew (E. Biancardi and R.T. Lewellen)

Downy mildew, caused by *Peronospora schachtii* (*farinosa*), attacks newly developed leaves in rainy or cool climates. Severe damage is quite frequent in northern Europe in both sugar and seed crops (Duffus and Ruppel, 1993). Crombie and O'Connor (1960) observed a correlation between ploidy level and disease severity. In particular, tetraploid genotypes appeared more resistant than 3n and 2n genotypes. Campbell and Russell (1964) listed six distinct independent resistance mechanisms, of which the most important is the hypersensitive reaction of host cells. In contrast to the hypersensitive response, the other factors are quantitative and relative severity is influenced by interactions with the environment.

According to Bogdanova and Shevstov (1995), resistance to downy mildew appears to be dominant. Downy mildew resistance is present in the sections *Vulgares* and *Corollinae* of the genus *Beta* (Frese et al., 2001; Asher et al., 2001a).

Predominant resistance breeding has been carried out in California (McFarlane, 1952, 1971), where susceptible varieties have been gradually replaced or integrated with moderately resistant materials (Duffus and Ruppel, 1993). McFarlane (1952) noted that the percentage of infected plants in inbred lines ranged from 3 to 85 and that greenhouse selection permitted rapid improvement. Selecting for downy mildew resistance in self-fertile lines was more successful than selection in cross-pollinated genotypes. The demand for monogerm hybrid varieties prompted the introduction of moderate resistance into CMS lines. Lewellen and Skoyen (1988) registered four monogerm, self-fertile O-type lines with moderate resistance to downy mildew.

2.10.4.4 *Fusarium* yellows (L. Panella and R.T. Lewellen)

Fusarium spp. can attack sugar beet as a seedling (seedling wilt), at the mature root stage (fusarium yellows), and while flowering during seed production (fusarium stalk rot). These vascular diseases of sugar beet are caused mainly by the fungus *Fusarium oxysporum* (Schneider and Whitney, 1986b; Duffus and Ruppel, 1993) but other *Fusarium* spp. can also cause them (Mukhopadhyay, 1987; Ruppel, 1991). Fusarium yellows is of particular importance in the Central High Plains of the USA (Montana, Wyoming, Colorado, Nebraska), but is found throughout the country (Harveson and Rush, 1997) and has also been found across Europe and Asia (Hillmann and Schlösser, 1987; Hull, 1960a; Mukhopadhyay and Thakur, 1970; Rudakou and Al'Akhovskaya, 1964). Although the pathogen is designated as *Fusarium oxysporum* f. sp. *betae*, spinach and pigweed were shown to host it, while some weeds (*Chenopodium album, Brassica nigra*, and *Anethum gravolens*) have been identified as symptomless carriers (Armstrong and Armstrong, 1976; MacDonald and Leach, 1976).

Stewart (1931) first reported and described the symptoms of fusarium yellows in the Arkansas valley of south-eastern Colorado. Early attempts to screen open-pollinated varieties, breeding lines, and inbred populations for resistance to the disease in the 1930s showed genetic variation for resistance present. The results of four years of selection indicated that progress could be made using mass selection in field plots with a history of the disease (Bockstahler, 1940).

In the early 1970s, fusarium stalk blight emerged as a problem in the US sugar beet seed fields (Gross and Leach, 1973) and screening of hybrid parents for stalk blight resistance was begun (MacDonald and Leach, 1976). It was observed that individual susceptible plants occurred throughout most breeding lines and populations, but that, inadvertently, curly top-resistant, monogerm O-type, inbred parental lines, such as widely used C563 and C563CMS, were highly susceptible leading to problems in seed production. It was noted that old lake bottom soils with high organic matter were the most conducive to the disease and, in part, seed production was improved by growing known susceptible inbreds on mineral soils. In 1976, J.S. McFarlane at the USDA-ARS station in Salinas CA began a program breeding for resistance to fusarium stalk blight that ultimately led to the release of the stalk blight-resistant parental lines, C566 and C566CMS (McFarlane, 1982), and germplasm, C554 (McFarlane et al., 1982b) and C859 (Lewellen, 1995c). From crosses made in his breeding program, McFarlane concluded that *Fusarium* resistance was dominant, not linked to the monogerm trait, and polygenic (McFarlane, 1981). Isolates causing fusarium stalk blight appeared to differ from those that caused fusarium yellows but strains of *Fusarium oxysporum* isolated from diseased sugar beet stalks have caused fusarium yellow symptoms in greenhouse tests (Hanson, pers. comm.).

Genetic variation in *Fusarium oxysporum* f. sp. *betae* is reportedly considerable (Harveson and Rush, 1997; Hanson et al., 2001). Martyn and coworkers (Martyn et al., 1989) found that symptoms caused by *F. oxysporum* isolates in Texas differed from those normally associated with fusarium yellows. Anecdotal evidence indicates resistant varieties performed well in some locations in the presence of the disease but proved susceptible in others. This has also been seen in preliminary greenhouse tests in which resistant varieties appear susceptible to other isolates of *F. oxysporum* (Hanson, pers. comm.). If races of *F. oxysporum* were present, this would have to be taken into account in any program breeding for resistance to fusarium yellows. Because other *Fusarium* spp. are able to cause yellow symptoms (Mukhopadhyay, 1987; Ruppel, 1991; Hanson, pers. comm.), possibily they are the culprits confounding efforts to develop fusarium yellow-resistant germplasm.

At present, there are no active public breeding programs for *Fusarium* resistance in the USA or western Europe, although most of the seed companies active in the USA have released *Fusarium*-resistant varieties. It is not known whether the material released from the USDA-ARS station in Salinas has been used in proprietary breeding programs but many seed companies screen and select in field nurseries in the USA. A search of the USDA-ARS *Beta* germplasm collection revealed that of the 80 accessions screened for resistance to *Fusarium*, three have been rated as *Fusarium* resistant (rating 3 on a scale of 0 = no disease to 9 = dead). Based on the experience of Bockstahler (1940) and McFarlane (McFarlane, 1982; McFarlane et al., 1982b), quite likely resistance to *Fusarium* is present in a number of sugar beet genetic backgrounds and field screening and selection is a feasible method for crop improvement.

2.10.4.5 Root rots (L. Panella)

Rhizoctonia solani is endemic in all sugar beet production areas and causes root and crown rot (AG-2-2) and seedling damping-off (primarily AG-4) (Herr, 1996). *Pythium* spp.—*Pythium aphanidermatum* (Hine and Ruppel, 1969; Schneider and Whitney, 1986d), *Pythium deliense* (Stanghellini et al., 1982), and *Pythium ultimum* (Abawi et al., 1986; Leach and Garber, 1970)—can harm sugar beet seedlings (damping-off) or cause root and tip rot in the mature root. *Phoma betae* can cause a seedling disease (damping-off) often referred to as "black leg", and a postharvest rot of roots in storage piles (Mukhopadhyay, 1987). Seedling damping-off and storage rot are the most damaging diseases caused by *Pythium betae* (Duffus and Ruppel, 1993). The seedling damping-off caused by these pathogens is generally controlled by chemical seed treatments and, although there have been some attempts to develop resistant germplasm (Bugbee et al., 1989), this is not a major breeding goal for commercial seed companies.

Rhizoctonia

Rhizoctonia root and crown rot of sugar beet which affects or threatens 5-10% of the US and European growing areas (Büttner et al., 2003; Herr, 1996), is present worldwide in sugar beet areas (Ahmadinejad and Okhovat, 1976; Mukhopadhyay, 1987; Ogata et al., 2000; Stevens and Dewar, 1996; Toporovskaya, 1975), and has become more of a problem wherever crop rotations are shortened. Although azoxystrobin fungicide effectively controls this disease, timing of application is critical (Stump et al., 2002) and the best control is an integrated management program including good cultural practices, timely fungicide application, and resistant germplasm (Herr, 1996).

Gaskill began a rhizoctonia root and crown rot resistance breeding program in the late 1950s at the USDA-ARS research station in Fort Collins CO. He released the first two resistant varieties in 1966 (Gaskill, 1968; Hecker and Gaskill, 1972) and releases of resistant germplasm from this program have continued since 1966 (**Table. 2.2**). Root yield of resistant germplasm is not reduced under heavy disease pressure and sucrose yield remains constant when disease pressure is mild or absent (Ruppel and Hecker, 1994). Gaskill and coworkers realized that natural field epiphytotics did not produce consistent, uniform disease pressure and he and others used mass selection or recurrent field selection under

Table 2.2 Rhizoctonia-resistant germplasm released by the USDA-ARS at Fort Collins CO. The FC number is given, the year the germplasm was registered in Crop Science (usually one to two years after release), the genetic source of germplasm, and the Crop Science volume and page of the registration with the germplasm number.

Release	*Year*	Source	*Crop Science Registration*
FC701	1972	Great Western (GW359-52R)	*CS* 12:400 B GP-1
FC701-4	1976	Great Western (GW359-52R)	*CS* 17:678 B GP11
FC701-4(4X)	1976	Great Western (GW359-52R) B Colchicine doubled	*CS* 17:678 B GP12
FC702	1972	Great Western (GW359-52R)	CS 12:400 B GP2
FC702-4	1978	Great Western (GW359-52R)	*CS* 19:934 935 B GP55
FC702-6	1981	Great Western (GW359-52R)	*CS* 22:454 B GP65
FC702-7	1983	Great Western (GW359-52R)	*CS* 25:374 B GP92
FC702-4(4X)	1978	Great Western (GW359-52R) B Colchicine doubled	*CS* 19:934 935 B GP56
FC703	1976	Great Western (GW359-52R)	*CS* 17:678 B GP13
FC703(4X)	1976	Great Western (GW359-52R) B Colchicine doubled	*CS* 17:678 B GP14
FC703-4	1981	Great Western (GW359-52R)	*CS* 22:1275 1276 B GP84
FC704	1978	German red beet - Great Western population	*CS* 19:934 935 B GP54
FC705	1978	Great Western (GW359-52R)	*CS* 19:934 935 B GP57
FC705-1	1983	Great Western (GW359-52R)	*CS* 25:374 B GP93
FC706	1978	GW359-52R, *B. v.* ssp. *maritima*	*CS* 19:934 935 B GP58
FC707	1978	Great Western (GW359-52R)	*CS* 19:934 935 B GP59
FC707(4X)	1986	Great Western (GW359-52R)B Colchicine doubled	*CS* 27:822 B GP119
FC708	1980	ARS-CTR[1], -LSR[2], *B. v.* ssp. *maritima,* GW359-52R	*CS* 21:802 B GP63
FC708CMS	1980	CMS equivalent of FC708	*CS* 21:802 B GP64

(Table 2.2 Contd.)

(Table 2.2 Contd.)

FC709	1987	Great Western (GW359-52R)	*CS* 28:1039 B GP128
FC709-2	1999	Great Western (GW359-52R)	*CS* 39:298-299 - GP200
FC710	1990	SP 5831-0 (some *B. v.* ssp. *maritima)*	*CS* 31:494 - GP133
FC711	1982	Heterogeneous Japanese lines (GW359-52R)	*CS* 23:601 602 - GP87
FC712	1985	Great Western (GW359-52R)	*CS* 26:213 214 B GP97
FC712(4X)	2001	Great Western (GW359-52R) Colchicine doubled	*CS* 41:1374 B GP217
FC715	1992	FC708 & Fort Collins LSR[2] Germplasm	*CS* 34:290 - GP148
FC715CMS	1992	CMS equivalent of FC715	*CS* 34:290 - GP149
FC716	1992	Several commercial hybrids	*CS* 35:291 - GP143
FC717	1992	Commercial hybrid (ACH14) & FC708	*CS* 35:291 - GP144
FC718	1992	Four Russian open-pollinated populations (PIs)	*CS* 35:291 - GP145
FC719	1992	Three high sucrose Polish lines & GW359-52R	*CS* 35:291 - GP146
FC721	1997	ARS-CTR[1], -LSR[2], GW359-52R, & C718 (USDA, Salinas, CA)	*CS* 37:1675-1676 - GP185
FC721CMS	1997	CMS equivalent of FC 721	*CS* 37:1675-1676 - GP186
FC724	2003	ARS-CTR[1], -LSR[2], & FC702	in press
FC725	1996	C37 (USDA, Salinas, CA) & FC707	*CS* 36:819-820 - GP167
FC726	1996	Peramano (fodder beet) & FC703	*CS* 36:819-820 - GP168
FC727	1999	FC703 & three high sucrose sources (Polish, Spanish, USA)	*CS* 39:298-299 - GP201
FC728	1996	FC708 & three commercial hybrids	*CS* 36:819-820 - GP169

[1]ARS-CTR: Early ARS curly top resistant (CTR) hybrids

[2]ARS-LSR: Early ARS leaf spot resistant (LSR) hybrids

artificially induced infection (Hecker and Ruppel, 1977b; Ruppel et al., 1979; Schneider et al., 1982), a method used by many breeding companies today (Panella, 1998). Greenhouse screening is also possible (Büttner et al., 2004; Campbell and Altman, 1976; Scholten et al., 2001; Weiland et al., 1999).

In sugar beet, resistance to *R. solani* is polygenic, reported to involve at least two loci, two or three alleles, and modifying genes in some populations (Hecker and Ruppel, 1975). Initially Gaskill reported resistance in FC702/3 to be almost completely dominant (Gaskill et al., 1970). Later broad-sense heritability was estimated at 0.65 with some dominance and nonadditive components of the variance (Hecker and Ruppel, 1975). Ogata reported mainly additive effects (Ogata et al., 2000). Resistance is equal in diploid and tetraploid lines but a dosage effect was observed in diploid and triploid resistant hybrids produced by crossing a susceptible CMS line with a resistant pollinator (Hecker and Ruppel, 1976). There have been continued efforts to screen genetic resources for new sources of resistance for over 40 years (Burenin, 2001; Luterbacher et al., 2000; Panella, 1998; Panella and Frese, 2003; Schneider and Gaskill, 1962), nonetheless, most commercial resistant germplasm can trace its parentage to the USDA-ARS program in Fort Collins CO. Today, resistance to *Rhizoctonia* has been incorporated by major seed companies into O-type maintainer, CMS female, and multigerm-pollinator lines and is currently also deployed in commercial hybrids combined with rhizomania resistance.

Phoma

Phoma betae has been reported throughout the world as causing not only seedling disease, but also leaf spot, stem rot, and crown and root rot (Mukhopadhyay, 1987) and, in areas where sugar beet is stored in piles after harvest, can cause a devastating storage rot (Bugbee and Cole, 1976; Burenin, 2001; Mukhopadhyay, 1987; Schneider and Whitney, 1986c; Sheikholeslami et al., 1998). In the USA, there has long been an interest in utilizing host-plant resistance to reduce the loss of beets to this disease while they are piled throughout the winter (Gaskill, 1950a,b, 1952; Nelson and Oldemeyer, 1952). This work was in some part based on the research done in the former Soviet Union (Campbell and Bugbee, 1993; Gaskill, 1952; Nelson and Oldemeyer, 1952) where susceptible commercial varieties suffered up to twice as much damage from storage rots as resistant commercial varieties (Popova, 1961).

Working at the USDA-ARS Research Station in Fargo ND, Bugbee utilized methods for selecting individual roots resistant to fungi causing storage rot (Bugbee, 1979a,b) and demonstrated that protection offered by genetic resistance was comparable to that from fungicide application

(Bugbee and Cole, 1979). Five germplasm with resistance to storage rot have been released from the USDA-ARS in Fargo ND (Bugbee, 1978; Campbell and Bugbee, 1985). One of these lines (F1006) is a polycross of resistant roots from a number of accessions from the USDA-ARS *Beta* collection, three germplasm were derived from Russian sources (F1001, F1004, and F1005), and one germplasm (F1002) was selected from a rhizoctonia resistant source, FC701/4 (Hecker and Ruppel, 1977a). Hybrids with a *Rhizoctonia* or *Aphanomyces*-resistant CMS parent were more resistant to storage rots than hybrids with susceptible CMS parents (Campbell and Bugbee, 1993). Further research demonstrated that selecting for storage rot resistance did not necessarily compromise agronomic performance (Campbell and Bugbee, 1988) and showed a relationship between rhizoctonia resistance and storage rot resistance (Bugbee and Campbell, 1990). However, although seedlings of germplasm selected as mature roots for storage rot resistance showed increased resistance to *P. betae*, they did not show increased resistance to *Rhizoctonia solani* (AG2-2 or AG4) (Bugbee et al., 1989). This USDA-ARS research has been reviewed by Campbell and Bugbee (1993).

Currently there are no major efforts in North America to screen the USDA-ARS National Plant Germplasm System's *Beta* collection for resistance to *Phoma betae* and the GRIN data base contains no evaluation results for either *Phoma* root rot or storage rot. There is some screening and breeding for *Phoma* resistance as a component of a blackleg resistance breeding program in Russia (Burenin, 1998, 2001).

Pythium

Although *Pythium* spp. are known to cause seedling damping-off, they can also cause root rot in the mature beet root (Ahmadinejad and Okhovat, 1976; Rush, 1987; Stanghellini et al., 1982; Stanghellini et al., 1983; von Bretzel et al., 1988), and *Pythium* spp. are commonly found with other pathogenic fungi in a "root rot complex" of diseases (Burenin, 1998; Harveson and Rush, 1994). There has been very little effort in breeding for resistance to the mature root rot or screening germplasm for resistance other than that reported in Russia as part of the blackleg resistance breeding program (Burenin, 2001). Recently, the EC/GRP collaborative project "Evaluation and Enhancement of *Beta* Collections for the Extension of Agricultural Production" (GENRES CT95 42) evaluated 597 accessions from a synthetic core collection for seedling resistance to *Pythium ultimum*, and these results are available on the World-Wide Web and can be accessed on the "*Beta*-version" of the redesigned International Data Base for Beta (IDBB) at: http://ice.zadi.de/idbbonline/beta.php (Asher et al., 2001a; Luterbacher et al., 2000).

2.10.4.6 Southern sclerotium root rot (R.T. Lewellen and E. Biancardi)

Southern sclerotium root rot, caused by the fungus *Sclerotium rolfsii,* is a problem in warm beet-growing areas, especially when sugar beet is grown as a winter crop. The effects of the disease on the plants are devastating: a blackish rot develops rapidly in the taproots, which become completely covered by thick strands of white mycelium. Several dark brown spherical sclerotia subsequently appear in this cottony layer (Whitney and Duffus, 1986). The disease can develop rapidly in piled beets, causing severe postharvest storage losses, and subsequently problems in processing (slicing, diffusion etc.). In some production areas, infection by *Sclerotium* may occur prior to seedling emergence (Fahim et al., 1981). Attempts to control the disease with chemicals, crop rotation, or biological control agents have not been successful (Duffus and Ruppel, 1993). As of now, no effective control method is available. Heavy applications of nitrogen fertilizer appear to reduce the damage caused by southern sclerotium root rot (Leach and Davey, 1942; Thakur and Mukhopadhyay, 1972).

The first attempt to select for resistance was initiated in 1946 in California (Lawlor and Doxtator, 1950). Differences in survival percentage were detected among and within the breeding lines. Coe and O'Neill (1983) described a greenhouse method for screening lines in early growth stages, when the hypocotyl diameter is about 6 mm. Most surviving seedlings in inoculated pots appeared to have a larger hypocotyl diameter. The correlation coefficient between hypocotyl diameter and disease severity was low but significant. Progeny of plants selected on the basis of both hypocotyl diameter and disease rating were evaluated in field nurseries. Heritability of hypocotyl diameter was low, but mass selection based on both traits resulted in reasonably resistant offspring. The resistance exhibited by the most resistant progeny (SP7822) provided adequate protection from the disease under low/moderate levels of infection.

Srivastava (1995) outlined procedures for obtaining varieties for cultivation in the presence of southern sclerotium root rot in India. Genotypes, previously selected for tolerance/resistance to high temperatures, were grown in fields infected with *Sclerotium rolfsii.* Modified mass selection of mother beets lead to the development of two varieties "LS6" and "IISR-Comp1". Farid and Sajjad (2001) evaluated nineteen varieties in a disease nursery: only the multigerm variety "Mezzano Au-Poly" displayed a useful level of resistance.

2.10.4.7 Polymyxa betae (E. Biancardi)

Polymyxa betae is a *plasmodiophoromycete* which hosts and transmits the rhizomania virus (BNYVV). It is present in most soils where sugar beet is grown. It appears that the mycete, *per se*, i.e. without the virus, does not cause damage (Winner and Schäufele, 1977; Brunt and Richard, 1989). Several researchers have hypothesized that rhizomania resistant varieties are able not only to delay the reproduction or spread of the virus inside the root (Poggi Pollini and Giunchedi, 1989; Tuitert, 1994; Delasalle and Dhellemmes, 1996), but also to impede or hinder transmission of the BNYVV inoculum from the vector. Reduced reproduction potential of *P. betae* in root tissue should reduce virus entry into the plant (Lewellen *et al.*, 1987; Gerik and Duffus, 1988; Whitney, 1986a; Abe and Ui, 1986).

According to Fujisawa and Sugimoto (1979) resistance to *P. betae* is present in sections *Corollinae* and *Procumbentes*. With monosomic additions to *B. procumbens,* Paul *et al.* (1993, 1994) established that genes for resistance to *P. betae* are located on chromosomes IV and VIII. Resistance could reduce the virus concentration in the roots and thereby delay both increase of the virus in the soil and infection inside the roots (van Geyt *et al.*, 1990; Paul *et al.*, 1992). Preliminary attempts to develop *P. betae* resistant varieties and lines were summarized by Peretyat *et al.* (1992).

2.10.4.8 Black root (L. Panella)

The *oomycete Aphanomyces cochlioides* is responsible for black root disease, which occurs as both an acute seedling disease (damping-off) and later as a chronic root rot of the mature root (Schneider and Whitney, 1986a; Mukhopadhyay, 1987). Black root (chronic rot in the mature root) is often part of a root rot complex that can include fusarium yellows and rhizoctonia root rot (Harveson and Rush, 2002). This disease is serious in parts of the United States, especially in eastern North Dakota, Minnesota, and Texas, but can be a problem in most growing areas. It has been reported in Canada, Chile, France, Germany, Hungary, the UK, Russia, eastern Europe and former Russia-allied states, and Japan (Schneider and Whitney, 1986a; Duffus and Ruppel, 1993).

In the early 1940s, when black root was increasing in significance in the Red River Valley of North America (Doxtator and Downie, 1947), the American Crystal Sugar Co. began a plant breeding program to develop *Aphanomyces* resistant sugar beet (Doxtator and Downie, 1948; Downie et al., 1952). Early research by the USDA indicated resistance was dominant, as expressed in an experimental hybrid (Bockstahler et al., 1950); however progress in breeding was slow when using mass selection in infested fields (Doxtator and Finkner, 1954). Coe and Schneider (1966) together developed a greenhouse screening method for determining seedling resistance to aphanomyces black root, which also correlated with resistance to rot of the mature root (Schneider, 1954, 1978). With this method, they screened breeding lines (Schneider and Gaskill, 1962;

Schneider and Hogaboam, 1983) and foreign introductions to the USA; much of the *Aphanomyces* resistance in current commercial hybrids is derived from these early USDA-ARS breeding programs (Coe and Hogaboam, 1971; Coe, 1974; Hogaboam et al., 1982). The *Aphanomyces* program is still active at the USDA-ARS station in East Lansing MI (McGrath et al., 2003).

Today, many commercial breeding companies have *Aphanomyces* resistance breeding programs, especially for varieties to be marketed in Minnesota and eastern North Dakota. This resistance is primarily against the chronic rot in the mature root because, in most cases, seedling damping-off is controlled by fungicide application on the seed at planting. *Aphanomyces*, as part of a larger root rot syndrome, has been an increasing problem in the Central High Plains of North America (Harveson, 2000; Harveson, pers. Comm.). In Europe, where *Aphanomyces* causing seedling damping-off has been the most important manifestation of the disease, the mature root rot has been a sporadic (local) problem, observed mainly under wet weather conditions, as seen in the last few years in the Netherlands, Moldavia, Poland, and Germany (Büttner, pers. Comm.). Outside the USA, there are also active breeding programs for *Aphanomyces* resistance in Russia (Burenin, 1998, 2001) and in Japan, where *Aphanomyces* is considered a very serious problem (Ohtake et al., 1996; Taguchi et al., 2000; Watanabe et al., 2000).

2.10.4.9 Minor fungal diseases (E. Biancardi)

Alternaria alternata, *Ramularia beticola,* and *Uromyces betae* are the causal agents of leaf diseases that usually induce limited damage to the beet crop. Climatic conditions favorable to any one of these pathogens are favorable for the other two (McFarlane et al., 1954). Fungicide treatments are occasionally required in the more serious cases. Chemicals provide satisfactory control in most situations. Few attempts at breeding for resistance have been reported.

Alternaria alternata incites the formation of dark brown to black leaf spots about 10 mm in diameter; the disease is infrequent in dry areas (Mukhopadhyay, 1987). El Kholi *et al.,* (1994) found some genetic variability and observed that the susceptibility of plants increased over time. Russell (1972) observed that resistance was partially dominant and associated with virus yellows tolerance/resistance. McFarlane et al., (1954) concluded that genes conditioning susceptibility were relatively widespread within US breeding stocks. Therefore, elimination of these resistance genes should be avoided in breeding programs.

Ramularia beticola is frequent in cool humid production areas. After infection, spots similar to those resulting from cercospora l.s. appear on the leaves (Mukhopadhyay, 1987). Adams et al., (1996) and Petersen et al., (2001) described a potentially useful inoculation method for screening for *Ramularia* resistance. The few attempts to improve resistance have not been successful. Adams *et al.,* (1996) found no interaction between resistances to *Cercospora* and to *Ramularia.*

The rust fungus *Uromyces betae* is common in some of the principal sugar beet-growing areas, but damaging rust attacks are quite rare. Another rust of minor importance is caused by *Puccinia subnitens,* present only in limited areas of the USA (Mukhopadhyay, 1987). *U. betae* produces reddish-brown pustules on the leaves. Genetic variability among breeding lines and commercial varieties has been observed during severe rust attacks (Kerr and Leaman, 1996; O'Sullivan, 1997).

Most sugar beet varieties developed for the coastal valleys of California appear to be rust resistant (Whitney and Duffus, 1986). Lewellen and Skoyen (1988) obtained the line C789/2 after five cycles of selection for improved sugar yield under infection of the prevalent diseases of central California. C789/2 displays a moderate degree of rust resistance. Significant progress through selection has not been reported.

2.10.5 Nematodes

2.10.5.1 Cyst nematode (M.H. Yu)

Sugar beet is an excellent host for numerous nematode species. The cyst nematode (*Heterodera schachtii*) is the most destructive plant-parasitic nematode of sugar beet. This nematode is widely distributed in sugar beet production areas and frequently occurs wherever species in the *Chenopodiaceae* and *Cruciferae* families grow. It damages the sugar beet primary root system and severely limits root yield, quality, and sucrose concentration. Typical symptoms are a whiskered appearance of the roots and wilted leaves in warm dry climates. Management of the nematode in sugar beet fields is challenging due to its wide host range and increasing restriction on nematicide utilization. The most promising means of nematode control is through planting resistant varieties. However, until very recently, heritable resistance to cyst nematode was not available in cultivated sugar beet.

In attempting to control the cyst nematode, Schneider (1937) was the first to cross sugar beet and *Beta procumbens.* In three successive decades numerous experiments on *Patellares* interspecific hybridization were attempted; however, none was successful in transferring resistance to sugar beet. Introgression of nematode resistance into sugar beet was not realized until: "Two diploid nematode-resistant plants were selected from 8,834 backcross plants in the progenies of trisomics, and resistance was transferred from both of these plants to F_1 hybrids" (Savitsky H., 1975). With her perseverance, Helen Savitsky was able to accomplish the tedious multiyear task in which cyst nematode resistance from a remote species, *B. procumbens,* was, for the first time, genuinely transferred to diploid sugar beet. It was a breakthrough that renewed interest in *Patellares* × sugar beet hybridization and cytogenetic research in Germany, the Netherlands, Poland, Sweden, Denmark, Japan, and the United States.

A milestone that signaled the beginning of a new era for sugar beet nematode resistance breeding was *Positional cloning of a gene for nematode resistance in sugar beet* published by Jung and a team of 12 others from Germany, Denmark, and the Netherlands (Cai et al., 1997). The cloning of $Hs1^{pro\text{-}1}$ gene, a locus that confers resistance to *H. schachtii* derived from *B. procumbens* chromosome 1, was a giant step forward. It opened the gate to the uncharted territory of sugar beet transformation, providing an avenue for development of productive cyst nematode resistant sugar beet varieties in the foreseeable future.

The life history of *H. schachtii* includes four molts and five stages. Typically, an infective second-stage juvenile (J2) hatches in the soil and is attracted to the elongation zone of the beet root. The J2 penetrates, establishes a sedentary feeding site, and forms a syncytial feeding structure in the root where the final three molts occur. Following the fourth molt, females grow larger and break through the root surface where their white, lemon-shaped bodies cling to the roots. Adult males are vermiform and migratory after emerging from roots. The female dies and her body wall eventually hardens and is transformed into a reddish-brown cyst containing from a few to more than 600 eggs (Raski, 1950; Steele, 1984; Baldwin and Mundo-Ocampo, 1991; Wyss, 1992; Wubben et al., 2001).

Host-plant resistance is currently the most effective and environmentally safe method to control plant-parasitic nematodes. However, selection for cyst nematode resistance within sugar beet varieties has been unfruitful because resistant genotypes are rare or nonexistent (Owen et al., 1938; Doney and Whitney, 1969). Use of nematode resistance from noncultivated *Beta* species via introgression has long been anticipated. Wild beets in section *Patellares* of genus *Beta* were known to be resistant to the cyst nematode (Hijner, 1952; Golden, 1959). Greenhouse inoculation tests demonstrated that all three *Patellares* species—*B. procumbens*, *B. webbiana* and *B. patellaris*—were highly resistant, but not immune, to *H. schachtii* (Yu, 1984a).

Breeding for nematode resistance in sugar beet was first attempted around 1912 (Koch, 1988). After Schneider's (1937) pioneer sugar beet and *Patellares* hybrid, there were numerous attempts to transfer nematode resistance to sugar beet (Coons, 1975). In the search for compatible interspecific matings and bridge hybrids, several *B. vulgaris* subspecies and varieties were examined. Genotypes of diploid and tetraploid sugar beet, red table beet, fodder beet, Swiss chard, *B. vulgaris* ssp. *maritima, B. macrocarpa,* and *B. atriplicifolia* were among the plants used as seed parents in crosses with *B. procumbens, B. webbiana,* and *B. patellaris* (Simon, 1947; Stewart, 1950; Gaskill, 1954; Coe, 1954; Oldemeyer, 1954; Oldemeyer and Brewbaker, 1956; Savitsky, 1960; Johnson and Wheatley, 1961; Ohta, 1961; Kuzdowicz and Brzeskwinska, 1967; Bosemark, 1969).

Despite the ambitious experiments in *Patellares* interspecific hybridization, transfer of a durable nematode resistance gene(s) to sugar beet was not achieved. Individual workers encountered one or more of the following challenges: i) almost all F_1 hybrid seedlings failed to grow secondary roots; ii) surviving and/or rescued plants hardly sustained a viable seed set; iii) meiotic pairing between homologous *Beta* chromosomes seldom occurred, and iv) the acquired "resistance" gene somehow got lost in the breeding process. In Stewart's (1950) example, the resultant *B. vulgaris* x *B. procumbens* diploid hybrids produced no resistant progeny after the second backcross.

The interspecific hybridization strategy was adopted by Savitsky (1975). She used F_1 triploid (2n = 27) progeny of tetraploid *B. vulgaris* (2n = 36) and diploid *B. procumbens* (2n = 18) crosses as bridge hybrids to transmit the resistance. The nonviable F_1 seedlings were grafted on young sugar beet seed stalks (Johnson, 1956). The hybrid population of over 60 grafted F_1 plants was pollinated by susceptible diploid sugar beet. The BC1 progeny and plants of all succeeding backcross generations were screened for nematode resistance. Resistant aneuploid progeny were backcrossed to diploid sugar beet pollinators. Four highly resistant alien monosomic addition plants (2n = 19) were selected from 6,750 BC1 progeny that originated from approximately 60,000 seeds (Savitsky, 1973). At this stage, a *B. procumbens* chromosome bearing the gene(s) for cyst nematode resistance had been added to the full complement of sugar beet chromosomes. In later generations of the alien monosomic addition backcrosses, three nematode-resistant diploid plants (2n = 18) were selected from 14,120 progenies (Savitsky, 1975, 1978). The diploid plants transmitted resistance to their F_1 hybrids. This initial success was soon followed by selection of additional resistant diploids (Nakamura, 1976; Yu, 1978).

Savitsky's successful transfer of *H. schachtii* resistance from *Patellares* to sugar beet stimulated a wave of research activities, especially in Europe. Programs on hybridization with all *Patellares* species were initiated by Speckmann and De Bock (1982) and Loptien (1984a), and with

B. patellaris by Heijbroek et al., (1983). Thus this strategy was seriously pursued by at least three groups at: i) the Institute of Applied Genetics, Hannover and the Institute of Crop Science and Plant Breeding, Kiel, Germany; ii) the Institute for Rationale Suikerproductie together with the CPRO-DLO, Wageningen, the Netherlands, in addition to iii) continued resistance breeding research at the USDA-ARS station in Salinas, California (Kleine et al., 1997). Various nematode-resistant monosomic additions in diploid *B. vulgaris* were established, each carrying an extra chromosome, or chromosome segment, from one of the *Patellares* species. These materials resulted in the successful transfer of nematode resistance to diploid cultivated sugar beet.

Occurrences of trivalent associations in pollen mother cells were rare; however, trivalent associations between the extra *B. procumbens* chromosomes and their homoeologous sugar beet chromosomes were evidenced cytologically at pachytene through metaphase I. Similarly, centromere associations involving two or three nonhomologous bivalents at prophase I of the *B. procumbens* monosomic additions were observed. These meiotic phenomena provide opportunities for crossovers between the remotely related homoeologous chromosomes to occur. Successful selection of resistant diploid plants in progenies of resistant parents with alien monosomic additions indicated that homoeologous pairing and crossing over did occur, permitting a transfer of the resistance gene(s) from *B. procumbens* to the sugar beet chromosome. Nearly all examined pollen mother cells of Savitsky's source resistant diploids contained one or two dicentric bridges associated with acentric fragments at anaphase I and/ or II. These abnormalities resulted from meiotic recombination between two heterozygous paracentric inversion chromosomes (Savitsky, 1975, 1978; Nakamura, 1976; Yu, 1978, 1982a; Nakamura and Tsuchiya, 1982, 1988; Loptien, 1984b; De Jong and Stam, 1985; De Jong et al., 1986; Tsuchiya, 1986).

It is expected that elimination of unwanted alien genes and reincorporation of sugar beet genes can be achieved by successive crossing over during meiosis in resistant heterozygous plants. Selection of favorable genetic recombinants is possible by recovering resistant haploid plants from heterozygous resistant diploids either via anther culture or unfertilized ovule culture. Recovery of resistant haploids from heterozygous diploid sources could provide a direct means for producing homozygous resistant diploids (Hosemans and Bossoutrot, 1983; Nakamura et al., 1992; Yu, 1992).

The transmission rate for nematode resistance in resistant heterozygous plants was lower than 50% in most cases. Female and male transmission rates of the alien monosomic addition chromosomes and nematode resistance averaged overall 17.9% and 1.7% respectively. Their transmission rates ranged from 1.5 to 47.6% for females and 0.1 to 5.1% for males, based on all accessible documented data of approximately 183,600 plants, progeny of over 50 *Patellares* monosomic addition lines (Savitsky, 1975; Nakamura, 1976; Yu, 1978; Loptien, 1984b; Speckmann et al., 1985; Jung and Wricks, 1987; Nakamura and Tsuchiya, 1988; Lange et al., 1990). In Savitsky's (1975) program, resistance was transferred to the eighth backcross generation through seed parents with an average transmission rate of 12.2%.

In diploid sugar beet, nematode resistance was also transmitted through the pollen at lower rates than through egg cells, as observed in alien monosomic additions. In resistant heterozygous plants, transmission through female gametes ranged from 11.0 to 31.4%, and through male gametes of the same plants from 0 to 19.7% (Savitsky, 1978). In diploid homozygous resistant sugar beet, resistance transmission was not complete in successive generations, being 97.5% through eggs and 95.7% through pollen of S_2 plants (Yu, 1981; 1984b). It seems that through traditional breeding approaches, insertion of the alien chromosomal material in the genome of the recipient sugar beet parent may cause some instability. This suggests that sugar beet hybrid varieties should be developed from an early generation resistant homozygote to maintain the highest level of nematode resistance.

Field infestations of *H. schachtii* may be localized or cover entire fields. The highest nematode population densities usually occur about 5-25 cm below the soil surface (Steele, 1986). The *Patellares* source of resistance caused high mortality to *H. schachtii* juveniles feeding on the beet plants. Resistance in these plants was not due to failure of nematode juveniles to enter roots of resistant hosts, but failure of the juveniles to achieve full development (Yu and Steele, 1981; Jung and Wyss, 1999; Holtmann et al., 2000). The mechanism of nematode resistance of this *Patellares* source was antibiosis, triggered by *de novo* transcription of the resistance gene (e.g., $Hs1^{pro\text{-}1}$) upon nematode infection (Yu, 1982b; Yu and Jones, 1983; Thurau et al., 2003).

The influence of resistant sugar beet crops on the *H. schachtii* population dynamics correlated with the initial nematode population density (Pi). Resistant sugar beet varieties released thus far are superior in yield only on heavily infested soils. At high Pi levels the yield of resistant hybrids exceeded control plants. However, at low Pi levels the susceptible control variety produced a higher yield than the resistant hybrids. Sugar beet varieties with *H. schachtii* resistance have been commercially used in Belgium and Germany since 1998. The *B. procumbens* resistant varieties showed approximately 90 to 95% resistant plants and the nematode multiplication factor was 5 to 10 times lower than with classic varieties, reducing nematode populations up to 80%. If the yield potential of resistant varieties in uninfested fields was 85 to 90% of susceptible varieties, a Pi of 500 to 1,500 eggs + juveniles per 100 mL soil would justify their use. In the Netherlands, similar findings were obtained; no substantial interactions between soil pathogens nor types of resistance were detected (Schlang and Muller, 1996; Wauters and Keleman, 1998; Muller, 1998a; Heinrichs, 2000; Heijbroek et al., 2002). The cropping sequence should include a rotation between resistant and susceptible varieties in combination with trap crops to avoid development of resistance-breaking pathotypes.

Planting trap crops or intercropping is an alternative nematode control measure. Viable nematode populations in soil from the root zones of resistant trap crops were lower than in adjacent susceptible crops (McFarlane et al., 1982a; Yu, 1994). Homozygous resistant diploid sugar beet lines have been developed from resistant heterozygotes (Yu, 1983; Heijbroek et al., 1988; Lewellen, 1995d). Sugar beet varieties resistant to nematode reproduction may be valuable as trap crops. Trap crops with resistance to cyst nematode include two varieties, "Pegletta" and "Nemex", of oleiferous radish (*Raphanus sativus* ssp. *oleiformis*) and two varieties, "Emergo" and "Maxi", of white mustard (*Sinapis alba*) (Behringer and Furst, 1985). The use of these crops as green manure to reduce soil infestation appears promising. Planting resistant *R. sativus* varieties after cereal crops reduced soil infestation by 60%, allowing three-year rotations (Caubel et al., 1985). Later experiments demonstrated that both the radish varieties and kenaf (*Hibiscus cannabinus*) cv. "Cuba 108" delay development of *H. schachtii* and provide effective control of this nematode (Tacconi and Pola, 1996; Held et al., 2000).

Cultivated sugar beet is also a host for another, less devastating cyst nematode, *H. trifolii* (the yellow beet cyst nematode). Plants homozygous or heterozygous for resistance to *H. schachtii* differed from one another in resistance to *H. trifolii*. Plants that had withstood *H. schachtii* infection showed increased resistance to *H. trifolii*. There was variation among nematode populations for fitness and virulence to sugar beet. Several strains of *H. schachtii* and *H. trifolii* were capable of overcoming resistance in sugar beet derived from *Patellares*. Populations Schach 0, Schach 1, and 129-SB of *H. schachtii* were able to produce a few cysts that multiplied for six generations on resistant plants. In general, the number of females that developed on resistant sugar beet lines was low (mostly less than 10 per plant), and the resultant cysts formed earlier and were smaller than those on susceptible sugar beet (Lange and De Bock, 1988; Lange et al., 1993; Porte et al., 1997; Muller, 1998b).

In addition to major nematode resistance genes from the *Patellares* species, so-called "partial resistance" has been observed in *B.v.* ssp. *maritima*. Partial resistance genotypes exhibit a wide range of resistance; cyst size also varies, with many of the cysts smaller than those on susceptible varieties. In the root system of partially resistant plants, larval penetration was equal to that in susceptible sugar beet, but the ratio of females to males lower (Heijbroek, 1977; Mesken and Lekkerkerker, 1988; Lange and De Bock, 1994). However, more recent studies indicate that sugar beet varieties with *B.v.* ssp. *maritima* resistance do not have reduced multiplication rates. The yield of sugar beet crops in uninfected soils was high, but profitability rapidly decreased with increasing nematode populations (Wauters and Keleman, 1998).

Several challenges encountered in sugar beet nematode resistance breeding through *Patellares* hybridizations, such as inadequate resistance transmission and unsatisfactory agronomic performance, could be overcome by molecular genetic engineering. A specific manipulation of resistance genes (e.g., *Hs1, Hs2,* etc.) and marker-assisted selection should hasten the development of nematode-resistant sugar beet via transformation (Kleine et al., 1997, 1998; Jung et al., 1998; Wozniak, 2001). Cooperation between breeders and biotechnologists should provide new approaches for genetic modification and accelerate resistance breeding. Extensive field trials are still required in varietal development programs.

2.10.5.2 Root-knot nematode (M.H. Yu)

Root-knot nematodes, *Meloidogyne* spp., cause extensive damage to a variety of crops, including sugar beet. The nematodes induce root galls that not only reduce yields, but also reduce quality. The extensive host range of root-knot nematodes precludes effective management through crop rotation and cultural practices. Fumigation was the most reliable means of control, but environmental concerns have restricted nematicide usage. Application of the soil fumigant Telone (1,3-dichloropropene) has been prohibited or limited in California since 1990, and methyl bromide (an ozone-depleting fumigant widely used for both preplant and postharvest pest and pathogen control) will be phased out in industrialized nations by 2005. Planting nematode-resistant sugar beet, therefore, would be the most economical and environmentally sound control strategy.

The life-cycle of *Meloidogyne* spp. is similar to that of *H. schachtii.* However, root-knot nematodes display a marked sexual dimorphism; females are pyriform, males vermiform. J2 juveniles invade the small fibrous roots, causing formation of galls that contain one to several nematodes. At optimum temperatures the life-cycle of *Meloidogyne* spp. is completed within 20-25 days. In warm climates, four or five generations develop in a growing season. Infected plants often have chlorotic leaves, stunted growth, or seedlings fail to emerge. In light, sandy soils or warm dry climates, severely infected plants may wilt or collapse and total loss of the crop may result (Steele, 1984; Eisenback and Triantaphyllou, 1991; Yu, 1995a).

Golden (1959) reported that all *Patellares* species-*B. patellaris, B. procumbens,* and *B. webbiana*, highly resistant to the cyst nematode, were susceptible to *Meloidogyne* spp.; however, Di Vito (1983) classified *B. procumbens* as partially resistant (two other species were susceptible) to root-knot nematodes. Resistance to root-knot nematode has been identified in rare strains of sea beet, *B. vulgaris* ssp. *maritima* (Yu, 1995b). Throughout monoxenic J2 inoculation tests with six *Meloidogyne* species and field trials with two species (*M. incognita* and *M. javanica*, most widespread), the sea beet source was resistant to all nematode species tested: *M. incognita, M javanica, M. arenaria, M. hapla, M. chitwoodi,* and *M. fallax* (Yu et al., 1999; Yu and Roberts, 2002). These six species comprise an estimated 98% of root-knot nematodes in agricultural soils of the world (Sasser, 1980). Host-plant resistance to such a wide range of *Meloidogyne* spp. is uncommon. Multispecific resistance should facilitate sugar beet breeding efficacy.

Resistance breeding was carried out progressively through controlled pollinations, eggs + juveniles inoculations, field nurseries, chromosomal examinations, isozyme and molecular marker analyses, and visual selections. In progeny populations, phenotypic expression of several undesirable wild beet characteristics remained. Nonetheless, the sprangled root structure and annual bolting trait have become less intense and sucrose concentration increased, in response to additional breeding efforts. Positive results were manifested in improved taproot conformation and root yield. Significant reductions in root weights occurred when susceptible plants were grown in infested soil. Susceptible sugar beet lines were markedly more sensitive to prolonged temperature stress (>38°C) and secondary pathogenic invasions than resistant counterparts (Yu and Roberts, 2002; Yu, 2003).

USDA has released six root-knot nematode-resistant germplasms (M66, M6-1, M6-2, Mi-1, M1-2, and M1-3; PI 586688, PI 613165, PI 632234, PI 593237, PI 614899, and PI 628749 respectively) (e.g., Yu, 2002; Yu and Lewellen, 2004). Among the six, M66, Mi-1 and M1-2 retained many wild *B.v.* ssp. *maritima* characteristics, whereas M1-3 and M6-2 are sugar beet lines homozygous for resistance to root-knot nematode derived from later generation progenies of backcrosses. There are two unique sources of resistance to *Meloidogyne* spp., Mi-1 and M66. The resistance derived from M66 (M66, M6-1, and M6-2) is identifiable by a molecular CAPS marker and resistance from Mi-1 (Mi-1, M1-2 and M1-3) by isozyme PGM banding patterns (Yu et al., 2001; Weiland and Yu, 2003). Homozygous resistant lines could be used by sugar beet growers as trap crops, and by researchers to facilitate resistance breeding efforts through biotechnological manipulations that eliminate the need for greenhouse or field testing. The availability of two, or more, sources of resistance minimizes genetic vulnerability to resistance-breaking biotypes (Kaloshian et al., 1996; Williamson, 1998; Castagnone-Sereno, 2002).

Beta vulgaris ssp. *maritima* is not only morphologically similar to *B. vulgaris*, but also phylogenetically closer than any species in the *Patellares* section (Ford-Lloyd et al., 1975). Introgression of nematode resistance and development of resistant sugar beet varieties should be facilitated by this close relationship. If the *B.v.* ssp. *maritima* source of resistance to multiple species of *Meloidogyne* nematode could be transgenically conferred (Boerma and Hussey, 1992; Jung et al., 1998) to other crops of agronomic importance, such as *Solanum, Phaseolus,* and *Lycopersicon*, it would contribute even more to agriculture.

2.10.6 Insects and Mites

2.10.6.1 Aphids (L.G. Campbell)

Although a number of leaf-feeding aphid species invade sugar beet, *Myzus persicae*, the green peach or peach-potato aphid, and *Aphis fabae*, the black or bean and beet aphid, are the most widely recognized as serious pests (Yun, 1986c). Losses of 4 to 9 tons per hectare have been attributed to *A. fabae* feeding (Jones, 1957); however, its ability to transmit viruses causing beet yellows (BYV), beet western yellows (BWYV), beet mosaic (BMV), and beet mild yellows (BMYV) (Lejealle and d'Aguilar, 1982) is often of greater significance. Populations of *Myzus persicae* are seldom large enough to cause direct injury (Jones, 1957) but because of its superiority as a virus vector (Bennet, 1960; Hull, 1960a), it is often the primary target of control efforts. Sugar beet root aphid *Pemphigus sp.* is a widespread pest of sugar beet in North America and also occurs in central and eastern Europe (Lejealle and d'Aguilar, 1982). Root aphids colonize roots, reducing both size and quality of the crop (Yun, 1986b).

Lowe and Russell (1969) found the inheritance of aphid resistance complex and noted that environmental influences on aphid behavior complicate breeding for resistance. Lines differed in settling resistance (nonpreference), resistance to multiplication (nonpreference or antibiosis), and tolerance. *Myzus persicae* and *A. fabae* responded differently to resistance mechanisms. Furthermore, the mechanism involved in settling of the apterous and alate aphid forms differed. Settling resistance may result in more rapid viral transmission (BYV or BMV) among non-aphid preferred resistant plants than between susceptible plants or slow the spread of viruses that require long feeding times for transmission (BMYV)(Russell, 1964; Lowe and Russell, 1969). Lowe and Singh (1985) documented the need for field trials when evaluating *M. persicae* resistance and formulated an index to facilitate selection when aphid populations are not uniform. Differences in vigor among *M. persicae* clones had no effect upon the relative resistance of sugar beet lines (Lowe, 1974). Plant maturity, infection with BMYV or BYV, (Kift et al., 1996), and aphid colony source (Lowe, 1974) should be considered when evaluating any aphid control measure because they affect survival and vigor of *M. persicae.*

Wallis and Gaskill (1963) identified GW674, a selection from GW359 (Lewellen, 1992), as root aphid resistant. Resistance in both GW674 and GW359 was confirmed by Harper (1964) in a comparison of 11 varieties. In a nursery of 31 lines and hybrids, Wallis and Turner (1968) found substantial differences in resistance to root aphid. Both antibiosis and antixenosis (nonpreference) have been shown to be important in determining resistance (Campbell and Hutchinson, 1995b). A root aphid damage index (Hutchinson and Campbell, 1994), inoculation techniques, and rearing methods have been developed (Campbell and Hutchinson, 1995a,b).

There appears to be sufficient genetic variability within available sugar beet stocks for the development of aphid-resistant varieties. Selection will be complicated by the complex inheritance of resistance and inconsistent environmental conditions. However, the occurrence of resistance in adapted germplasm and the low probability of aphids overcoming multigene inheritance and multiple-mechanism resistance make resistance, combined with natural enemies and viral resistance, an attractive alternative to the introduction of new chemicals and corresponding development of insecticide-insensitive aphid strains (Lowe, 1978; de Pointi and Mollema, 1991). Virus resistance appears to be independent of resistance to aphids and it seems likely the primary emphasis will continue to be on improving resistance to viruses (Russell, 1964, 1966b.c; McFarlane et al., 1969; Lewellen, 1973). However, control measures that reduce viruliferous aphid populations will reduce viral damage (Harrington et al., 1989). Root aphid-resistant commercial hybrids are available and development of additional resistant parental lines should be routine. Due to absence of reliable insecticides, resistance is the primary control method for root aphid. General techniques for evaluating insect resistance have been summarized by Painter (1951) and Smith et al., (1994).

2.10.6.2 Spinach leaf miner (E. Biancardi)

Larvae of *Pegomya betae* feed and develop in the mesophyll of beet leaves, causing formation of characteristic large blotches. Heavy damage is rather rare. Genetic resistance has been identified in sections *Corollinae* and *Procumbentes* (Frese et al., 2001). Luczak (1996) observed useful levels of genetic variability in a few commercial varieties. Resistance to *Pegomya betae* appeared to be associated with poor attractiveness for ovipositing, and/or inhibition of larval development.

2.10.6.3 Sugar beet root maggot (L.G. Campbell)

Sugar beet root maggot, *Tetanops myopaeformis* (*Diptera*: *Otitidae*) adults are shiny black flies (about 6 mm long) with transparent wings that have small dark areas on the front margins (Gojmerac, 1956). The larvae are white, legless maggots without distinct heads (Bjerke et al., 1992). Mature larvae overwinter in soil. Pupation occurs in early spring. Adult females migrate to recently emerged sugar beet fields and lay eggs in the soil near seedlings. Developing larvae feed on the root by tunneling along the surface. Mature larvae cease feeding and enter diapause in late summer (Harper, 1962; Yun, 1986a). Yield reductions (Campbell et al., 1998) may be the result of stand loss early in the season, but damage occurs primarily from larval feeding throughout the growing season.

The root maggot, a native of North America, was first recognized as a pest of sugar beet in the 1920s (Hawley, 1922). Since then it has become a destructive pest of sugar beet in large areas of North America. The primary control is chemical insecticides that reduce larval populations in sugar beet fields. Currently, all selection for resistance and evaluations are dependent upon natural infestations at sites where root maggot populations are usually high. Damage, root scarring, is assessed in mid to late summer after roots are carefully dug up and washed. A 0 to 5 damage rating scale (Blickenstaff et al., 1977), widely used to assess insecticide effectiveness, did not suffice to differentiate between lines and was expanded to the following 0 to 9 scale: 0 = no visible root maggot feeding scars; 1 = 1 to 4 small (pinhead size) scars; 2 = 5 to 10 small scars; 3 = up to 3 large scars or scattered small scars; 4 = a few large scars and/or numerous small scars; 5 = several large scars and/or heavy feeding on lateral roots; 6 = numerous scars with up to ¼ of root scarred; 7 = ¼ to ½ of root blackened by feeding scars; 8 = ½ to ¾ of root blackened; 9 = more than ¾ of root surface blackened.

Theurer et al., (1982) reported a linear trend in increasing and decreasing maggot damage in response to selection for high- and low-maggot damage respectively. Both tolerance and antibiosis appeared to contribute to resistance. Two root maggot-resistant germplasm lines, F1015 (PI 605413) and F1016 (PI 608437), have been developed (Campbell et al., 2000). In a two-year trial, root maggot damage ratings for F1015 and F1016 were 3.3 and 2.2 respectively, compared to 5.5 for commercial hybrids. Two globe-shaped red beet accessions from the USDA *Beta* collection, PI 179180 and PI 181718, have also been identified as resistant. Populations resulting from 3 to 4 cycles of mass selection typically have significantly lower damage ratings (4.3 - 4.5) than their heterogeneous parental population or commercial hybrids (6.0). Replicated progeny tests are usually required to identify the most resistant plants. Immunity to sugar beet root maggot feeding has not been observed.

Within field variation and year-to-year differences in the severity of root maggot damage complicate selection for resistance, hamper inheritance studies, and make it difficult to associate resistance with molecular markers. Even under the best conditions, the labor required for damage assessment may restrict the volume of material evaluated. Attempts to develop mass rearing techniques and efficient greenhouse or laboratory screening procedures have not been productive. Transgenic sugar beet plants that produce inhibitors that bind to the digestive proteinases in the root maggot midgut may have promise (Wilhite et al., 1998; Smigocki et al., 2003). Also, the introduction of genes that enhance cytokinin production may reduce root maggot damage (Snyder et al., 1999) but engender detrimental effects that have to be overcome before these genes can the employed as a source of resistance. If the root maggot were to develop resistance to the few insecticides currently used or these insecticides become unavailable because of environmental concerns, sugar beet production in some areas would be threatened until alternatives such as host-plant resistance are developed.

2.10.6.4 Flea beetle (E. Biancardi)

The insect *Chaetocnema concinna* chews small circular holes in the leaves of seedlings and can be particularly damaging in the cotyledon stage, i.e. the initial emergence of monogerm varieties. Insecticides incorporated in seed pellets or distributed along the rows at seeding time have made control easier and minimized damage. Selections to improve resistance to the flea beetle have been carried out on genotypes subjected to chemical mutagenesis. This induced resistance appeared to be multigenic and recessive (Golev et al., 1984). No other reports are available in the literature.

2.10.6.5 Spider mite (E. Biancardi)

The spider mite *Tetranychus urticae* frequently causes significant damages in the UK and the USA (Whitney and Duffus, 1986; Haylock and Dewar, 2003). During a serious attack of spider mites on a variety trial, Bush and Brewbaker (1956) observed that infestation intensities differed among genotypes. In particular, variety GW359 was less damaged than GW59. Unlike GW59, GW359 was resistant to cercospora l.s. This association between spider mite resistance and cercospora l.s. resistance was observed in other entries in the trials, with just one exception in which a genotype susceptible to cercospora l.s. and resistant to curly top was not heavily damaged by spider mites.

Dewar *et al.*, (2000) reported resistance in the commercial variety "Roberta". This observation was confirmed by Haylock and Dewar (2003), and resistance in two additional genotypes reported. The trait reduced the number of mites and eggs on the leaves compared to the susceptible variety "Access". Reduced infestations corresponded with increased sugar yield. Effective control of the pest is achieved with a combination of acaricide treatments and resistant varieties.

2.10.7 Multiple Disease Resistance (E. Biancardi)

Sugar beet diseases sometimes appear alone but frequently more than one disease must be dealt with. In many situations, genotypes with resistance to two or more diseases would be useful (McFarlane, 1971). An example of combining resistance to several pathogens in a genotype is cited by Coons *et al.*, (1955). To satisfy this need in areas of the USA affected by cercospora l.s. and curly top (BCTV), a program was established to intercross inbred lines resistant to a single disease. Backcrossing and selection in environments affected alternatively by the two diseases, resulted in the release of US104, resistant to both.

During the 1980s, rhizomania spread rapidly in parts of Europe where cercospora l.s. is prevalent, creating the need for varieties resistant to both diseases. The multigerm variety "Alba P" was the first to be identified with resistance to both diseases. The availability of monogenic sources of resistance to rhizomania ("Rizor type" and "Holly type") and CMS lines, such as FC504, FC506, FC606, FC607 etc., resistant to cercospora l.s., assure improving the yield potential of double-resistant varieties in the USA (Smith and Ruppel, 1986).

In the 1990s, Panella et al., (1995) and Ruppel et al., (1995) released several multigerm (FC716, FC717, FC718, etc.) and monogerm O-type (FC715) *Rhizoctonia solani*-resistant lines. These lines were crossed with lines bearing monogenic resistance to rhizomania. These hybridizations led to the production of several varieties, now being tested in Europe, which are resistant to both diseases.

Examples of multiple resistance to cercospora l.s., virus yellows, mosaic, black leg, were identified in commercial and other diseases varieties by Kovbasyuk and Bulin (1983) and Burenin and Timoshenko (1989). Introduction of resistance traits into susceptible varieties was always accompanied by a reduction in sugar yield, as has been documented with cercospora l.s. (Shane and Teng, 1992). This negative relationship is a major obstacle to combining high yield with resistance to two or more pests; however, the availability of commercial varieties with resistance to both cercospora l.s. and rhizomania are encouraging. Double, triple, and quadruple resistances to some foliar diseases (BMYV, BMV, powdery mildew, and cercospora l.s.) are relatively frequent in section *Procumbentes* (Luterbacher, pers. comm.), suggesting that species from this section may provide unique sources of multiple disease resistance in future.

Multiple resistance, according to Scholten *et al.,* (1999) can also refer to combining different sources of resistance to a single disease in the same genotype. At present, this is only practical with rhizomania, in which different types of resistance have been identified and used in commercial production. Combining the "Alba P" source with "Rizor" or "Holly" resistance genes does indeed increase the plant's ability to combat rhizomania (Lewellen and Biancardi, 1990). A significant improvement in rhizomania resistance was not observed in lines with both the "Rizor" and "Holly" resistance genes (Biancardi et al., 2002) however, a further indication of their similarity.

2.11 Resistance to Abiotic Stresses (P. Stevanato)

Realized productivity of sugar beet is considerably less than its potential because of abiotic stresses caused by physical (temperature, light) and/or physicochemical (water and nutrient availability) factors. Hence, there is considerable interest in selecting genotypes resistant to adverse environmental conditions with the objective of improving sugar yield stability in current production areas and expanding the areas where sugar beet growing is profitable.

2.11.1 Cold Stress

During the vegetative cycle of sugar beet low temperatures are frequently encountered in the early developmental stages (for autumnal sowing) or during germination (for the usual spring sowing). In some production areas, there is a tendency to plant early in order to allow plant development during a time of low evapotranspiration, to harvest earlier, and to extend the campaign.

Sugar beet germination requires a minimum temperature of 3°C and plant emergence is related to the heat units received and to seed quality (Scott and Jaggard, 1993). Quality seed germinates faster, even at low temperatures. Chemical or physical treatments such as priming (Murray et al., 1993) can improve germinability under low temperatures. Cold resistance, initially at least, increases as the plant develops: a temperature of about −2°C is sufficient to destroy all seedlings at the cotyledon stage, −10°C is required for the same effect at the 2-3 true leaf stage (Biancardi, 1999). Apart from the early phases of vegetative growth, cold resistance is quite useful during postharvest storage: temperatures lower than 0°C damage root tissue, creating conditions for rot development with a consequent reduction in sugar yield. Cold resistance is also necessary for the survival of plants in winter nurseries or for plants remaining in the field during winter before they are brought to the sugar factories. Maintaining a regular plant population is necessary for the production of uniform roots (Slavov, 1984).

Earlier studies have shown differences in frost resistance among breeding lines. However, except for one reported case in which cold resistance was improved in two varieties (Dix et al., 1994), progress in genetic improvement has not been encouraging. In other species, control of cold resistance is multigenic (Thomashow, 1993) and its expression can be influenced by both low temperatures and hormonal factors (Palva and Heino, 1997). Coons (1936) found differences in cold resistance at early growth stages among inbred lines obtained from fodder beet or fodder beet by sugar beet crosses and the character seemed to be heritable. Additional research in the USA revealed genotypic differences in response to naturally or artificially imposed low temperatures (Wood et al., 1950; Wood, 1952). Hence the existence of sufficient genetic variability to allow for effective selection has been verified. An increase in sucrose content was observed in the progenies of plants (1%) that survived frost. In this study however, no information on the developmental stage of the plants exposed to low temperatures was provided. Genetic variability was also observed for frost damage caused by temperatures lower than -5°C on mature plants close to harvest (Wood, 1952). In this case, a high negative correlation between severity of damage and sucrose content of the root was found. It was also noted that cold sensitivity and susceptibility to cercospora l.s. correlated, suggesting a common mechanism for resistance to both stresses (Wood et al., 1950).

2.11.2 Heat and UV Stresses

In countries with a subtropical climate, e.g. India, Pakistan, Egypt, and Morocco, sugar beet is frequently subjected to thermal stress. High temperatures (35-45°C) coupled with dry winds for long periods impose stress conditions on the crop with subsequent reduction in productivity. Srivastava (1996) demonstrated the existence of sufficient genetic variation for improvement of resistance to thermal stress. A test based on chlorophyll fluorescence measurement was suggested for the selection of high temperature resistant genotypes (Clarke et al., 1995).

The electromagnetic spectrum of solar light reaching the ground is enriched with the UV-B component (280-320 nm). Reduction in growth and biomass production is among the effects of UV-B radiation on plants. It has been shown in sugar beet, that UV-B radiation can decrease photosynthetic efficiency (Panagopoulos et al., 1990). Levall and Borman (1993) demonstrated the effectiveness of *in vitro* selection for resistance to UV-B radiation stress; chlorophyll fluorescence and carotenoid content appeared useful as predictive indicators of the trait.

2.11.3 Water Stress

The possibility of enhancing resistance to drought was investigated as early as 1936 in eastern Europe and Russia but no positive results were obtained from these initial studies (Knapp, 1958). Only recently has the problem come to the attention of researchers again. Variability in response to water stress has been demonstrated in sugar beet by Sadeghian et al., (2000) and Ober and Luterbacher (2002). The structural and morphophysiological adaptive responses of sugar beet to water deficiency however are not understood. If the stress is imposed during the early growth stages, root development proceeds at a slower rate, leading to a final root yield reduction of up to 46% (Noghabi and Williams, 2000). Late water stress on the other hand, reduces the unfolding and expansion of leaves and consequently the interception of light radiation.

This response is partly explained by the limited availability of nitrogen in the soil and, as a matter of fact, water deficiency can slow or even completely block ion uptake by the roots, thus creating strong nutrient fluctuations in the plant and eventually causing nutrient deficiencies. The negative effect of late water stress is less pronounced because the plant has already developed its root system and the leaf canopy fully covers the soil. A well-developed root system increases the plant's water extraction efficiency and hence its capacity to tolerate drought. Variation in response to water stress observed among commercial varieties might be due to differences in the development of deep root systems (Ober and Luterbacher, 2002).

The effects of water stress are first manifested in the leaves. A small decrease in leaf water potential suffices to diminish development of total leaf area. A low potential favors development of new leaves and concomitantly accelerates the senescence of older leaves. Sugar beet leaf stomata are very sensitive to changes in water status, closing in response to a drop in water potential from –5 to –15 bars (Milford and Lawlor, 1976). Low stomatal density in sugar beet seemed to correlate with water stress resistance. Thomas and Clarke (1995), confirming results from the studies of Visser (1951), found that stomatal densities ranged from 70 to 150 stomata per square millimeter and that varieties more capable of tolerating water deficiency had lower stomatal densities. Given that stomatal density correlates positively with sugar yield (Thomas and Clarke, 1995), attempting to improve resistance to drought by selecting for fewer stomata might lead to a reduction in sugar yield.

It has been reported that water stress in sugar beet provokes accumulation of osmoregulatory compounds such as betaine, proline and fructans. The evidence suggests that such an accumulation is an adaptive response and correlates with a reduction in the cell's osmotic potential. It is not clear however, whether plants accumulating a high concentration of these compounds as a consequence of water stress are drought resistant (Stajner et al., 1995; Ghoulam et al., 2002).

Water stress causes a significant increase in potassium and sodium concentration and, as a consequence, a reduction in sugar extractability during processing. Marschner et al., (1981) observed a positive correlation between the accumulation of sodium and chlorine in the leaves and salt resistance.

Acknowledgments. The author thanks Prof. Massimo Saccomani, Dr. Marco Bertaggia and Mr. Gianpaolo Fama for their collaboration in editing this section.

2.12 Sugar Content, Root Weight, and Sugar Yield (M. McGrath)

Sugar yield is of utmost importance to sugar beet growers. Gross sugar yield is the product of the weight of the roots per unit area and the sugar percentage. Extractable or recoverable white sugar indicates how much sugar is produced at the factory. Ideally, gross sugar yield and extracted sugar yield should be equal, however the presence of nonsucrose bearing materials (beet tops, harvested weeds, soil tare) and naturally occurring nonsucrose constituents that impede sucrose crystallization invariably reduce actual sugar yield.

The inheritance of "sugar yield" is quantitative and affected by environment (Powers et al., 1963). Nonadditive gene action is prevalent in controlling root yield (Campbell, 2002b), while sucrose concentration inheritance is additive, without expression of heterosis or dominance (Oldemayer and Rush, 1960; Smith et al., 1973). Evaluation of the inheritance of these traits demonstrates the difficulty, not so much in improving sugar concentration, but rather in increasing root yield and sucrose concentration simultaneously (Hecker, 1967; MacLachlan, 1972). Campbell (2002b) concluded that improvement of root weight had a larger impact on improving sugar yield than increasing sugar content alone. Schneider et al., (2002) showed that sugar yield correlated highly with root yield and not so well with sugar concentration. When root weight was increased, sucrose concentration (polarimetric degree) and juice quality decreased and *vice versa*. The most problematic relationship for the breeder is the negative correlation between root yield and sugar concentration. Improvements in sugar yield are obtained by controlling both characters. This justifies the high number of experimental crosses and trials that seed companies must examine when attempting to improve production traits.

Root yield has not been as amenable to selection as sucrose concentration (Savitsky, 1940; Bosemark, 1993), most likely because of the influence of nonadditive genetic variation on root yield. Testing combining ability of sugar beet breeding lines is a time-consuming and expensive, but essential task for improving root yield and thus sugar yield (Smith et al., 1973, Lasa et al., 1989). **Table 2.3** shows relationships among traits evaluated in multiyear agronomic trials. If the same characteristics are measured on individual roots, variability noticeably increases. In this case, correlation coefficients may assume very different values but tend to be similar to the plot results.

Snyder and Smith (1986) found a high heritability (0.90) following two cycles of selection for high or low root weight/leaf weight ratio. Selecting for

Table 2.3 Simple correlations between production traits in sugar beet (After Biancardi, 1999). The symbols *, ** and *** show significance at the P = 0.05, 0.01 and 0.001 values, respectively.

	Sugar content	*Sugar yield*	*Potassium*	*Sodium*	*Amino nitrogen*	*Purity*
Root yield	-0.225***	0.830***	–0.048	–0.235***	–0.100**	0.217***
Sugar content		0.328***	–0.407***	–0.573***	–0.122***	0.558**
Sugar yield			-0.234***	–0.511***	–0.194***	0.496***
Potassium				0.452***	0.067	–0.608***
Sodium					0.202***	–0.902***
Amino nitrogen						–0.536***

high root weight/leaf weight ratios increased root yield and thus sugar yield. Redirecting photosynthate accumulation toward the roots shows promise as a selection criterion. Assimilate partitioning is receiving renewed interest with the cloning of sucrose-proton antiporter protein genes (Bush, 1999). Since sucrose is the predominate form of chemical energy transported between plant organs, altered partitioning favoring roots should increase root growth and thus increase yield.

The sucrose concentration in mature sugar beet roots is about 75% of dry weight (15-20% fresh weight). Sucrose biosynthesis likely occurs by mechanisms similar to those in other plants, although some key control points remain to be specifically elucidated for beets (Avigad and Dey, 1997; Winter and Huber, 2000). While sucrose metabolism is paramount to sucrose accumulation, other processes may be equally important for sugar beet economic value, in particular the transport of sucrose and its regulation (Kock, 1996; Bush, 1999). Sucrose is transported from sites of synthesis (leaves) to storage and utilization sites (roots, seeds, stems). Sucrose appears to have a direct role in the regulation of gene expression (Smeekens, 2000) but it is not known how sucrose concentrations are regulated in sugar beet or what role sucrose plays in sensing the metabolic status of sugar beet cells.

Sucrose is not distributed uniformly throughout the root; it is concentrated in the vascular tissue and adjacent parenchyma cells of the innermost five of the 10 to 12 concentric cortical rings, and is highest around the point of maximum root girth (Elliott and Weston, 1993). Sucrose concentration in root cells is proportional to cell volume up to a point, beyond which sucrose concentration decreases (Milford, 1973) while water content and nonsucrose dry matter remain constant (Milford, 1973; Milford et al., 1988), suggesting a physiological and genetic association between cell size and sucrose concentration. Watson and Goldman (1997) described a red beet mutant with a blotchy color distribution in the root. Since pigments and sucrose may share similar sites of accumulation, similar irregular distribution of sucrose may go unnoticed.

Varieties have improved only slightly in sucrose concentration since the early 20th century, mainly as a result of repeatedly selecting high sucrose segregants in heterogeneous breeding populations. Savitsky demonstrated that sucrose concentration depended on four to five genes (Coons, 1936; Savitsky, 1940) in crosses among divergent types such as sugar (15-20% sucrose), fodder and red beet (3-12% sucrose), but curiously, one or more of these genes was present in the phylogenetically more distant Swiss chard types (12-15% sucrose). In the sugar beet x Swiss chard populations, segregation into high and low sucrose types could be explained by as few as two major genes. Further experiments have confirmed that sucrose percentage in sugar beet is a quantitatively controlled trait with high heritability (Culbertson, 1942; Powers et al., 1963; Zhao et al., 1997). Schneider et al., (2002) identified five quantitative trait loci (QTL) associated with sucrose concentration in a multilocation, replicated field trial using a single molecularly mapped population. Interestingly, only two QTLs with effects on root yield were detected.

Theurer (unpubl.) suggested sucrose concentration in dried sugar beet and fodder beet roots was similar. The implication is that water concentration influences sucrose concentration on a fresh-weight basis, and root yield as well. Variability of water content in fresh roots has not been subjected to genetic characterization, but one of the sugar concentration QTL regions identified by Schneider et al., (2002) contains a gene involved in cellular water relations. Percent sucrose correlated negatively with percent water in roots from multiple environments and representing diverse genotypes. If the suggestion of relative constancy of sucrose as a percentage of root dry weight proves valid (Wyse, 1980), the most progress would result from increasing total dry weight (biomass) of the root. However, this strategy would not necessarily produce roots with acceptable processing quality.

2.13 Ethanol Production (E. Biancardi)

Breeding genotypes with enhanced ethyl alcohol (ethanol) production potential requires increasing synthesis and accumulation of sugars, the raw product from which alcohol is obtained after fermentation and distillation. Traditional quality characteristics of sugar beet can be disregarded because they are only of importance in the conventional extraction of sucrose (Biancardi and Pavarin, 1993).

Since traits related to processing quality are unimportant, genotypes selected for ethanol production will not be suitable for processing in sugar factories. Therefore, it is important to differentiate ethanol beets from sugar beet. The "red skin" trait derived from garden beet would provide a reliable means of marking varieties selected for ethanol production. Sugar beet x garden beet crosses have low sugar content and yield, but this could be overcome with backcrossing (Biancardi and Pavarin, 1993). Theurer et al., (1987) reported that ethanol production potential is lower in fodder beet and their hybrids than in sugar beet.

2.14 Postharvest Storage Traits (L.G. Campbell)

After harvest, most of the sugar beet crop is stored in exposed piles for up to 200 days, during which respiration, rotting, accumulation of impurities, and physical deterioration decrease extractable sucrose. Respiration is often responsible for 50 to 70% of the loss (Wyse and Dexter, 1971b), while losses attributable to storage rots can be considerable. Three fungi - *Phoma betae*, *Botrytis cinerea*, and *Penicillium claviforme* - are widely recognized as storage-rot pathogens (Bugbee, 1986, 1993). Fungi of local or lesser importance include other *Penicillium* species (Bugbee, 1975a; Bugbee and Nielsen, 1978) and species of *Aspergillus* (Halloin and Roberts, 1995), *Alternaria*, *Chaetomium*, *Fusarium*, *Mucor*, *Rhizopus*, *Sclerotinia*, *Stemphylium*, (Lejealle and d'Aguilar, 1982), and *Pythium* (Bugbee, 1986). Once storage piles have cooled, bacteria and yeast seldom cause storage diseases unless oxygen is depleted (Bugbee, 1982). The decrease in recoverable sucrose during storage is not only the result of sucrose lost to respiration or consumed by pathogens but due to an increase in impurities (McCready and Goodwin, 1966; Hobbis and Batterman, 1975; Tungland et al., 1998; Martin et al., 2001a,b). Endophytic bacteria in healthy sugar beet tissue may contribute to the formation of impurities in some situations (Bugbee et al., 1975; Cole and Bugbee, 1976).

Differences in storage respiration rates among sugar beet lines have been documented (Nelson and Wood, 1950; Cole, 1977; Akeson and Widner, 1981). No adverse relationship between low respiration rate and other desirable characters was observed by Nelson (1954). Wyse et al., (1978) found ample genetic diversity for development of low storage respiration genotypes. Contrary to Stout's (1954) conclusion that low respiration is associated with larger roots, correlations between respiration rate and root size or sucrose concentration at harvest were nonsignificant. The respiration rate of hybrids was lower than the midparent, and in several cases lower than both parents. Male-fertile parents respired at higher rates than CMS lines. In another study (Theurer et al., 1978), hybrids also had lower respiration rates than either parent and relative respiration rates of inbreds and hybrids were similar in two years. Both additive and nonadditive gene action conditioned respiration rate. Differences among reciprocal crosses were small. The inbred line with the highest respiration rate also showed the greatest heterosis for low respiration rate in hybrids, suggesting selection for low respiration on the basis of combining ability may be necessary. Stout (1954) observed that polyploid lines respired more slowly than diploid varieties.

Gaskill (1950a, 1952) demonstrated the possibility of selecting for resistance to *Phoma* and *Botrytis.* Nelson and Oldemeyer (1952) suggested that selecting lines with resistance to *Phoma* and low respiration would be feasible. Cormack and Moffatt (1961) discovered differences in virulence among isolates of *Phoma.* Bugbee (1979a) cautioned that the presence of the sexual stage of *Phoma* could lead to the development of more virulent *Phoma* strains. Bugbee (1973a) found no commercial variety with resistance to *P. betae*; however, a few sugar beet, mangel, and table beet accessions from the USDA *Beta* collection with relatively high frequencies of roots resistant to *P. betae* were identified. Bugbee (1979c) selected lines with combined resistance to *Phoma* and *Botrytis*. Resistance to the two organisms appeared to be genetically independent. Bugbee (1978) and Campbell and Bugbee (1985) developed five germplasm lines with resistance to *P. betae, B. cinerea*, and/or *P. claviforme*. Three of these lines were selected from germplasm obtained from Russia, likely a reflection of Russian efforts to develop varieties with improved storage characteristics (Bugbee, 1993). Bugbee and Campbell (1990) noted an association between resistance to crown and root rot caused by *Rhizoctonia solani* and resistance to *P. betae* and *B. cinerea*, suggesting a common biochemical mechanism for resistance. Storage rot control with the use of resistant varieties would likely be equal to that obtained with fungicides (Bugbee and Cole, 1979).

Respiration rate is determined by measuring oxygen consumption (Nelson and Wood, 1950) or carbon dioxide (CO_2) production (Burke et al., 1979; Cole, 1980a, 1986; Wyse and Dexter, 1971b). A widely accepted method is to measure the CO_2 produced by roots, commonly 10 to 12, in sealed containers. Outside air (Cole, 1977; Campbell and Seiler, 1994) or air with a known CO_2 concentration (Wyse et al., 1978; Akeson, 1981) is allowed to flow through the containers at a controlled rate. The CO_2 concentration of air from the exit tube is determined, often with an infrared analyzer. The CO_2 concentration from the exit tube of a container without roots or the concentration of the background air supplied to the containers is subtracted from this measurement and respiration rate expressed as CO_2 produced per kilogram of roots over time (Dilley et al., 1969). Respiration rates stabilize after approximately 20 days of storage and reflect the potential of a variety. In general, relative respiration rates recorded early in the storage season correspond with those obtained after a longer storage time (Cole, 1977; Wyse et al., 1978). Disease-free roots must be used to measure inherent differences in respiration rates (Mumford and Wyse, 1976) since respiration resulting from fungal activity can overshadow the respiration required to maintain the integrity of the root.

Resistance breeding programs need rapid dependable screening methods to identify valuable germplasm. Gaskill (1950b, 1952) evaluated a number of inoculation techniques, some involved injuring intact roots while others used tissue samples. Halloin and Roberts (1995) placed millet caryopses carrying *Aspergillus fumigatus* in a small hole drilled in roots to inoculate them. Bugbee (1973b) used toothpicks that were autoclaved, soaked in agar, and inoculated with *P. betae* to inoculate roots. Inoculated roots were incubated at 10°C for four weeks and the resultant lesions measured. Another method (Bugbee, 1976) examined rot progression along an 18 × 60 mm core removed from healthy roots with a cork bore. Cores were placed on-end in contact with cultures of a single storage rot fungus. In a modification of this technique (Bugbee, 1979b; Campbell and Bugbee, 1985) that allowed for simultaneous screening of many roots to multiple storage rot fungi, 1 cm^3 blocks were cut from the smooth surface of each root and placed in petri dishes in contact with the rot fungi. After incubation, each block was cut down the center and rated on a scale of 0 (no rot) to 5 (completely rotted). Roots corresponding to cubes with low ratings were subsequently planted to produce seed. Cultures of a single fungus should be used when evaluating germplasm because antagonisms between fungi (Bugbee, 1976) may influence results. Roots are usually stored for approximately 80 days prior to determining response to storage rots. Roots should not be allowed to lose excessive moisture prior to evaluation (Bugbee, 1979c) and genotypes to be compared should be grown under uniform conditions (Bugbee, 1979b; Bugbee and Cole, 1986).

Even under favorable storage conditions, sucrose losses due to accumulation of reducing sugars, raffinose, and amino acids may be substantial (Martin et al., 2001a,b; Wyse and Dexter, 1971a; Cole, 1977). Wood (1954) used mass selection to produce genotypes with low raffinose concentration. Raffinose concentration appeared to be controlled by five unlinked genes in another population (Wood et al., 1956). Akeson and Widner (1981) reported that varieties differed in raffinose concentration after storage but not at harvest, confirming the Finkner et al., (1959) suggestion that selection for low raffinose should begin after several weeks of storage. Akeson and Widner (1981) concluded differences among varieties for invert sugar concentration at harvest were not a reliable indicator of relative invert sugar accumulation. Differences in variety rankings among impurity components and respiration rate, and storage time by variety interactions complicate development of varieties with improved storage characteristics (Akeson and Widner, 1981; Wyse and Dexter, 1971a; Martin et al., 2001a,b). Developing varieties with relatively low respiration rates and resistance to prevalent storage rot fungi ought not to present unusual problems. However, incorporating additional traits into a breeding program slows progress toward other goals or requires additional resources. Damage caused by diseases that occur during the growing season often is easier to quantify with yield trials and more apparent than storage losses. Grower payment is often based on yield and quality at harvest; therefore, growers will not plant varieties with improved storage characteristics unless they are more productive than other varieties or processors incorporate storage traits into their payment system or criteria for approving varieties. A better understanding of physiological processes (Tungland et al., 1998; Klotz and Finger, 2001) and disease progression (Bugbee, 1975b) would facilitate selection for productive varieties with improved storage characteristics.

2.15 Processing Quality (L.G. Campbell)

Processing quality encompasses a number of complex characters that affect the quantity and ease with which processors can extract sugar from roots. Many of the traits that influence quality are under genetic control but the impact of cultural practices, harvest method, and postharvest storage conditions frequently overshadow inherent differences. Breeding programs often accentuate traits related to root composition or internal quality, while external quality is dependent on those who manage harvest operations and storage sites and on environmental conditions (Harvey and Dutton, 1993).

Payment to growers for sugar beet constitutes the major expense of processors, and most of their revenue is realized from the sale of crystallized sucrose. Therefore, sugar beet quality generally refers to sucrose concentration and the concentrations of naturally occurring nonsucrose constituents that impede sucrose crystallization. Purity, the ratio of sucrose to total dry substance expressed as a percent, is a widely accepted measure of processing quality (Milliken and Hecker, 1967; Hartman, 1977). Attached leaf material and soil, mechanical damage, freezing, and dehydration may cause storage problems (Bugbee, 1993) and hamper sucrose extraction but are minimized through proper handling and cultural practices and are generally not responsive to selection.

Attempts to improve quality while maintaining or improving root yields have been complicated by a negative relationship between root yield and sucrose concentration (Wyse, 1979; Campbell and Kern, 1983; Campbell and Cole, 1986). At harvest, 73 to 77% of the weight of a typical root is water and 16 to 22% soluble solids, of which approximately 80% is sucrose (Bichsel, 1988; Bohn et al., 1998). Therefore, any increase in sucrose concentration resulting from a decrease in water content will reduce root yield. In general, genetic variance for sucrose concentration is predominately additive. In contrast, nonadditive genetic variance and specific combining ability components are significant in determining root yield and hence play a role in determining sucrose yield per hectare (MacLachlan, 1972; Smith et al., 1973; Hecker, 1991; Ahmadi and Assad, 1998). Availability of both diploid and triploid commercial hybrids implies that the addition of a genome has neither a clear beneficial or detrimental effect on yield or sucrose. However, in instances wherein triploids enhanced sugar production, the benefit has almost always been related to an increase in root yield, with little affect on sucrose concentration (Hecker et al., 1970; McFarlane et al., 1972; Smith et al., 1979).

Among the soluble nonsucrose components or impurity components, sodium, potassium, and amino-nitrogen have received the most attention in variety development programs (Carruthers et al., 1962; Last and Draycott, 1977; Smith et al., 1977). Sodium, potassium, and amino-nitrogen levels can be shifted substantially with only a few selection cycles (Powers et al., 1963; Coe, 1987; Smith and Martin, 1989), suggesting that additive genetic variance is important in determining the relative levels of these traits (Smith et al., 1973). Interactions among impurity components, sucrose concentration, and root yield complicate selection for optimum levels of yield and quality traits. Coe (1987) noted an association between low concentrations of nonsucrose solubles and small root size. Selection for low sodium concentration was accompanied by an increase in extractable sugar per ton (Smith and Martin, 1989). However, selection for low amino-nitrogen concentration was associated with a decrease in extractable sugar and selection for low potassium had no effect on extractable sugar per ton. Heritability estimates for sodium, potassium, and amino-nitrogen were 0.23, 0.66, and 0.81 respectively. Finkner and Bauserman (1956) found that selection for sodium, *per se*, had little effect on sucrose concentration and concluded that selecting for low sodium was generally of little value. In contrast, Wood et al., (1958) presented evidence that selecting for both high sucrose and low sodium was more effective than selecting for sucrose concentration alone. Dudley and Powers (1960) concluded that lines with both low sodium and potassium could be developed.

Other compounds that interfere with sucrose extraction have received less attention. Smith et al., (1977) found that betaine concentration had a significant effect on purity. Heterosis for low betaine was observed in an eight-parent diallel cross (Smith et al., 1973). Wood et al., (1956) demonstrated that raffinose concentration was heritable and Finkner and Bauserman (1956) observed a negative correlation between raffinose and sucrose concentration. Akeson and Widner (1981) found no differences among varieties for raffinose content at harvest. Varieties differed in invert sugar concentration at harvest but these differences were not predictive of relative invert sugar accumulation during storage.

Some physiological and anatomical characteristics have been associated with processing quality and, on occasion, selection for these may facilitate efforts to enhance sucrose concentration and extractability. Artschwager (1930) concluded there was no relationship between cambial ring number and structure and sucrose concentration. Pack (1930) found sucrose concentration to be positively associated with high density, high dry matter content, shoulder extension, firmness of tissue, root length, roughness of skin, and darkness of leaf color. Milford (1973) reported an association between small cell size and increased sucrose concentration. Up to 20% of sugar beet is crown tissue (Cole, 1980b) and relationships among yield and quality traits are strongly influenced by crown size (Campbell and Cole, 1986). Crowns have lower sucrose concentration than roots (Cole and Seiler, 1976); however, complete removal of the crown would substantially decrease recoverable sucrose yield per hectare (Zielke, 1973). Zielke and Snyder (1974) reported that varietal rankings for impurity concentration in the crown differed from rankings based on concentration in the root. The concentration of impurities in the crowns averaged 70% more than in the roots. Doney and Theurer (1990) found that the osmolality of root extracts of L19, an inbred line noted for its high sucrose concentration and ability to increase sugar concentration in hybrids in which it is a parent (Theurer, 1981), was higher than other lines; they also observed a unique seasonal sucrose accumulation pattern for L19. In comparisons of diploids, triploids, and tetraploids, ploidy status influenced the activity of several important enzymes (Spettoli et al., 1976). Klotz and Finger (2002) suggested that sucrose synthase might be important in controlling root size and acknowledged the need for a better understanding of relationships between enzymes, yield, and quality. Production of smooth-root hybrids (sugar beet roots without the two vertical groves and associated mass of fibrous roots characteristic of most current varieties) would reduce the amount of soil in storage piles and could be considered a quality factor if premiums were paid for clean roots (Theurer, 1993). Smooth roots would also facilitate peeling or removing the skin to reduce the impurity load in the factory, if processors were to adopt this practice (Edwards et al., 1989).

Environmental variation (Milford and Thorne, 1973; Burley, 1990; Scott and Jaggard, 1993; Campbell, 2002a) and variety x environment interactions (Ulrich, 1961) must be considered in any plant breeding program. Relatively small variety x location, variety x year, and variety x location x year variance components for sucrose, sodium, and potassium concentration, compared to the varietal variance component, indicated consistency in the relative expression of these characters (Campbell and Kern, 1982). Significant variety x planting date interactions for yield and sucrose concentration indicated that planting date affects the relative yield and quality of hybrids (Yonts et al., 1999). Lasa et al., (1989) found sucrose yield of triploids to be more stable across environments than that of diploids.

Production of quality sugar beet will allow the beet sugar industry to remain competitive in a sweetener market that includes cane sugar and high-fructose corn sweeteners. Enhancing quality through breeding is slow and laborious. Negative correlation between root yield and sucrose concentration, interactions among quality traits and yield, and the requirement that commercial hybrids have resistance to prevalent pests compound the task. Processors often pay premiums for higher extractable sucrose concentrations; however, nonuniformity of these premiums complicate breeders' attempts to maximize economic return to growers. Increasing root yields has often been more productive than attempts to increase sucrose concentration and enhance quality; hence, future improvements in quality will require increased emphasis on quality traits when developing elite breeding populations.

References

Abawi, G.S., D.C. Croseir, A.C. Cobb, and R.F. Becker. 1986. Root rot of table beets in New York state. New York's Food and Life Sci. Bull. 115. Cornell Univ. New York, NY.

Abe, H. and T. Ui. 1986. Host range of *Polymyxa betae* strains in rhizomania infested soils of sugar beset fields in Japan. Ann. Pytopath. Soc. Japan 52:394-403.

Abe, J., G.P. Guan, and Y. Shimamoto. 1997. A gene complex for annual habit in sugar beet (*Beta vulgaris* L.). Euphytica 94:129-135.

Abegg, F.A. 1936. A genetic factor for the annual habit in beets and linkage relationship. J. Agric. Res. 53:493-511.

Abegg, F.A. 1938. The inheritance of curly-top resistance in beets. Proc. ASSBT 1:72.

Abegg, F.A. 1940. List of characters and gene symbols reported for the species *Beta vulgaris* L. Proc. ASSBT 2:109-113.

Abegg, F.A. and F.V. Owen. 1936. A genetic factor for curly-top disease resistance in beets (*Beta vulgaris* L.) and linkage relationships. Amer. Nat. 70:36.

Adams, H., W.R. Schäufele, and B. Märländer. 1996. A method for the artificial inoculation of sugar beet with *Ramularia beticola* in field conditions. Zeitschrift Pflanzenkrankheiten Pflanzenschutz 103:321-323.

Agrios, G.N. 1988. Plant Pathology. Academic Press, San Diego, CA.

Ahmadi, M. and M.T. Assad. 1998. Estimating genetic parameters of agronomic and quality traits in a diallel cross of sugar beet. Iran Agric. Res. 17:19-34.

Ahmadinejad, A. and M. Okhovat. 1976. Pathogenicity test of some soil-borne fungi on important field crops. Iran. J. Plant Pathol. 12:10-18.

Akeson, W.R. 1981. Methods for estimating sucrose loss in laboratory storage tests. J. ASSBT 21:56-73.

Akeson, W.R. and J.N. Widner. 1981. Differences among sugar beet cultivars in sucrose loss during storage. J. ASSBT 21:80-91.

Allison, M.F. and G.R.G. Clover. 1996. Nutrient uptake by the sugar beet fibrous root system. Proc. IIRB 59:165-172.

Armstrong, G.M. and J.K. Armstrong. 1976. Common hosts for *Fusarium oxysporum formae speciales Spinaciae* and *Betae*. Phytopathology 66:542-545.

Artschwager, E. 1930. A study of the structure of sugar beets in relation to sugar content and type. J. Agric. Res. 40:867-915.

Asher, M.J.C. 1993. Rhizomania. *In:* D.A. Cooke and R.K. Scott (eds.). The Sugar Beet Crop. Chapman & Hall, London, UK, pp. 311-346.

Asher, M.J.C., L.C. Luterbacher, and L. Frese. 2001a. Wild *Beta* species as a source of resistance to sugar-beet pests and diseases. Proc. IIRB 64:141-152.

Asher, M.J.C., M.C. Luterbacher, and L. Frese. 2001b. Wild *Beta* species as a source of resistance to sugar-beet pests and diseases. Intl. Sugar J. 103:447-456.

Avigad, G. and P.M. Dey. 1997. Carbohydrate metabolism: storage carbohydrates. *In:* P.M. Dey and J.B. Harborne (eds.). Plant Biochemistry. Academic Press, New York, NY, pp. 143-204.

Baldwin, J.G. and M. Mundo-Ocampo. 1991. Heteroderinae, cyst- and non-cyst-forming nematodes. *In:* W.R. Nickle (ed.). Manual of Agricultural Nematology. Marcel Dekker Inc., New York, NY, pp. 275-362.

Barocka, K.H. 1985. Zucker und Futterrüben (*Beta vulgaris L.*). *In:* W. Hoffmann, A. Mudra, and W. Plarre. Lehrbuch der Züchtung landwirtschaftlicher Kulturpflanzen. Verlag Paul Parey, Berlin, Germany, vol. 2.

Barzen, E., R. Stahl, E. Fuchs, D.C. Borchardt, and F. Salamini. 1997. Development of coupling-repulsion-phase SCAR markers diagnostic for the sugar beet Rr1 allele conferring resistance to rhizomania. Molec. Breed. 3:231-238.

Battey, N.H. 2000. Aspects of seasonality. J. Exper. Bot. 51:1769-1780.

Battle, J.P. and W.J. Whittington. 1969. The relation between inhibitory substances and variability in time to germination of sugar beet clusters. J. Agric. Sci. 73:337-346.

Battle, J.P. and W.J. Wittington. 1971. Genetic variability in time to germination of sugar beet clusters. J. Agric. Sci. Camb. 76:27-32.

Bayford, W.J. 1978. Field experiments on sugar beet powdery mildew, *Erysiphe betae*. Ann. Appl. Biol. 88:377-382.

Behringer, P. and L. Furst. 1985. Biological control of *Heterodera schachtii* with resistant radish or mustard. Bayerisches Landwirtschaftliches Jahrbuch 62:35-38.

Bennett, C.W. 1960. Sugar beet yellows disease in the United States. USDA Tech. Bull. 1218.

Bennett, C.W. 1971. The Curly Top Disease of Sugarbeet and Other Plants. Amer. Phytopath. Soc., St. Paul, MN, Monog. No. 7.

Bennett, C.W. and C. Munch. 1946. Yellow wilt of sugar beet in Argentina. J. Agric. Res. 73:45-64.

Bennett, C.W. and A. Tanrisever. 1958. Curly top disease in Turkey and its relationship to curly top in North America. J. ASSBT 10:189-211.

Bennett, C.W. and J.S. McFarlane. 1967. Damage produced by beet yellows and beet western yellows under greenhouse and field conditions. J. ASSBT 14:619-636.

Bennett, C.W. and L.D. Leach. 1971. Diseases and their control. *In:* R.T. Johnson, J.T. Alexander, G.E. Bush, and G.R. Hawkes (eds.). Advances in Sugarbeet Production: Principles and Practices. Iowa State Univ. Press, Ames, IA, pp. 223-285.

Bennett, C.W., F.J. Hills, K. Ehrenfeld, J. Valenzuela, and C.K. Klein. 1967. Yellow wilt of sugar beet. J. ASSBT 14:480-510.

Bernier, G., A. Havelange, C. Houssa, A. Petitjean, and P. Lejeune. 1993. Physiological signals that induce flowering. Plant Cell 5:1147-1155.

Biancardi, E. 1999. Miglioramento genetico. *In:* B. Casarini, E. Biancardi, and P. Ranalli (eds.). La barbabietola negli ambienti mediterranei. Edagricole. Bologna, Italy, pp. 57-145.

Biancardi, E. and A. Graf. 1984. Cercospora e varietà. Sementi Elette 1-2:31-38.

Biancardi, E. and G. Pavarin. 1993. La barbabietola per impieghi energetici. Informatore Agrario 5:87-89.

Biancardi, E., R.T. Lewellen, M. De Biaggi, A.W. Ericsen, and P. Stevanato. 2002. The origin of rhizomania resistance in sugar beet. Euphytica 127:383-397.

Bichsel, S.E. 1988. An overview of the U.S. beet sugar industry. *In:* M.A. Clarke and M.A. Godshall (eds.). Chemistry and Processing of Sugar Beet and Sugar Cane. Elsevier Sci. Publ., Amsterdam, Netherlands, pp. 1-8.

Bjerke, J.M., A.W. Anderson, and T.P. Freeman. 1992. Morphology of the larval stages of *Tetanops myopaeformis* (Röder) (*Diptera*: *Otitidae*). J. Kansas Ent. Soc. 65:59-65.

Blickenstaff, C.C., R.E. Peckenpaugh, and G.C. Mahrt. 1977. Rating sugar beets for damage by the sugar beet root maggot. J. ASSBT 19:188-191.

Bockstahler, H.W. 1940. Resistance to Fusarium yellows in sugar beets. Proc. ASSBT 3:191-198.

Bockstahler, H.W. G.J Hogaboam, and C.L. Schneider. 1950. Further studies on the inheritance of black root resistance in sugar beets. Proc. ASSBT. 6:104-107.

Boerma, H.R. and R.S. Hussey. 1992. Breeding plants for resistance to nematodes. J. Nematol. 24:242-252.

Bogdancva, E.Y. and I.A. Shevestov. 1995. Phenotypic variation in resistance to downy mildew in plants of sugar beet in the event of genotype x environment interaction. Tsitol. Genet. 1:37-41.

Bohn, K., M.A. Clark, K. Buchholz, K.M. Bliesiner, R. Buczys, and K. Thielicki. 1998. Composition of sugar beet and sugarcane and chemical behavior of constituents in processing. *In:* P.W. van der Poel, H. Schiweck, and T. Schwartz (eds.). Sugar Technology, Beet and Cane Manufacture. Verlag Dr. Albert Bartens, Berlin, Germany, pp. 115-208.

Bongiovanni, G.C. and L. Lanzoni. 1964. La rizomania della bietola. Progr. Agric. 2:209-220.

Bordonos, M.G. 1939. A study on the inheritance of the single germ character in beets. USSR. All Union Res. Inst. Sugar Beet Industry, Kiev, USSR, pp. 357-359.

Bosemark, N.O. 1967. Genoypic competition in anisoploid sugar beet. Hereditas 58:111-134.

Bosemark, N.O. 1969. Interspecific hybridization in *Beta* L.: prospects and value in sugar beet breeding. Proc. IIRB 4:112-121.

Bosemark, N.O. 1971. Use of Mendelian male-sterility in recurrent selection and hybrid breeding in beets. Rept. Meet. Eucarpia Fodder Crops Section. Lusignan, France, 15-17 Sept. 1970, pp. 127-136.

Bosemark, N.O. 1993. Genetics and breeding. *In:* D.A. Cooke and R.K. Scott (eds.). The Sugar Beet Crop. Chapman & Hall, London, UK, pp. 59-119.

Boudry, P., H. McCombie, and H. Van Dijk. 2002. Vernalisation requirement of wild beet *Beta vulgaris* ssp. *maritima*: among-population variation and its adaptive significance. J. Ecol. 90:693-703.

Boudry, P., R. Wieber, P. Saumitou-Laprade, K. Pillen, H. Van Dijk, and C. Jung. 1994. Identification of RFLP markers closely linked to the bolting gene *B* and their significance for the study of the annual habit in beet (*Beta vulgaris* L). Theor. Appl. Genet. 88:852-858.

Brants, I., P. Steen, S. Bisgaard, H.C. Pederson. 1995. Roundup ReadyTM sugar beet. Proc. IIRB. 58:557-559.

Brewbaker, H.E., R.R. Wood, and H.L. Bush. 1946. Single germ seed. Proc. ASSBT 4:259-262.

Brunt, A.A. and K.E. Richard. 1989. Biology and molecular biology of furoviruses. Adv. Virus Res. 36:1-32.

Bugbee, W.M. 1973a. Resistance in *Beta vulgaris* to *Phoma* storage rot in the North Central Region. Pl. Dis. Rept. 57:204-207.

Bugbee, W.M. 1973b. Sucrose and cell walls as factors affecting *Phoma* storage rot of sugar beet. Phytopathology 63:480-484.

Bugbee, W.M. 1975a. *Penicillium claviforme* and *Penicillium variabile*: pathogens of stored sugar beet. Phytopathology 65:926-927.

Bugbee, W.M. 1975b. Peroxidase, polyphenoloxidase, and endopolygalacturonate activity in different tissues of sugar beet infected with *Phoma betae*. Can. J. Bot. 53:1347-1351.

Bugbee, W.M. 1976. *Penicillium claviforme*: sugar beet pathogen and antagonist of *Botrytis cinerea*. Can. J. Plant Sci. 56:647-649.

Bugbee, W.M. 1978. Registration of F1001 and F1002 sugar beet germplasm. Crop Sci. 18:358.

Bugbee, W.M. 1979a. *Pleospora bjoerlingii* in the USA. Phytopathology 69:277-278.

Bugbee, W.M. 1979b. Resistance to sugar beet storage rot pathogens. Phytopathology 69:1250-1252.

Bugbee, W.M. 1979c. The effect of plant age, storage, moisture, and genotype on storage rot evaluation of sugar beet. Phytopathology 69:414-416.

Bugbee, W.M. 1982. Storage rot of sugar beet. Plant Dis. 66:871-873.

Bugbee, W.M. 1986. Storage rot of sugar beet. *In:* E.D. Whitney and J.E. Duffus (eds.). Compendium of Beet Diseases and Insects. Amer. Phytopath. Soc., St. Paul MN, pp. 37-39.

Bugbee, W.M. 1993. Storage. *In:* D.A. Cooke and R.K. Scott (eds.). The Sugar Beet Crop. Chapman & Hall, London, UK, pp. 551-570.

Bugbee, W.M. and D.F. Cole. 1976. Sugarbeet storage rot in the Red River Valley, 1974-75. J. ASSBT 19:19-24.

Bugbee, W.M. and G.E. Nielsen. 1978. *Penicillium cyclopium* and *Penicillium funiculosum* as sugar beet storage rot pathogens. Plant Dis. Rep. 62:953-954.

Bugbee, W.M. and D.F. Cole. 1979. Comparison of thiabendazole and genetic resistance for control of sugar beet storage rot. Phytopathology 69:1230-1232.

Bugbee, W.M. and D.F. Cole. 1986. Sucrose content, clear juice purity and storage rot of sugar beet. J. ASSBT 23:154-161.

Bugbee, W.M. and L.G. Campbell. 1990. Combined resistance in sugar beet to *Rhizoctonia solani*, *Phoma betae*, and *Botrytis cinerea*. Plant Dis. 74:353-355.

Bugbee, W.M., D.F. Cole, and G. Nielsen. 1975. Microflora and invert sugars in juice from healthy tissue of stored sugar beets. Appl. Microbiol. 29:780-781.

Bugbee, W.M., L.G. Campbell, and M. El-Kholi. 1989. Seedling response of storage-rot-resistant sugar beets to *Phoma betae* and *Rhizoctonia solani*. J. Sugar Beet Res. 26:33-39.

Bulder, H.A.M., W.R. van der Leij, E.J. Speek, P.R. van Hasselt, and P.J.C. Kuiper. 1989. Interactions of drought and low temperature stress on lipid and fatty acid composition of cucumber genotypes differing in growth response at suboptimal temperature. Physiol. Plant 75:362-368.

Burenin, V.I. 1998. Resistance to blackleg of beet: adaptive potential of *Beta* L. genetic resources. *In:* L. Frese, L. Panella, H.M. Srivastava, and W. Lange (eds.). Rept. 4th Intl. *Beta* Genetic Resources Workshop and World *Beta* Network Conf. Izmir, Turkey, 28 February - 3 March 1996. IPGRI, Rome, Italy, pp. 78-79.

Burenin, V.I. 2001. Genetic resources of sugar beet and resistance to diseases. Sakharnaya Svekla 7:25-29.

Burenin V.I. and Z.V. Timoshenko. 1989. Multiple disease resistance in sugar beet. Nauchno Tekhn. Bull. 194:36-39.

Burke, J.I., B. Rice, and V.A. Dodd. 1979. Measurement of respiration rate of stored sugar beet. Irish J. Agric. Res. 18:305-313.

Burky, K. and G. Büttner. 1989. Rizomania tolerante Zuckerrübensorten: II. BNYVV-Gehalt junger Testpflanzen und Ertrag in Feld unter Rizomaniabefall. Zuckerindustrie 114:225-228.

Burley, J.B. 1990. Sugar beet productivity model for Clay County Minnesota. J. Sugar Beet Res. 27:50-57.

Bush, D.R. 1999. Sugar transporters in plant biology. Curr. Opin. in Plant Biol. 2:187-191.

Bush, H.L. and H.E. Brewbaker. 1956. Genetic variability of sugar beets in relation to spider mite injury. J. ASSBT 9:275-276.

Büttner, G. 2002. The pathogenicity of beet soil-borne virus (BSBV) and its possible role in rhizomania disease of sugar beet. Proc. IIRB 65:413-418.

Büttner, G., B. Pfähler, and J. Petersen. 2003. Rhizoctonia root rot in Europe - incidence, economic importance and concept for integrated control. Proc. ASSBT/IIRB 1:897-901.

Büttner, G., B. Pfähler, and B. Märländer. 2004. Greenhouse and field techniques for testing sugar beet (*Beta vulgaris* L.) for resistance to Rhizoctonia root and crown rot (*Rhizoctonia solani* Kühn). Plant Breed. (in press).

Bychkova, V.A. and N.M. Krivoklyakin. 1993. Quality of inbred lines and F1 hybrids. Sakharnaya Svekla 5:27-28.

Byford, W.J. and K. Bentley. 1976. Cleistothecia of *Erysiphe betae* in England. Trans. British Mycol. Soc. 67:544-545.

Cai, D., M. Kleine, S. Kifle, et al., 1997. Positional cloning of a gene for nematode resistance in sugar beet. Science 275:832-834.

Campbell, C.D. and W.D. Hutchinson. 1995a. Rearing methods and demographic statistics for a subterranean morphology of the sugar beet root aphid (*Homoptera*: *Aphididae*). Can. Ent. 127:65-77.

Campbell, C.D. and W.D. Hutchinson. 1995b. Sugar beet resistance to Minnesota populations of sugar beet root aphid (*Homoptera*: *Aphididae*). J. Sugar Beet Res. 32:37-46.

Campbell, C.L. and J. Altman. 1976. Rapid laboratory screening of sugar beet cultivars for resistance to *Rhizoctonia solani*. Phytopathology 66:1373-1374.

Campbell, G.K.G. and G.E. Russell. 1964. Breeding in sugar beet. Ann. Rept. Plant Breed. Inst. Cambridge pp. 1-31.

Campbell, L.G. 2002a. Sugar beet quality improvement. *In:* A.S. Basra and L.S. Randhawa (eds.). Quality Improvement in Field Crops. Food Products Press, Binghamton, NY, pp. 395-413.

Campbell, L.G. 2002b. Sugar beet breeding and improvement. *In:* M.S. Kang (ed.) Crop Improvement in the Twenty-first Century. Food Products Press, Binghamton, NY, pp. 193-221.

Campbell, L.G. and J.J. Kern. 1982. Cultivar x environment interactions in sugar beet yield trials. Crop Sci. 22:932-935.

Campbell, L.G. and J.J. Kern. 1983. Relationships among components of yield and quality of sugar beets. J. ASSBT 22:135-145.

Campbell, L.G. and W.M. Bugbee. 1985. Registration of storage rot resistant sugar beet germplasms F1004, F1005, and F1006. Crop Sci. 25:577.

Campbell, L.G. and D.F. Cole. 1986. Relationships between taproot and crown characteristics and yield and quality traits in sugar beets. Agron. J. 78:971-973.

Campbell, L.G. and W.M. Bugbee. 1988. Selection for improved sugar beet storability. Crop Sci. 28:33-36.

Campbell, L.G. and W.M. Bugbee. 1993. Pre-breeding for root-rot resistance. J. Sugar Beet Res. 30:241-252.

Campbell, L.G. and G.J. Seiler. 1994. Internal CO_2 concentration as a selection criterion for storage respiration rate in sugar beet. Plant Breed. 112:96-101.

Campbell, L.G., A.W. Anderson, and R.J. Dregseth. 2000. Registration of F1015 and F1016 sugar beet germplasms with resistance to the sugar beet root maggot. Crop Sci. 40:867-868.

Campbell, L.G., A.W. Anderson, R. Dregseth, and L.J. Smith, 1998. Association between sugar beet root yield and sugar beet root maggot (*Diptera: Otitidae*) damage. J. Econ. Ent. 91:522-527.

Canova, A. 1959. Ricerche sulla biologia e l'epidemiologia della *Cercospora beticola* Sacc. Ann. Sper. Agric. 13:1-46; 157-203; 477-497; 685-767; 855-897.

Carruthers, A., J.F.T. Oldfield, and H.J. Teague. 1962. Assessment of beet quality. Proc. 15th Tech. Conf. British Sugar Corp. Peterborough, UK, pp. 28.

Carsner, E. 1926a. Resistance in Sugar Beets to Curly Top. USDA Circular 388.

Carsner, E. 1926b. Seasonal and regional variations in curly-top of sugar beets. Science 63:213-214.

Carsner, E. 1933. Curly-top Resistance in Sugar Beets and Test of the Resistant Variety US1. USDA Tech. Bull., no. 360.

Casarini, B. 1999. Le avversità: loro natura, prevenzione e lotta. *In:* B. Casarini, E. Biancardi, and P. Ranalli (eds.). La barbabietola negli ambienti mediterranei. Edagricole. Bologna, Italy, pp. 273-421.

Casarini B., E. Biancardi, and P. Ranalli. 1999. La barbabietola negli ambienti mediterranei. Edagricole, Bologna, Italy.

Casarini-Camangi, P. and A. Canova. 1987. Infezione ed incidenza sulla produttività del virus della rizomania in alcuni genotipi di barbabietola da zucchero. Phytopath. Medit. 26:91-99.

Castagnone-Sereno, P. 2002. Genetic variability of nematodes: a threat to the durability of plant resistance genes? Euphytica 124:193-199.

Caubel, G., F. Le Cochec, and M. Renard. 1985. Resistance against *Heterodera schachtii* in beet and some crucifers: practical use. C. R. Acad. Agric. France, 71:731-740.

Christmann, J. 1983. Les varietes sauvages et la selection des types resistants — Stude du cas de la betterave a sucre. Proc. IIRB 46:91-100.

Chroboczek, E. 1934. A study of some ecological factors influencing seed-stalk development in beets (*Beta vulgaris* L.). Cornell Univ. Memoir 154:3-83.

Clark, M.F. and A.N. Adams. 1977. Characteristics of the microtiter plate method of enzyme-linked immunosorbent assay for the detection of plant viruses. J. Gen. Virol. 34:475-483.

Clarke, N.A., H.M. Hetschkun, M. Redfearn, T.H. Thomas, and V. Naumenko. 1995. Response and tolerance of sugar beet to stress. J. ASSBT 28:189-197.

Coe, G.E. 1954. A grafting technique enabling an unthrifty interspecific hybrid of *Beta* to survive. Proc. ASSBT 8, Part 2:157-160.

Coe, G.E. 1974. Registration of sugar beet parental lines SP 69550-0 and SP69550-01. Crop Sci. 14:343.

Coe, G.E. 1987. Selecting sugar beets for low content of nonsucrose solubles. J. ASSBT 24:41-48.

Coe, G.E. and C.L. Schneider. 1966. Selecting sugar beet seedlings for resistance to *Aphanomyces cochlioides*. J. ASSBT 14:164-167.

Coe, G.E. and G.J. Hogaboam. 1971. Registration of sugar beet parental line SP 6322-0. Crop Sci. 11:947.

Coe, G.E. and N.R. O'Neill. 1983. Selecting sugar beet seedlings in the greenhouse for resistance to *Sclerotium rolfsii*. J. ASSBT 22:35-45.

Cole, D.F. 1977. Effect of cultivar and mechanical damage on respiration and storability of sugar beet roots. J. ASSBT 19:240-245.

Cole, D.F. 1980a. Effect of complete crown removal on quality of sugar beets. J. ASSBT 20:449-454.

Cole, D.F. 1980b. Post-harvest respiration rates and internal CO_2 concentration in sugar beet roots. Can. J. Plant Sci. 60:1489-1491.

Cole, D.F. 1986. Relationship between internal CO_2 and respiration in selected *Beta vulgaris* L. genotypes. J. ASSBT 23:116-120.

Cole, D.F. and W.M. Bugbee. 1976. Changes in resident bacteria, pH, sucrose, and invert sugar levels in sugar beet roots during storage. Appl. Envir. Microbiol. 31:754-757.

Cole, D.F. and G.J. Seiler. 1976. Effect of crown material on yield and quality of sugar beet roots: A grower survey. J. ASSBT 19:131-137.

Coons, G.H. 1936. Improvement of the sugar beet. USDA Yearbook of Agriculture, Washington, DC, pp. 625-657.

Coons, G.H. 1975. Interspecific hybrids between *Beta vulgaris* L. and the wild species of *Beta*. J. ASSBT 18:281-306.

Coons, G.H., D. Stewart, and H.A. Elcock. 1931. Sugar beet strains resistant to leaf spot and curly-top. USDA Yearbook of Agriculture, Washington, DC, pp. 493-496.

Coons, G.H., F.V. Owen, and D. Stewart. 1955. Improvement of the sugar beet in the United States. Adv. Agron. 7:89-139.

Cormack, M.W. and J.F. Moffatt. 1961. Factors influencing storage decay of sugar beets by *Phoma betae* and other fungi. Phytopathology 51:3-5.

Crespi, P., C. Penel, M. Crèvecoeur, and H. Greppin. 1993. Plasma membrane sterols and flowering induction. Plant Sci. 89:153-160.

Crombie, B. and L.J. O'Connor. 1960. The variable incidence of downy mildew *(Peronospora schachtii)* on a series of sugarbeet varieties, strains and family groups. Proc. IIRB 23:453-457.

Crosthwaite, S.K. and G.I. Jenkins. 1993. The role of leaves in the perception of vernalizing temperatures in sugar beet. J. Exper. Bot. 44:801-806.

Culbertson, J.O. 1942. Inheritance of factors influencing sucrose percentage in *Beta vulgaris*. J. Agric. Res. 64: 153-172.

Curth, P. 1955. Temperatur und Licht als blüheninduzierende Faktoren bei der Zuckerrübe. *In:* G. Becker, H. Kappert, H. Kuckuck, K. Pätau, R. Schick, and H. Stubbe (eds.). Der Züchter. Springer-Verlag, Berlin, Germany, pp. 176-181.

Curth, P. 1960. Der Übergang in die reproduktive Phase bei der Zuckerrübe in Abhängigkeit von verschiedenen Umweltfaktoren. Beiträge zur Rübenforschung 4:7-80.

Curth, P. 1963. Über die Modifikation des kälte- und lichtinduzierten Blüheffektes bei Zuckerrüben durch Anwendung des Wirkstoffes AMO-1618. Flora 153:87-89.

De Biaggi, M. 1987. Methodes de selection—Un cas concret. Proc. IIRB 50:157-161.

De Biaggi, M., A.W. Ericsen, R.T. Lewellen, and E. Biancardi. 2003. The discovery of the rhizomania resistance traits in sugar beet. Proc. ASSBT/IIRB 1:131-147.

De Jong, J.H. and P. Stam. 1985. The association of centromeres of nonhomologous chromosomes at meiotic prophase in *Beta vulgaris* L. Can. J. Genet. Cytol. 27:165-171.

De Jong, J.H., G.J. Speckmann, T.S.M. De Bock, W. Lange, and A. van Voorst. 1986. Alien chromosome fragments conditioning resistance to beet cyst nematode in diploid descendants from monosomic additions of *Beta procumbens* to *B. vulgaris*. Can. J. Genet. Cytol. 28:439-443.

De Kock, P.C., R.F. Hunter, and I.R. McDonald. 1953. A germination inhibitor from sugar beet. J. Expl. Bot. 4:272-282.

De Pointi, O.M.B. and C. Mollema. 1991. Emerging breeding strategies for insect resistance. *In:* H.T. Stalker and J.P. Murphy (eds.). Plant Breeding in the 1990s. CAB Int. Oxon, UK, pp. 323-346.

Debenham, G.B. 1999. Bolting and flowering mechanisms in sugar beet, *Beta vulgaris* ssp *vulgaris* L. PhD diss. IACR Broom´s Barn, UK.

Decoux, L. and A. Weenen. 1962. Factors affecting viability of sugar beet seed. Proc. IIRB 25:397-400.

Delasalle, L. and C. Dhellemmes. 1996. Barbabietola da zucchero: attuali esigenze e richieste, metodologie genetiche per risposte possibili. Sementi Elette 42:65-72.

Desprez, M. 1993. Evolution des methodes de selection de la betterave sucriere des origines a nos jours. C.R. Acad. Agric. France 79:71-84.

Dewar, A.M., L.A. Haylock, K.M. Bean, B.H. Garner, and R. Boyce 2000. The ecology and control of two-spotted spider mite, *Tetranychus urticae*. Proc. BCPC Conf., Brighton, UK, pp. 913-918.

Di Vito, M. 1983. Reaction of *Beta* spp. to root-knot nematodes. J. Nematol. 15:144-145.

Dilley, D.R., D.H. Dewey, and D.D. Dedolph. 1969. Automated system for determining respiratory gas exchange of plant materials. J. Amer. Soc. Hort. Sci. 94:138-141.

Dix, P., F. Iseult, and J.I. Burke. 1994. Genotypic differences in cold tolerance are masked by high sucrose and cytokinin in shoot cultures of sugarbeet. Plant Cell Tiss. Org. Cult. 36:285-290.

Doney, D.L. and E.D. Whitney. 1969. Screening sugar beet for resistance to *Heterodera schachtii* Schm. J. ASSBT 15:546-552.

Doney, D.L. and J.C. Theurer. 1978. A modified reciprocal recurrent selection breeding method for sugarbeet. *In:* Agronomy Abstracts. ASA, Madison, WI, p. 49.

Doney, D.L. and J.C. Theurer. 1990. Osmolality of L19 type sugar beet germplasm. J. Sugar Beet Res. 27:81-89.

Downie, A.R., C.W. Doxtator, J.F. Swink, and J.C. Tanner. 1952. Reaction of sugar beet strains to *Aphanomyces cochlioides* at three different locations. Proc. ASSBT 8:393-395.

Doxtator, C.W. and A.R. Downie. 1947. Breeding for resistance to Aphanomyces root rot. Proc. ASSBT 4:34-39.

Doxtator, C.W. and A.R. Downie. 1948. Progress in breeding sugar beets for resistance to Aphanomyces root rot. Proc. ASSBT 5:130-136.

Doxtator, C.W. and R.E. Finkner. 1954. A summary of results in the breeding for resistance to *Aphanomyces cochlioides* (Drecks) by the American Crystal Sugar Company since 1942. Proc. ASSBT 8:94-98.

Doxtator, C.W. and R.E. Finkner. 1958. Sugar beet germination selection results in osmotic pressure selection: I. Germination methodology and osmotic selection effects on germination of four varieties. J. ASSBT 10:237-246.

Dudley, J.W. and L. Powers. 1960. Population genetic studies on sodium and potassium in sugar beets (*Beta vulgaris* L.). J. ASSBT 9:97-127.

Duffus, J.E. 1960. Radish yellow, a disease of radish, sugar beet, and other crops. Phytopathology 50:389-394.

Duffus, J.E. 1964. Host relationships of beet western yellows virus strains. Phytopathology 54:736-738.

Duffus, J.E. 1978. The impact of yellows control on California sugarbeets. J. ASSBT 20:1-5.

Duffus, J.E. 1986. Beet curly top. *In* E.D. Whitney, and J.E. Duffus (ed.) Compendium of Sugar Beet Diseases and Insects. Am. Phytopath. Soc., St. Paul, MN, pp. 31-32.

Duffus, J.E. and G.E. Russell. 1975. Serological relationship between beet western yellows and beet mild yellowing viruses. Phytopathology 65:811-815.

Duffus, J.E. and E.G. Ruppel. 1993. Diseases. *In:* D.A. Cooke and R.K. Scott (eds.). The Sugar Beet Crop. Chapman & Hall, London, UK, pp. 346-427.

East, E.M. and A.J. Mangelsdorf. 1925. A new interpretation of hereditary behavior of self-sterile plants. Proc. Natl. Acad. Sci., Washington, DC, 11:166-171.

Edwards, R.H., J.M. Randall, and L.W. Rodel. 1989. Peeling of sugar beets by use of high pressure steam. J. Sugar Beet Res. 26:63-76.

Eisenback, J.D. and H.H. Triantaphyllou. 1991. Root-knot nematodes: *Meloidogyne* species and races. *In:* W.R. Nickle (ed.) Manual of Agricultural Nematology. Mercel Dekker, Inc., New York, NY, pp. 191-274.

El Kholi, M.M. A.M. Ragab, M.Y. Hussein, and M.M. Ragab. 1994. Alternaria leaf spot of sugar beet in Egypt. Egyptian J. Phytopath. 22:179-193.

Elliott, M.C. and G.D. Weston. 1993. Biology and physiology of the sugar-beet plant. *In:* D.A. Cooke and R.K. Scott (eds.). The Sugar Beet Crop. Chapman & Hall, London, UK, pp. 37-66.

El-Mezawy, A., F. Dreyer, G. Jacobs, and C. Jung. 2002. High-resolution mapping of the bolting gene *B* of sugar beet. Theor. Appl. Genet. 105:100-105.

Esau, K. 1930. Studies of the breeding of sugar beets for resistance to curly top. Hilgardia 4:415-441.

Ezra, D., I. Barash, L. Valinsky, and S. Manulis. 2000. The dual function in virulence and host range restriction of a gene isolated from the pPATHEhg plasmid of *Erwinia herbicola* pv. *gypsophilae*. Molec. Plant-Microbe Interact. 13:683-692.

Fahim, M.M., M.A. Kararah, A.A. El-Gharabawi, and K.A.M. Abada. 1981. Studies on fungi causing root rot of sugar beet with special reference to *Sclerotium rolfsii*. Egyptian J. Phytopath. 13:1-10.

Farid, G. and K. Sajjad. 2001. Evaluation of candidate varieties of sugarbeet against sclerotial rot (*Sclerotium rolfsii* Sacc). Proc. 2nd Workshop R&D Activities, Shakarganj Sugar Inst. 16: 106-107.

Fauchere J. 2001. Influence de la qualite des graines sur le varietes experimentees. Proc. IIRB 64:257-268.

Feltz, H. and R. Marx, 1960. Zur Mosaikresistenz bei Zuckerrüben. Zucker 13:165-168.

Finkner, R.E. and H.M. Bauserman. 1956. Breeding sugar beets with reference to sodium, sucrose, and raffinose content. J. ASSBT 9:170-177.

Finkner, R.E., M.D. Finkner, and R.F. Olson. 1959. Effects of storage on raffinose content of sugar beets I. Varietal changes occurring during storage. J. ASSBT 10:481-488.

Fischer, H.E. 1956. Untersuchungen an Zwilligen von *Beta vulgaris* L. Züchter 26:136-152.

Ford-Lloyd, B.V., A.L.S. Williams, and J.T. Williams. 1975. A revision of *Beta* section *Vulgares* (*Chenopodiaceae*), with new light on the origin of cultivated beets. Bot. J. Linn. Soc. 71:89-102.

Francis, S. 2002. Sugar-beet powdery mildew (*Erysiphe betae*). Molec. Plant Pathol. 3:119-124.

Frese, L., B. Desprez, and D. Ziegler. 2001. Potential of genetic resources and breeding strategies for base-broadening. *In:* H.D. Cooper, C. Spillane, and T. Hodgkin (eds.). Broadening the Genetic Base of Crop Production. IPGRI-FAO, Rome, Italy, pp. 295-309.

Fujisawa, I. and T. Sugimoto. 1979. The reaction of some beet species of Section *Patellares*, *Corollinae* and *Vulgares* to rhizomania of sugar beet. Proc. Sugar Beet Res. Assoc. Japan 21:31-38.

Gaskill, J.O. 1950a. Drying after harvest increases storage decay of sugar beet roots. Phytopathology 40:483-486.

Gaskill, J.O. 1950b. Possibilities for improving storage-rot resistance of sugar beets through breeding. Proc. ASSBT 6:664-669.

Gaskill, J.O. 1952. A study of two methods of testing individual sugarbeet roots for storage pathogens. Proc. ASSBT 7:575-580.

Gaskill, J.O. 1954. Viable hybrids from matings of chard with *Beta procumbens* and *B. webbiana*. Proc. ASSBT 8, Part 2:148-152.

Gaskill, J.O. 1968. Breeding for rhizoctonia resistance in sugar beet. J. ASSBT 15:105-119.

Gaskill, J.O. and R.K. Ehrenfeld. 1976. Breeding sugar beet for resistance to yellow wilt. J. ASSBT 19:25-44.

Gaskill, J.O., D.L. Mumford, and E.G. Ruppel. 1970. Preliminary report on breeding sugarbeet for combined resistance to leaf spot, curly top and *Rhizoctonia*. J. ASSBT 16:207-213.

Gaskill, J.O., C.L. Schneider, A.M. Murphy, and G.E. Coe. 1967. Breeding for combined resistance to leaf spot and curly top in sugar beet. J. ASSBT 14:518-537.

Gerik, J.S. and J.E. Duffus. 1988. Differences in vectoring ability and aggressiveness of isolates of *Polymyxa betae*. Pytpathology 78:1340-1341.

Ghoulam, C., A. Foursy, and K. Fares. 2002. Effects of salt stress on growth, inorganic ions and proline accumulation in relation to osmotic adjustment in five sugarbeet cultivars. Envir. Exper. Bot. 47:39-50.

Giunchedi, L., M. De Biaggi, and C. Poggi Pollini. 1985 Evaluation of ELISA technique for the screening of rhizomania tolerant sugar beet genotypes. Proc. IIRB 48:385-390.

Giunchedi, L., M. De Biaggi, and C. Poggi Pollini. 1987. Correlation between tolerance and BNYVV in sugar beet genotypes. Phytopath. Medit. 26:23-28.

Gojmerac, W.L. 1956. Description of the sugar beet root maggot, *Tetanops myopaeformis* (von Röder), with observations on reproductive capacity. Entomol. News 6: 203-210.

Golden, A.M. 1959. Susceptibility of several *Beta* species to the sugar-beet nematode (*Heterodera schachtii*) and root-knot nematodes (*Meloydogyne* spp.). J. ASSBT 10:444-447.

Golev, I.F., K.P. Goleva, and L.K. Lunin. 1984. The possibilities of selecting for resistance to flea beetles in sugar beet. Zashchita Rastenii 4:22.

Goodshall, M.A., M.A. Clarke, C.D. Dooley, and R.S. Bianco. 1991. Progresses in beet sugar colorant research. J. Sugar Beet Res. 3-4:155-165.

Grassi, G., R. Fantini, and E. Biancardi. 1988. Prospective method of selecting sugar beet for resistance to rhizomania virus BNYVV. Zuckerindustrie 113:594-596.

Grassi, G., R. Fantini, and E. Biancardi. 1989. A new approach to selecting sugar beet for resistance to rhizomania virus (BNYVV). Phytopath. Medit. 28:131-139.

Gross, D.C. and L.D. Leach. 1973. Stalk blight of sugar beet seed crops caused by *Fusarium oxysporum* f. sp. *betae*. Phytopathology 63:1216.

Gummerson, R.J. 1986. The effects of constant temperatures and osmotic potential on the germination of sugar beet. J. Exper. Bot. 37:729-741.

Halloin, J.M. and D.L. Roberts. 1995. Temperature as a determining factor in the storage rot of sugar beet caused by *Aspergillus fumigatus*. J. Sugar Beet Res. 32:59-67.

Hanson, L.E., A. Hill, and L. Panella. 2001. Variability in *Fusarium oxysporum* isolates from sugar beet. Phytopathology 91:S36.

Harju, V. and M. Richard-Molard. 2002. Rhizomania P type: A new threat to growers? British Sugar Beet Rev. 3:22-27.

Harju, V.A., R.A. Mumford, A. Bockley, et al., 2002. Occurrence in the United Kingdom of BNYVV isolates which contain RNA 5. Plant Path. 51:811.

Harper, A.M. 1962. Life history of the sugar beet root maggot *Tetanops myopaeformis* (Röder) (*Diptera*: *Otitidae*) in southern Alberta. Can. Entomol. 94:1334-1340.

Harper, A.M. 1964. Varietal resistance of sugar beets to the sugar beet root aphid *Pemphigus betae* Doane (*Homoptera*: *Aphididae*). Can. Entomol. 96:520-522.

Harrington, R., A.M. Dewar, and B. George. 1989. Forecasting the incidence of virus yellows in sugar beet in England. Ann. Appl. Biol. 114:459-469.

Hartman, E.M. 1977. Technical vocabulary for American beet-sugar processors. J. ASSBT 19:345-364.

Harveson, R.M. 2000. First report of aphanomyces root rot of sugar beet in Nebraska and Wyoming. Plant Dis. 84:596.

Harveson, R.M. and C.M. Rush. 1994. Evaluation of fumigation and rhizomania-tolerant cultivars for control of a root disease complex of sugar beets. Plant Dis. 78:1197-1202.

Harveson, R.M. and C.M. Rush. 1997. Genetic variation among *Fusarium oxysporum* isolates from sugar beet as determined by vegetative compatibility. Plant Dis. 81:85-88.

Harveson, R.M. and C.M. Rush. 2002. The influence of irrigation frequency and cultivar blends on the severity of multiple root diseases in sugar beets. Plant Dis. 86:901-908.

Harvey, C.W. and J.W. Dutton. 1993. Root quality and processing. *In:* D.A. Cooke and R.K. Scott (eds.) The Sugar Beet Crop. Chapmann & Hall, London, UK, pp. 571-617.

Hasegawa, T., T. Takeda, K. Sekimura, and M. Hachinohe. 1977. Studies on back crossing in monogerm sugar beet: II. Back crossing European multigerm strains. J. ASSBT 19:109-122.

Hawley, I.M. 1922. The sugar beet root maggot (*Tetanops aldrichi* Hendel), a new pest of sugar beets. J. Econ. Entomol. 15:388-391.

Haylock, L. and A. Dewar. 2003. Might mite be a future sugar beet problem? British Sugar Beet Rev. 3:20-25.

Hecker, R.J. 1967. Evaluation of three sugar beet breeding methods. J. ASSBT 14:309-318.

Hecker, R.J. 1985. Reciprocal recurrent selection for the development of improved sugar beet hybrids. J. ASSBT 23:47-57.

Hecker, R.J. 1991. Effect of sugar beet root size on combining ability of sucrose yield components. J. Sugar Beet Res. 28:41-48.

Hecker, R.J. and J.O. Gaskill. 1972. Registration of FC 701 and FC 702 sugarbeet germplasm. Crop Sci. 12:400.

Hecker, R.J. and E.G. Ruppel. 1975. Inheritance of resistance to rhizoctonia root rot in sugar beet. Crop Sci. 15:487-490.

Hecker, R.J. and E.G. Ruppel. 1976. Polyploid and maternal effects on rhizoctonia root rot resistance in sugarbeet. Euphytica 25:419-423.

Hecker, R.J. and E.G. Ruppel. 1977a. Registration of diploid and tetraploid FC 701/4 and FC 703 sugarbeet germplasm. Crop Sci. 17:678.

Hecker, R.J. and E.G. Ruppel. 1977b. Rhizoctonia root-rot resistance in sugarbeet: breeding and related research. J. ASSBT 19:246-256.

Hecker, R.J. and R.M. Helmerich. 1985. Sugar beet breeding in the United States. *In:* G.E. Russell (ed.) Progress in Plant Breeding. Butterworths, London, UK, pp. 37-61.

Hecker, R.J., R.E. Stafford, R.H. Helmerick, and G.W. Maag. 1970. Comparison of the same sugar beet F_1 hybrids as diploids, triploids, and tetraploids. J. ASSBT 16:106-116.

Hecker, R.J., E.G. Ruppel, G.W. Maag, and D.M. Rasmuson. 1975. Amino acids associated with *Cercospora* leaf spot resistance in sugarbeet. *Phytopathologische Zeitschrift,* 82:175-181.

Heide, O.M. 1973. Environmental control of bolting and flowering in red garden beets. Sci. Rept. Agric. Univ. Norway 52:1-17.

Heijbroek, W. 1977. Partial resistance of sugar beet to beet cyst eelworm (*Heterodera schachtii* Schm.). Euphytica 26:257-262.

Heijbroek, W., A.J. Roelands, and J.H. De Jong. 1983. Transfer of resistance to beet cyst nematode from *Beta patellaris* to sugar beet. Euphytica 32:287-298.

Heijbroek, W., R.G. Munning, and A.C.P.M. van Swaaij. 2002. The effect of different levels of beet cyst nematodes (*Heterodera schachtii*) and beet necrotic yellow vein virus (BNYVV) on single and double resistant sugar beet cultivars. Eur. J. Plant Path. 108:735-744.

Heijbroek, W., A.J. Roelands, J.H. De Jong, C. van Hulst, A.H.L. Schoone, and R.G. Munning. 1988. Sugar beets homozygous for resistance to beet cyst nematode (*Heterodera schachtii* Schm.), developed from monosomic additions of *Beta procumbens* to *B. vulgaris*. Euphytica 38:121-131.

Heinrichs, V.C. 2000. Solutions for the control of the sugar beet cyst nematode *Heterodera schachtii*: Experiences with nematode resistant sugar beets in the Rhineland. Gesunde Pflanzen 52:67-70.

Held, L.J., J.W. Jennings, D.W. Koch, and F.A. Gray. 2000. Trap drop radish: a sustainable alternative for nematoide in sugar beets. J. Amer. Soc. Farm Manag. Apprais. pp. 118-126.

Herr, L.J. 1996. Sugar beet diseases incited by *Rhizoctonia* spp. *In:* B. Sneh (ed.). *Rhizoctonia* Species: Taxonomy, Molecular Biology, Ecology, Pathology and Disease Control. Kluwer Acad. Publ., Dordrecht, Netherlands, pp. 341-350.

Hijner, J.A. 1952. The susceptibility of wild beets for *Heterodera schachtii*. Med. Inst. Rat. Suikerprod. 21:1-13.

Hillmann, U. and E.Schlösser. 1987. *Fusarium oxysporum* f. sp. *betae* als Erreger von Umfallkrankheit und Hypocotylfäule an Zuckerrüben. Gesunde Pflanzen 39:78-83.

Hills, F.J., D.H. Hall, and D.G. Kontaxis. 1975. Effect of powdery mildew on sugarbeet production. Plant Dis. Rept. 59:513-515.

Hills, F.J., L. Chiarappa, and S. Geng. 1980. Powdery mildew of sugar beet: disease and crop loss assessment. Phytopathology 70:680-683.

Hine, R.B. and E.G. Ruppel. 1969. Relationship of soil temperature and moisture to sugarbeet root rot caused by *Pythium apanidermatum* in Arizona. Plant Dis. Rept. 59:989-991.

Hobbis, J. and L. Batterman. 1975. Storability characteristics of sugar beet varieties. Sugar J. 12:32-35.

Hogaboam, G.J. 1957. Factors influencing phenotypic expression of cytoplasmic male sterility in the sugar beet (*Beta vulgaris* L.). J. ASSBT 9:457-465.

Hogaboam, G.J. 1961. Radiographing as a method of observing some seed characters in monogerm sugar beet fruit. J. ASSBT 11:605-609.

Hogaboam, G.J., R.C. Zielke, and C.L. Schneider. 1982. Registration of EL40 sugar beet parental line. Crop Sci. 22:700.

Holtmann, B., M. Kleine, and F.M.W. Grundler. 2000. Ultrastructure and anatomy of nematode-induced syncytia in roots of susceptible and resistant sugar beet. Protoplasma 211:39-50.

Holtschulte, B. 2000. *Cercospora beticola* – worldwide distribution and incidence. *In: Cercospora beticola* Sacc. biology, agronomic influence and control measures in sugar beet. Advances in Sugar Beet Research. IIRB, Brussels, Belgium, vol. 2, pp. 5-16.

Hosemans, D. and D. Bossoutrot. 1983. Induction of haploid plants from in vitro culture of unpollinated beet ovules (*Beta vulgaris* L.). Z. Pflanzenzucht. 91:74-77.

Hull, R. 1960a. Sugar Beet Diseases, bull. no. 142. Ministry of Agriculture, Fisheries and Food, London, UK.

Hull, R. 1960b. The selection of sugar beet varieties for tolerance to virus yellows. Proc. IIRB 23:407-417.

Hutchinson, W.D. and C.D. Campbell. 1994. Economic impact of sugar beet root aphid (*Homoptera*: *Aphididae*) on sugar beet yield and quality in southern Minnesota. J. Econ. Entomol. 87:465-475.

Institut Technique Français de la Betterave Industrielle (ITB). 1995. Qualité de la récolte. Reduire la tare terre. Betteravier Français 660:15-18.

Institut Technique Français de la Betterave Industrielle (ITB). 1997. Varietes tolerantes a la rhizomanie. Betteravier Français 703:29-32.

Janssen, G.J.W., M. Nihlgaard, and T. Kraft. 2003. Mapping of resistance genes to powdery mildew (*Erysiphe betae*) in sugar beet. Proc. ASSBT/IIRB 1:175-180.

Jassem, M. 1976. Über die Verbesserung der Keimfähigkeit monokarper Zuckerrüben. Tag. Ber. Akad. Landwirtsch. Wiss. 147:41-49.

Johnson, R.T. 1956. A grafting method to increase survival of seedlings of interspecific hybrids within the genus *Beta*. J. ASSBT 9:25-31.

Johnson, R.T. and G.W. Wheatley. 1961. Studies on backcross generations and advanced generations of interspecific hybrids between *B. vulgaris* and *B. webbiana*. J. ASSBT 11:429-435.

Jones, F.G.W. 1957. Sugar Beet Pests. Bull. No. 162. Ministry of Agriculture, Fisheries and Food. London, UK.

Jung, C. and G. Wricks. 1987. Selection of diploid nematode-resistant sugar beet from monosomic addition lines. Plant Breed. 98:205-214.

Jung, C. and U. Wyss. 1999. New approaches to control plant parasitic nematodes. Appl. Microbiol. Biotech. 51:439-446.

Jung, C., D. Cai, and M. Kleine. 1998. Engineering nematode resistance in crop species. Trends Plant Sci. 3:266-271.

Kafka, S.R., W.M. Wintermantel, and R.T. Lewellen. 2002. Comparisons of soil and seed applied systemic insecticides to control Beet Curly Top Virus in the San Joaquin Valley. J. Sugar Beet Res. 39:59-74.

Kajanus, B. 1913. Über die Vererbungsweise gewisser Merkmale der *Beta* und *Brassica* Rüben. I.Z. Pflanzenzüchtung 1:125-186.

Kajanus, B. 1917. On the color variation of *Beta*. Z. Pflanzenzücht. 5:357-372.

Kaloshian, I., V.M. Williamson, G. Miyao, D.A. Lawn, and B.B. Westerdahl. 1996. "Resistance-breaking" nematode identified in California tomatoes. Calif. Agric. 50:18-19.

Kästner, B. 1990. Zur Optimierung von Pflanzenzahl und Verteilung sowie Ertrags und Qualitätsmerkmalen in der Zuckerrübenproduction. Zuckerindustrie 115:458-465.

Keller, W. 1936. Inheritance of some major color types in beets. J. Agric. Res. 52:27-38.

Kerr, S.P. and M. Leaman. 1996. Varieties for 1997. British Sugar J. 2:9-13.

Kift, N.B., A.M. Dewar, A.W. Werker, and A.F.G. Dixon. 1996. The effect of plant age and infection with virus yellows on the survival of *Myzus persicae* on sugar beet. Ann. Appl. Biol. 129:371-378.

Kleine, M., H. Voss, D. Cai, and C. Jung. 1998. Evaluation of nematode-resistant sugar beet (*Beta vulgaris* L.) lines by molecular analysis. Theor. Appl. Genet. 97:896-904.

Kleine, M, D. Cai, R.M. Klein-Lankhorst, et al., 1997. Breeding for nematode resistance in sugar beet: a molecular approach. *In:* C. Fenoll and F.M.W. Grundler (eds.). Cellular and Molecular Aspects of Plant-Nematode Interactions. Kluwer Acad. Publ., Dondrecht, Netherlands, pp. 176-190.

Klotz, K.L. and F.L. Finger. 2001. Activity of a soluble acid invertase from sugar beet roots. J. Sugar Beet Res. 38:121-137.

Klotz, K.L., and F.L. Finger. 2002. Contribution of invertase and sucrose synthase isoforms to sucrose catabolism in developing sugar beet roots. J. Sugar Beet Res. 39:1-24.

Knapp, E. 1958. *Beta* Rüben. *In:* T. Roemer and W. Rudorf (eds.). Handbuch der Pflanzenzüchtung. Paul Parey, Berlin, vol. 3.

Knapp, E. 1967. Die genetischen grundlagen der Einzelfrüchtigkeit (monocarpie) bei *Beta vulgaris*. Tagungsberichte Deutsche Akademie der Landwirtsch. Wiss. Berlin 89:189-235.

Koch, F. 1970. Resistenz und Resistenzzüchtung bei Zuckerrüben gegen *Cercospora beticola*-eine Literaturübersicht. J. IIRB 4:237-247.

Koch, F. 1988. Breeding work for the control of *Heterodera schachtii*. Sementi Elette 34:23-26.

Koch, G. 1997. Genetische Untersuchungen zur *Cercospora beticola* -- Resistenz in Zuckerrüben. Vorts. Pflanzenzüchtung 37:54-64.

Koch, H.J. and B. Jung. 1995. Zum Einfluss pflanzenbaulicher Massnahmen auf den Erdanhang von Zuckerrüben. Proc. IIRB 58:447-451.

Kock, K.E. 1996. Carbohydrate-modulated gene expression in plants. Ann. Rev. Physiol. Plant Molec. Biol. 47:509-540.

Koenig, R. and W. Burgermeister. 1989. Mechanical inoculation of sugarbeet roots with isolates of BNYVV having different RNA compositions. J. Phytopathology 124:249-255.

Koenig, R. and B.L. Lennefors. 2000. Molecular analyses of European A, B and P type sources of BNYVV and detection of the rare P type in Kazakhstan. Arch. Virol. 145:1561-1570.

Koenig, R., A.M. Haeberle, and U. Commandeur. 1997. Detection and characterization of a distinct type of BNYVV RNA-5 in a sugar beet growing area in Europe. Arch. Virol. 142:1499-1504.

Koenig, R., P. Lüddecke, and A.M. Haeberlé. 1995. Genome difference between beet necrotic yellow vein virus (BNYVV) sources from different parts of the world. Proc. IIRB 58:271-278.

Koenig, R., W. Burgermeister, H. Weich, W. Sebald, and C. Kothe. 1986. Uniform RNA pattern of BNYVV in sugar beet roots, but not in leaves from several plant species. J. Gen. Virol. 67:2043-2046.

Kontaxis, D.G., I. Meister, and R.K. Sharma. 1974. Powdery mildew epiphytotic on sugarbeets. Plant Dis. Rept. 58:904-905.

Kovbasyuk, E.P. and V.N. Bulin. 1983. Methods of increasing the resistance of sugar beet varieties to a number of diseases. Sekekt Semenov 9:25-26.

Kuzdowicz, A. and M. Brzeskwinska. 1967. Cytological research on F1 hybrids of sugar beet with *Beta webbiana* Moq. and *Beta procumbens* Chr. Biul. Inst. Hodowli Aklimat. Rosl. 3-4:3-8.

Lane, H.C., H.M. Cathey, and L.T. Evans. 1965. The dependence of flowering in several long-day plants on the spectral composition of light extending the photoperiod. Amer. J. Bot. 52:1006-1014.

Lange, W. and T.S.M. De Bock. 1988. Is resistance to beet cyst nematode colour-blind? Prophyta, Bijlage Januari, pp. 72-75.

Lange, W. and T.S.M. De Bock. 1994. Prebreeding for nematode resistance in beet. J. Sugar Beet Res. 31:13-26.

Lange, W., C. Jung, and W. Heijbroek. 1990. Transfer of beet cyst nematode resistance from *Beta* species of the section *Patellares* to cultivated beet. Proc. IIRB 53:89-102.

Lange, W., J. Muller, and T.S.M. De Bock. 1993. Virulence in the beet cyst nematode (*Heterodera schachtii*) versus some alien genes for resistance in beet. Fund. Appl. Nematol. 16:447-454.

Larsen, K. 1977. Self incompatibility in *Beta vulgaris* L. Four gametophytic complementary S-loci in sugar beet. Hereditas 85:227-248.

Larsen, K. 1978a. Four S-genes in one linkage group in *Beta vulgaris* L. Incompat. Newsl. 9:78-82.

Larsen, K. 1978b. Oligoallelism in the multigenic incompatibility system of *Beta vulgaris*. Incompat. Newsl. 10:23-28.

Lasa, J.M. and I. Romagosa. 1992. Mejora Genetica de la Ramolacha Azucarera. AIMCRA, Valladolid, Spain.

Lasa, J.M. I. Romagosa, R.J. Hecker, and J.M. Sanz. 1989. Combining ability in diploid and triploid sugar beet hybrids from diverse parents. J. Sugar Beet Res. 26:10-18.

Last, P.J. and A.P. Draycott. 1977. Relationships between clarified beet juice purity and easily-measured impurities. Intl. Sugar J. 79:183-185.

Lawlor, N. and C.W. Doxtator. 1950. Breeding for resistance to root rot caused by *Sclerotium rolfsii*. Proc. ASSBT 6:108-110.

Lawrence, W.J.C. 1930. Incompatibility in polyploids. Genetica 12:269-296.

Le Cochec, F. 1989. Les variétiés monogermes de betterave sucrière. Suggestions pour la sélection d'un autre type de variétetes: les hybrides F_1 ou hybrides entre deux lignées fixées. Sélect. Français 30:45-8.

Leach, L.D. and A.E. Davey. 1942. Reducing southern sclerotium rot of sugar beet with nitrogenous fertilizers. J. Agric. Res. 64:1-18.

Leach, L.D. and R.H. Garber. 1970. Control of Rhizoctonia. *In* J. Parmeter (ed.) *Rhizoctonia solani*: Biology and Pathology. Univ. California Press, Berkeley, CA, pp. 189-199.

Lejealle, F. and J. d'Aguilar. 1982. Pests, Diseases, and Disorders of Sugar Beet. Deleplanque, Maisons Lafitte, Paris.

Lemaire, O., M. Beuve, C. Weber, et al., 2003. Etiology and molecular epidemiology of a severe rhizomania disease occurring in confined locations in Europe: hypothesis for the implication of RNA-3 and/or RNA-5 of BNYVV P pathotype. Proc. ASSBT/IIRB 1:40-41.

Letschert, J.P.W. 1993. *Beta* section *Beta*; biogeographical patterns of variation, and taxonomy. PhD diss., Wageningen Agric. Univ., Wageningen, Netherlands.

Levall, M.W. and F.J. Bornman. 1993. Selection in vitro for UV-tolerant sugar beet (*Beta vulgaris*) somaclones. Physiol. Plant 88:37-43.

Levy, Y.Y. and C. Dean. 1998. Control of flowering time. Curr. Opin. Plant Biol. 1:49-54.

Lewellen, R.T. 1973. Inheritance of beet mosaic virus resistance in sugar beet. Phytopathology 63:877-881.

Lewellen, R.T. 1988. Selection for resistance to rhizomania in sugar beet. Abstract Proc. 5th Int. Congr. Plant Path. Kyoto, Japan, p. 445.

Lewellen, R.T. 1989. Registration of cytoplasmic male-sterile sugarbeet germplasm C600CMS. Crop Sci. 1:246.

Lewellen, R.T. 1992. Use of plant introductions to improve populations and hybrids of sugarbeet. *In:* Use of plant Introductions in Cultivar Development, Part 2. CSSA Spec. Publ. no. 20. Crop Sci. Soc. of Amer., Madison, WI, pp. 117-135.

Lewellen, R.T. 1994a. Registration of C762-17, a parental line of sugarbeet. Crop Sci. 34:319.

Lewellen, R.T. 1994b. Registration of C790-6, C790-15, and C790-54 parental lines of sugarbeet. Crop Sci. 34:319-320.

Lewellen, R.T. 1995a. Performance of near-isolines of sugarbeet with resistance to rhizomania from different sources. Proc. IIRB 58:83-89.

Lewellen, R.T. 1995b. Registration of C859 germplasm of sugar beet resistant to Rhizomania. Crop Sci. 35:289-290.

Lewellen, R.T. 1995c. Registration of sugar beet germplasm lines with multiple disease resistance: C39, C39R, C39R-6, C47, C47R, C93, and C94. Crop Sci. 35:596-597.

Lewellen, R.T. 1995d. Registration of three cyst nematode resistant sugar beet germplasms: C603, C603-1, and C604. Crop Sci. 35:1129-1130.

Lewellen, R.T. 1997. Registration of 11 sugar beet germplasm C79 with resistance to rhizomania. Crop Sci. 37:1026.

Lewellen, R.T. 2000. Registration of powdery mildew resistant sugarbeet germplasms CP01 and CP0. Crop Sci. 40:1515.

Lewellen, R.T. and E.D. Whitney. 1976. Inheritance of resistance to race C2 of *Cercospora beticola* in sugar-beet. Crop Sci. 16:558-561.

Lewellen, R.T. and I.O. Skoyen. 1988. Registration of four monogerm, self-fertile, random-mated sugar beet germplasm. Crop Sci. 5:873-874.

Lewellen, R.T. and E. Biancardi. 1990. Breeding and performance of rhizomania resistant sugar beet. Proc. IIRB 53:69-87.

Lewellen, R.T. and E.D. Whitney. 1993 Registration of germplasm lines developed from composite crosses of sugar beet x *Beta maritima*. Crop Sci. 33:882-883.

Lewellen, R.T. and J.K. Schrandt. 2001. Inheritance of powdery mildew resistance in sugar beet derived from *Beta vulgaris* subsp. *maritima*. Plant Dis. 85:627-631.

Lewellen, R.T., J.S. McFarlane, and I.O. Skoyen. 1978a. Registration of 11 germplasm lines of sugar beet. Crop Sci. 18:1100-1101.

Lewellen, R.T., E.D. Whitney, and C.K. Goulas. 1978b. Inheritance of resistance to Erwinia root rot in sugar beet. Phytopathology 68:947-950.

Lewellen, R.T., I.O. Skoyen, and E.D. Whitney. 1985a. Registration of C46 sugarbeet parental line. Crop Sci. 25:376.

Lewellen, R.T., E.D. Whitney, and I.O. Skoyen. 1985b. Registration of C37 sugarbeet parental line. Crop Sci. 25:375.

Lewellen, R.T., I.O. Skoyen, and A.W. Erichsen. 1987. Breeding sugar beet for resistance to rhizomania: evaluation of host-plant reaction and selection for inheritance of resistance. Proc. IIRB 50:139-156.

Lexander, K. 1969. Increase in bolting as an effect of low temperature on unripe sugar beet seed. 32nd IIRB Winter Cong., Paper no. 1.8.

Lexander, K. 1975. Bolting susceptibility of sugar beet (*Beta vulgaris*) in relation to contents of sulfhydryls and disulfides and to protein composition of membranes. Physiol. Plant. 33:142-150.

Lexander, K. 1985. *Beta vulgaris*, beet crops (sugar, forage, vegetable). *In:* A.H. Halevy (ed.). Handbook of Flowering. CRC Press Inc., Boca Raton, FL, vol. II., pp. 24-32.

Lexander, K. 1987. Characters related to the vernalization requirement of sugar beet. *In:* J.G. Atherton (ed.) Manipulation of Flowering. Butterworths, Bodmin, UK, pp. 147-158.

Lexander, K., 1993. Present understanding of the physiology of seed germination. Proc. IIRB 56:387-394.

Lichter, R. and G.H. Vieweg. 1969. Die Vervendung annueller Zuckerrüben in Rückkreuzungsprogrammen, dargestellt am Beispiel der Züchtung auf Einzelfrüchtigkeit (Monokarpie). J. IIRB 3:182-194.

Linde-Laursen, I. 1972. A new locus for colour formation in beet, *Beta vulgaris* L. Hereditas 70: 105-112.

Liu, H.Y., J.L. Sears, and R.T. Lewellen. 2003. A new beny-like sugar beet virus emerging in the United States. Proc. ASSBT/IIRB 1:53.

Longden, P.C. 1990. Seed quality research for improved establishment. Proc. IIRB 53:63-68.

Loptien, H. 1984a. Breeding nematode-resistant beets. I. Development of resistant alien additions by crosses between *Beta vulgaris* L. and wild species of the section *Patellares*. Z. Pflanzenzucht. 92:208-220.

Loptien, H. 1984b. Breeding nematode-resistant beets. II. Investigations into the inheritance of resistance to *Heterodera schachtii* Schm. in wild species of the section *Patellares*. Z. Pflanzenzucht. 93:237-245.

Lowe, H.J.B. 1974. Intraspecific variation of *Myzus persicae* on sugar beet (*Beta vulgaris*). Ann. Appl. Biol. 78:15-26.

Lowe, H.J.B. 1978. Possible advantages of small differences in resistance to aphids. *In:* Plant Breeding for Resistance to Insect Pests: Considerations about the Use of Induced Mutations. Intl. Atomic Energy Agency Tech. Doc. 215:91-95.

Lowe, H.J.B. and G.E. Russell. 1969. Inherited resistance of sugar beet to aphid colonization. Ann. Appl. Biol. 63:337-343.

Lowe, H.J.B., and M. Singh. 1985. Glasshouse tests and field selection for heritable resistance to *Myzus persicae* in sugar beet. Ann. Appl. Biol. 107:109-116.

Luczak, I. 1996. Mechanism of resistance to the beet fly *Pegomya betae* in sugar beet. Plant Breed. Seed Sci. 1-2:109-116.

Luterbacher, M.C., J.M. Smith, and M.J.C. Asher. 1998. Sources of disease resistance in wild *Beta* germplasm. Aspects Appl. Biol. 52:423-430.

Luterbacher, M.C., J.M. Smith, M.J.C. Asher, and L. Frese. 2000. Disease resistance in collections of *Beta* species. J. Sugar Beet Res. 37:39-47.

Lysgaard, C.P. 1978. Selection for reduced bolting susceptibility in beet and swedes, and the influence of environmental factors on bolting. Kgl. Veterinär og Landbohöjskole pp. 138-158.

MacDonald, J.D. and L.D. Leach. 1976. Evidence for an expanded host range of *Fusarium oxysporum* f. sp. *Beta*. Phytopathology 66:822-827.

MacLachlan, J.B. 1972. Estimation of genetic parameters in a population of monogerm sugar beet (*Beta vulgaris*): Sib-analysis of mother-line progeny: 1. Offspring/parent regression analysis of mother-line progeny: 2. Analysis of a diallel set of crosses among heterozygous populations: 3. Irish J. Agric. Res. 11:237-246, 319-325, 327-338.

Magyarosy, A.C., P. Schurmann, and B.B. Buchanan. 1976. Effect of powdery mildew infection on photosynthesis by leaves and chloroplasts of sugar beets. Plant Physiol. 57:486-489.

Maletskii, C.I., Y.N. Shavrukov and A.V. Mglinets. 1989. Linkage of separate flowering and incompatibility genes in sugar beet. Doklady Akademii Nauk USSR 308:203-206.

Maletsky, S.I., S.G. Veprev, and Y.N. Shavrukov. 1991. Le controle genetique dans la reproduction de la betterave à sucre. Nauka 168.

Marcum, W.B. 1948. Inheritance of bolting resistance. Proc. ASSBT 5:154-5.

Margara, J. 1960. Recherches sur le déterminisme de l'elongation et de la floraison dans le genre *Beta*. PhD diss., Faculté des Sciences de l'Université de Paris, Paris.

Marschner, H., A. Kylin, and P.J.C. Kuiper. 1981. Differences in salt tolerance of three sugar beet genotypes. Physiol. Plant. 51:234-238.

Martin, S.S. 1977. Accumulation of flavonoids Betagarin and Betavulgarin in *Beta vulgaris* infected by the fungus *Cercospora beticola*. Physiol. Plant Pathol. 11:297-303.

Martin, S.S. 1990. Flavonoid phytoalexins are synthetized by diverse *Beta vulgaris*. J. Sugar Beet Res. 27:72-80.

Martin, S.S., J.A. Narum, and K.H. Chambers. 2001a. Sugar beet biochemical quality changes during pile storage. Part I. Sugars. J. Sugar Beet Res. 38:35-53.

Martin, S.S., J.A. Narum, and K.H. Chambers. 2001b. Sugar beet biochemical quality changes during pile storage. Part II. Nonsugars. J. Sugar Beet Res. 38:173-188.

Martinez-Zapater, J.M., G. Coupland, C. Dean, and M. Koornneef. 1994. The transition of flowering in *Arabidopsis*. *In:* E.M. Meyerowitz, and C.R. Sommerville (eds.) *Arabidopsis*. Cold Spring Harbour Laboratory Press, Cold Spring Harbour, NY, pp. 403-433.

Martyn, R.D., C.M. Rush, C.L. Biles, and E.H. Baker. 1989. Etiology of a root rot disease of sugar beet in Texas. Phytopathology 73:879-884.

McCready, R.M. and J.C. Goodwin. 1966. Sugar transformations in stored sugar beets. J. ASSBT 14:197-205.

McFarlane, J.S. 1952. Breeding for resistance to downy mildew in sugar beet. Proc. ASSBT 8:415-420.

McFarlane, J.S. 1969. Breeding for resistance to curly top. J. IIRB 4:73-83.

McFarlane, J.S. 1971. Variety development. *In:* R.T. Jonhson, J.T. Alexander, G.E. Bush, and G.R. Hawkes (eds.). Advances in Sugar Beet Production. Iowa State Univ. Press, Ames, IA, pp. 402-435.

McFarlane, J.S. 1981. Fusarium stalk blight resistance in sugar beet. J. ASSBT 21:175-183.

McFarlane, J.S. 1982. Registration of two sugar beet parental lines. Crop Sci. 22:454.

McFarlane, J.S. 1993. The Savitsky story. J. Sugar Beet Res. 30:1-36; 125-137.

McFarlane, J.S., C. Price, and F.V. Owen. 1948. Strains of sugar beets extremely resistant to bolting. Proc. ASSBT 5:151-3.

McFarlane, J.S., R. Bardin, and W.C. Snyder. 1954. An alternaria leaf spot of the sugar beet. Proc. ASSBT 8:241-246.

McFarlane, J.S., I.O. Skoyen, and R.T. Lewellen. 1969. Development of sugarbeet breeding lines and varieties resistant to yellows. J. ASSBT 15:347-360.

McFarlane, J.S., I.O. Skoyen, and R.T. Lewellen. 1972. Performance of sugar beet hybrids as diploids and triploids. Crop Sci. 12:118-119.

McFarlane, J.S., H. Savitsky, and A. Steele. 1982a. Breeding for resistance to the sugar beet nematode. J. ASSBT 21:311-323.

McFarlane, J.S., I.O. Skoyen, and R.T. Lewellen. 1982b. Registration of four sugar beet germplasm lines. Crop Sci. 22:698.

McGrath, J. M., B. de los Reyes, and J.W. Saunders. 2003. *Beta* breeding and genetics at East Lansing, Michigan: Molecular methods, genetic diversity, and trait elucidation. J. Sugar Beet Res. 37:97-106.

Meriggi, P., F. Rosso, P.M. Joannides, and J. Ayala-Garcia. 2000. Fungicide treatments against cercospora leaf spot in sugar beet (*Beta vulgaris* L.). *In: Cercospora beticola* Sacc.: Biology, Agronomic Influence and Control Measures in Sugar Beet. Advances in Sugar Beet Research. IIRB. Brussels, vol. 2, pp. 77-102.

Mesken, M. 1987a. Mass selection for crown height in sugar beets (*Beta vulgaris* L.): I. Divergent selection in diploids. Euphytica 36:129-145.

Mesken, M. 1987b. Mass selection for crown height in sugar beets (*Beta vulgaris* L.): II. Divergent selection in tetraploids. Euphytica 36:433-440.

Mesken, M. and J. Dieleman. 1988. Breeding sugar beets with globe shaped roots: selection and agronomical performances. Euphytica Suppl. pp. 37-44.

Mesken, M. and B. Lekkerkerker. 1988. Selection for partial resistance to beet cyst nematode in crosses of sugar beet and fodder beet with *B. maritima.* Bijlage Januari, pp. 68-71.

Milford, G.F.J. 1973. The growth and development of the storage root of sugar beet. Ann. Appl. Biol. 75:427-438.

Milford, G.F.J. and G.N. Thorne. 1973. The effect of light and temperature late in the season on the growth of sugar beet. Ann. Appl. Biol. 75:419-425.

Milford, G.F.J. and D.V. Lawlor. 1976. Water and physiology of sugar beet. Proc. IIRB 39:95-108.

Milford, G.F.J., K.Z. Travis, T.O. Pocock, K.W. Jaggard, and W. Day. 1988. Growth and dry-matter partitioning in sugar beet. J. Agric. Sci. Camb. 110:301-308.

Miller, J., M. Rekoske, and A. Quinn. 1994. Genetic resistance, fungicide protection and variety approval policies for controlling yield losses from cercospora leaf spot infections. J. Sugar Beet Res. 31:7-12.

Milliken, G.A. and R. J. Hecker. 1967. Computer method for calculating percentage apparent purity of sugar beet thin juice. J. ASSBT 14:674-681.

Mouradov, A., F. Cremer, and G. Coupland. 2002. Control of flowering time: Interacting pathways as a basis for diversity. Plant Cell, Suppl. Pp. 111-130.

Mukhopadhyay, A.N. 1987. CRC Handbook on Diseases of Sugar Beet. CRC Press Inc., Boca Raton, FL.

Mukhopadhyay, A.N. and R.P Thakur. 1970. Fusarium root rot of sugar beet. Bull. no. 18. FAO Plant Protection, Rome, Italy.

Muller, J. 1998a. Resistance and tolerance to beet cyst nematode (*Heterodera schachtii*) in sugar beet cultivars. Zuckerindustrie 123: 688-693.

Muller, J. 1998b. New pathotypes of the beet cyst nematode *(Heterodera schachtii)* differentiated on alien genes for resistance in beet (*Beta vulgaris*). Fund. Appl. Nematol. 21:519-526.

Mumford, D.L. 1974. Procedure for inducing curly top epidemics in field plots. J. ASSBT 18:20-23.

Mumford, D.L. and R.E. Wyse. 1976. Effect of fungus infection on respiration and reducing sugar accumulation of sugar beet roots and use of fungicides to reduce infection. J. ASSBT 19:157-162.

Munerati, O. 1931. L'eredità della tendenza all'annualità nella comune barbabietola coltivata. Zeitschrift Pflanzenzücht. 17:84-9.

Munerati, O. and T. Costa. 1930. Osservazioni sulla trasmissione del carattere "pelle nera" nella barbabietola. Reprinted in: Sulla barbabietola da zucchero. ISCI, Rovigo, Italy, pp. 1347-1348.

Murphy, A.M. 1942. Production of heavy curly-top exposures in sugarbeet breeding fields. Proc. ASSBT 3:459-462.

Murphy, A.M. 1946. Sugar beet and curly-top history in Southern Idaho 1912-1945. Proc. ASSBT 4:408-412.

Murphy, A.M., and V.F. Savitsky. 1952. Breeding for resistance to curly top in hybrids with monogerm beets. Proc. ASSBT 7:387-389.

Murphy, A.M., G.K. Ryser, and F.V. Owen. 1952. Performance of F1 hybrids between curly-top resistant and curly-top susceptible sugar beets. Proc. ASSBT 7:390-392.

Murray, G., J.B. Swensen, and J.J. Gallian. 1993. Emergence of sugar beet seedlings at low soil temperature following seed soaking and priming. Hort. Sci. 28:31-32.

Nakamura, C. 1976. Cytogenetics of alien addition trisomics in sugar beet. MS thesis, Colorado State Univ., Fort Collins, CO.

Nakamura, C. and T. Tsuchiya. 1982. Cytogenetics of alien addition trisomics in sugar beet. I. Meiotic chromosome behavior in nematode-resistant trisomics. Biol. Zbl. 101:227-240.

Nakamura, C., and T. Tsuchiya. 1988. Nematode resistance and meiotic abnormality in diploid sugar beet selected from interspecific *Beta vulgaris* x *B. procumbens* hybrids. Plant Breed. 100:41-53.

Nakamura, C., G.N. Skaracis, and I. Romagosa. 1992. Cytogenetics and breeding in sugar beet. *In:* T. Tsuchiya and P.K. Gupta (eds.). Chromosome Engineering in Plants: Genetics, Breeding, Evolution, Part B. Elsevier, Amsterdam, Netherlands, pp. 295-314.

Nelson, R.T. 1954. Progeny test of sugar beet roots selected for low respiration rate. Proc. ASSBT 8:134-136.

Nelson, R.T. and R.R. Wood. 1950. Respiration and spoilage studies employing a modification of the method described by Stout and Fort. Proc. ASSBT 6:660-663.

Nelson, R.T. and R.K. Oldemeyer. 1952. Preliminary studies applicable to selection for low respiration and resistance to storage rots of sugar beets. Proc. ASSBT 7:400-406.

Nemazi, J. and Nielson K. 1967. Occurrence of double ovules in sugar beets. J. ASSBT 5:389-395.

Noe, B. and G. Mantovani. 1996. Resultats de l'enquete sur la qualité interne de la betterave sucriere. Proc. IIRB 59:329-342.

Noghabi, A.M. and B.F. Williams. 2000. Drought stress and weed competition in sugar beet. British Sugar Beet Rev. 1:47-49.

Nozaws, Y., H. Tida, H. Fukushima, K. Ohki, and S. Ohnishi. 1974. Studies on *Tetrahymena* membranes: Temperature induced alterations in fatty acid composition of various membrane fractions of *Tetrahymena pyriformis* and its effect on membrane fluidity as inferred by spin-label study. Biochem. Biophys. Acta 368:134-147.

Ober, E.S. and M.C. Luterbacher. 2002. Genotypic variation for drought tolerance in *Beta vulgaris*. Ann. Bot. 89:917-924.

Ogata, N., K. Taguchi, T. Kuranouchi, and M. Tanaka. 2000. Breeding for root rot resistant varieties. 3. Resistance of seed parents bred in NARCH. Proc. Jpn. Soc. Sugar Beet Tech. 42:13-19.

Ohta, T. 1961. On the F1 hybrid between sugar beet and *Beta webbiana*. Seiken Ziho 12:21-28.

Ohtake, M., H. Watanabe, and K. Kanzawa. 1996. Selection for varieties tolerant to Aphanomyces root rot in hydroponic gravel culture. Proc. Jpn. Soc. Sugar Beet Tech. 38:92-97.

Olak, L. 1950. La fecondation de *Beta vulgaris* L. Rev. Invest. Agric. 4:427-436.

Oldemeyer, R.K. 1954. Viable interspecific hybrids between wild species in the section *Vulgares* and species in the section *Patellares* in the genus *Beta*. Proc. ASSBT 8:153-156.

Oldemeyer, R.K. and H.E. Brewbaker. 1956. Interspecific hybrids in the genus *Beta*. J. ASSBT 9:15-18.

Oldemeyer, R.K. and G.E. Rush. 1960. Evaluation of combining ability in self-fertile lines of sugar beets using male sterile testers. J. ASSBT 11:175-184.

O'Sullivan, E. 1997. Responses of sugar beet cultivars to the control of rust (*Uromyces betae*). Irish J. Agric. Food Res. 2:175-184.

Owen, F.V. 1942. Inheritance of cross and self-sterility and self-fertility in *Beta vulgaris*. J. Agric. Res. 64:679-698.

Owen, F.V. 1945. Cytoplasmically inherited male-sterility in sugar beet. J. Agrc. Res. 71:423-440.

Owen, F.V. 1948. Utilization of male sterility in breeding superior yielding sugar beets. Proc. ASSBT 5:156-161.

Owen, F.V. 1952. Mendelian male sterility in sugar beet. Proc. ASSBT 7:371-376.

Owen, F.V. 1954. The significance of single gene reactions in sugar beets. Proc. ASSBT 8:392-8.

Owen, F.V. and G.K. Ryser. 1942. Some mendelian characters in *Beta vulgaris* and linkages observed in the Y-R-B group. J. Agric. Res. 65:155-171.

Owen, F.V., E. Carsner, and M. Stout. 1940. Photothermal induction of flowering in sugar beets. J. Agric. Res. 61:101-124.

Owen, F.V., G. Thorne, and C.W. McBeth. 1938. Varietal tests for resistance to sugar beet nematodes. Proc. ASSBT 1:71.

Pack, D.A. 1930. Selection characters as correlated with percent of sucrose, weight, and sucrose content of sugar beet. J. Agric. Res. 40:523-546.

Painter, R.H. 1951. Insect resistance in crop plants. MacMillan Co., New York, NY.

Palva, E.T., and P. Heino. 1997. Molecular mechanism of plant cold acclimation and freezing tolerance. *In:* Li, Paul H., Tony H.H. Chen. (eds.). Plant Cold Hardiness-molecular Biology, Biochemistry and Physiology. Plenum Press, New York, NY.

Panagopoulos, I., J.F. Bornman, and L.O. Bjorn. 1990. Effect of ultraviolet radiation and visible light on growth, fluorescence induction, ultraweak luminescence and peroxidase activity in sugar beet plants. J. Photochem. Photobiol. 8:73-87.

Panella, L. 1998. Screening and utilizing *Beta* genetic resources with resistance to Rhizoctonia root rot and Cercospora leaf spot in a sugar beet breeding program. p. 62-72. *In:* L.Frese, L.Panella, H.M.Srivastava, and W.Lange (eds.). Report of *Beta* Genetic Resources Workshop and 4th World *Beta* Network Conf. Izmir, Turkey, 28 Feb.- 3 Mar. 1996. IPGRI, Rome, Italy.

Panella, L. and E.G. Ruppel. 1997. Registration of sugarbeet germplasms FC721 and FC721CMS resistant to Rhizoctonia root rot and moderately resistant to the beet curly top virus. Crop Sci. 37:1675-1676.

Panella, L. and L. Frese. 2003. *Beta* germplasm evaluation data in two databases, GRIN & IDBB. Proc. ASSBT/IIRB. 1:233-241.

Panella, L.W., E.G. Ruppel, and R.J. Hecker. 1995. Registration of four multigerm sugar beet germplasm resistant to rhizoctonia root rot. Crop Sci. 1:291.

Paolini, R., M. Principi, R.J. Froud-Williams, and E. Biancardi. 1999. Competition between sugarbeet and *Sinapis arvensis* and *Chenopodium album* as affected by timing of nitrogen fertilization. Weed Res. 39:425-440.

Paul, H. 1993. Quantitative studies on resistance to *Polymyxa betae* and beet nerotic yellow vein virus in beet. PhD diss., Univ. of Wageningen, Wageningen, Netherlands.

Paul, H., B. Henken, M.F.J. Alderlieste. 1992. A greenhouse test for screening sugar beet for resistance to BNYVV. Netherlands J. Plant Path. 98:65-75.

Paul, H., B. Henken, T.S.M. De Boch, and W. Lange. 1993. Variation in the level of infection with *Polymyxa betae* and its effect on infection with BNYVV in beet accessions of the sections *Beta* and *Corollinae*. Proc. 2nd Symp. Intl. Working Group on Plant Viruses with Fungal Vectors, Montreal, Canada, pp. 133-136.

Paul, H., B. Henken, O.E. Scholten, T.S.M. De Boch, and W. Lange. 1994. Resistance to *Polymyxa betae* and Beet Necrotic Yellow Vein Virus in *Beta* species of the section *Corollinae*. J. Sugar Beet Res. 31:1-6.

Pedersen, A. 1944. Our Bederoermes farver. Arsskript Koengelije Veterinaer Landbohjskole, pp. 60-111.

Peretyat, V.G., L.V. Kehl, G.S. Stephanenko. 1992. Search for resistant varieties. Sakharnaya Svekla 2:23-27.

Petersen, J., H. Adams, W.R Schäufele, and G. Buttner. 2001. Untersuchungen zur Schadwirkung von *Ramularia beticola* in Zuckerruben und Moglichkeiten zur Differenzierung der Sortenfälligkeiten nach kunstlicher Inokulazion. Gesunde Pflanzen 53:141-147.

Peto, F.H. 1964. Method of loosening tight seed caps in monogerm seed to improve germination. J. ASSBT 13:281-286.

Peto, F.H. and J.M. Boyes. 1940. Comparison of diploid and triploid sugar beet. Can. J. Res. 18:273-282.

Poggi Pollini, C. and L. Giunchedi. 1989. Comparative histopathology of sugar beets that are susceptible and partially resistant to rhizomania. Phytopath. Medit. 28:16-21.

Popova, I.V. 1961. Selection of sugarbeet for resistance to storage rots. Agrobiologiya 5:762-763.

Porte, C., A. Mahfoud, and G. Caubel. 1997. Reaction of geographically dispersed populations of *Heterodera schachtii* and *H. trifolii* to resistant lines of sugarbeet. Nematol. Medit. 25:113-120.

Powers, L., W.R. Schmehl, W.T. Federer, and M.G. Payne. 1963. Chemical genetic and soils studies involving thirteen characters in sugar beet. J. ASSBT 12:393-448.

Prell, H. 1921. Das Problem der Unfrüchtbarkeit. Naturwiss. Wochenschr. 20:440-446.

Prince, J. and M. Durrant. 1990. Progress in seed quality and seed treatments. British Sugar Beet Rev. 4:4-6.

Radley, M.E. 1975. Sugar Beet. Growth Substances and Bolting. Rept. Rothamsted Exper. Stn. Ed. Whittingham, Harpenden, UK.

Raski, D.J. 1950. The life history and morphology of the sugar-beet nematode, *Heterodera schachtii* Schmidt. Phytopathology 40:135-152.

Rasmusson, J. and A. Levan. 1939. Tetraploid sugar beets from colchicine treatments. Hereditas 25:97-102.

Richard-Molard, M. 2002. Rhizomanie: interactions varities x lieux et consequences. Proc. IIRB 65:239-245.

Rietberg, H. 1959. Selection sur la tolerance à la jaunisse. Proc. IIRB 22:50-58.

Roland, G. 1936. Recherches sur la jaunisse de la betterave et quelques observations sur la mosique de cette plant. Sucrerie Belge 55:219-293.

Rosnowsky, S., S. Bardzinsky, and H. Lachowska. 1961. The disclosure of the degree of defect of sugar beet fibrosity caused by pre-winter sowing. Proc. IIRB 24:275-279.

Rossi, V., P. Battilani, G. Chiesa, L. Languasco, and P. Racca. 1997. Components of rate reducing resistance to cercospora leaf spot in sugar beet. Proc. IIRB 60:203-213.

Roundy, T.E. and J.C. Theurer. 1974. Inheritance of a yellow-leaf mutant and a pollen fertility restorer in sugar beet. Crop Sci. 14:62–63.

Röver, A. 1994. Lichtabsorption und Ertrag in Abhängigkeit vom Blattflächenindex bei Zuckerrüben. Zuckerindustrie 119:664-670.

Rubies-Autonell, C., R. Resca, and M. Turina. 1996. Survey of soil-borne virus diseases of sugar beet in Italy. Proc. 3rd Symp. Intl. Working Group: Plant Viruses with Fungal Vectors. Dundee, Scotland, 6-7 August 1996 pp. 121-124.

Rudakou, O.L., and T.F. Al'Akhovskaya. 1964. Fusarium wilt of sugar beet. Sakhav. Svekla. 9:34.

Ruppel, E.G. 1972. Negative relationship of stomatal size and density with resistance in sugar beet to *Cercospora beticola*. Phytopathology 62:1095-1096.

Ruppel, E.G. 1986. Powdery Mildew. *In:* E.D. Whitney and J.E. Duffus (eds.). Compendium of Beet Diseases and Insects. Amer. Phytopath. Soc., St. Paul, MN, pp. 13-15.

Ruppel, E.G. 1991. Pathogenicity of *Fusarium* spp. from diseased sugar beets and variation among sugar beet isolates of *F. oxysporum.* Plant Dis. 75:486-489.

Ruppel, E.G. and B.J. Tomasovic. 1977. Epidemiological factors of sugar beet powdery mildew. Phytopathology 67:619-621.

Ruppel, E.G. and R.J. Hecker. 1994. Rhizoctonia root rot on sugarbeet cultivars having varied degrees of resistance. J. Sugar Beet Res. 31:135-142.

Ruppel, E.G., F.J. Hills, and D.L. Muford. 1975a. Epidemiological observations on the sugarbeet powdery mildew epiphytotic in western U.S.A. in 1974. Plant Dis. Rept. 59:283-286.

Ruppel, E.G., M.D. Harrison, and A.K. Nielson. 1975b. Occurrence and cause of bacterial vascular necrosis and soft rot of sugar beet in Washington. Plant Dis. Rept. 59:837-840.

Ruppel, E.G., R.J. Hecker, and L.W. Panalla. 1995. Registration of two sugar beet germplasms resistant to rhizoctonia root rot. Crop Sci. 1:290.

Ruppel, E.G., C.L. Schneider, R.J. Hecker, and G.J. Hogaboam. 1979. Creating epiphytotics of Rhizoctonia root rot and evaluating for resistance to *Rhizoctonia solani* in sugarbeet field plots. Plant Dis. Rep. 63:518-522.

Rush, C.M. 1987. Root rot of sugar beet caused by *Pythium deliense* in the Texas panhandle. Plant Dis. 71:469.

Russell, G.E. 1958. Sugar beet yellows: a preliminary study of the distribution and interrelationships of viruses and virus strains found in East Anglia, 1955-57. Ann. Appl. Biol. 46:393-398.

Russell, G.E. 1960. Sugar beet yellows: further studies on viruses and virus strains and their distribution in East Anglia, 1958-59. Ann. Appl. Biol. 48:721-728.

Russell, G.E. 1964. Breeding for tolerance to beet yellows virus and beet mild yellowing virus in sugar beet: I. Selection and breeding methods. Ann. Appl. Biol. 53:363-376.

Russell, G.E. 1965. Breeding for resistance to sugar beet diseases. British Sugar Beet Rev. 1:19-22.

Russell, G.E. 1966a. Some effects of inoculation with yellowing viruses on the susceptibility of sugar beet to fungal pathogens, II. Susceptibility to *Erysiphe* species. Trans. Brit. Mycol. Soc. 49:621-628.

Russell, G.E. 1966b. Observations on the settling behaviour of aphids on sugar beet plants in the glasshouse. J. Agric. Sci. Camb. 67:405-409.

Russell, G.E. 1966c. Breeding for resistance to infection with yellowing viruses in sugar beet: I. Resistance in virus-tolerant breeding material. Ann. Appl. Biol. 57:311-320.

Russell, G.E. 1969. Different forms of inherited resistance to virus yellows in sugar beet. J. IIRB 4:131-142.

Russell, G.E. 1972. Components of resistance to diseases in sugar beet. Proc. 6th Cong. Eucarpia, pp. 99-107.

Sadeghian, S.Y. 1993. Bolting in Sugar Beet, Genetics and Physiological Aspects. Swedish Univ. Agric. Sci., Dept. Plant Breed. Res., Svalöv, Sweden.

Sadeghian, S.Y. and E. Johansson. 1993. Genetic study of bolting and stem length in sugar beet (*Beta vulgaris* L.) using a factorial cross design. Euphytica 65:177-185.

Sadeghian, S.Y. and K. Lexander. 1993. Fatty acid composition of plasma membranes in relation to bolting in sugar beet. Flowering Newsl. 16: 39-43.

Sadeghian, S.Y. and H. Khodaii. 1998. Diallel cross analysis of seed germination traits in sugar beet. Euphytica 103: 259-263.

Sadeghian, S.Y., H.C. Becker, and E. Johansson. 1993. Inheritance of bolting in three sugar beet crosses after different periods of vernalization. Plant Breed. 110:328-333.

Sadeghian, S.Y., H. Fazli, R. Mohammadian, D.F. Taleghani, and M. Mesbah. 2000. Genetic variation for drought stress in sugarbeet. J. Sugar Beet Res. 37:55-77.

Sasser, J.N. 1980. Root-knot nematodes: a global menace to crop production. Plant Dis. 64:36-41.

Savitsky, H. 1950. A method of determining self-fertility and self-sterility in sugar beets, based upon the stage of ovule development shortly after flowering. Proc. ASSBT 6:198-201.

Savitsky, H. 1960. Viable diploid, triploid, and tetraploid hybrids between *Beta vulgaris* and species of the section *Patellares*. J. ASSBT 11:215-235.

Savitsky, H. 1973. Meiosis in hybrids between *Beta vulgaris* L. and *Beta procumbens* Chr. Sm. and transmission of sugar beet nematode resistance. Genetics 74:241.

Savitsky, H. 1975. Hybridization between *Beta vulgaris* and *B. procumbens* and transmission of nematode (*Heterodera schachtii*) resistance to sugar beet. Can. J. Genet. Cytol. 17:197-209.

Savitsky, H. 1978. Nematode (*Heterodera schachtii*) resistance and meiosis in diploid plants from interspecific *Beta vulgaris* x *B. procumbens* hybrids. Can. J. Genet. Cytol. 20:177-186.

Savitsky, V.F. 1940. Genetics of Sugar Beets. USDA: National Agricultural Library Translation 5825.

Savitsky, V.F. 1950. Monogerm sugar beets in the United States. Proc. ASSBT 6:156-159.

Savitsky, V.F. 1952. A genetic study of monogerm and multigerm characters in beets. Proc. ASSBT 7:331-338.

Savitsky, V.F. 1954. Inheritance of the number of flowers in flower cluster of *Beta vulgaris* L. Proc. ASSBT 8:3-15.

Savitsky, V.F. 1957. Inheritance of chlorophyll deficiencies in *Beta vulgaris* L. J. ASSBT 9:321-336.

Savitsky, V.F. and A.M. Murphy. 1954. Study of inheritance for curly top resistance in hybrids between mono- and multigerm beets. Proc. ASSBT 8:34-44.

Schlang, J. and J. Muller. 1996. Resistance in sugar beet to *Heterodera schachtii*: nematode population dynamics and sugar beet yield under field conditions. Mitteilungen aus der Biologischen Bundesanstalt fur Land und Forstwirtschaft Berlin 317:129-140.

Schlösser, L.A. 1961. Breeding of nonfibrous sugar beet. Proc. IIRB 24:229-233.

Schmid, M.G. 1974a. Stoffwechselphysiologische Untersuchungen an vernalisierten und nicht vernalisierten Zuckerrüben, I. Enzymaktivitäten, Kohlenhydrate und stickstoffhaltige Verbindungen im Epikotyl, Hypokotyl und Wurzelteil von vegetativen und blühinduzierten Zuckerrüben. Angew. Botanik 48:221-237.

Schmid, M.G. 1974b. Stoffwechselphysiologische Untersuchungen an vernalisierten und nicht vernalisierten Zuckerrüben, II. Enzymaktivitäten in drei Zuckerrübenpopulationen mit unterschiedlichem Vernalisationsbedarf vor, während und nach einer Kaltebehandlung. Angew. Botanik 48:339-352.

Schneider, C.L. 1954. Methods of inoculating sugar beets with *Aphanomyces cochlioides* Drechs. Proc. ASSBT 8:247-251.

Schneider, C.L. 1978. Use of oospore inoculum of *Aphanomyces cochlioides* to initiate blackroot disease in sugar beet seedlings. J. ASSBT 20:55-62.

Schneider, C. L. and J.O. Gaskill. 1962. Tests of foreign introductions of *Beta vulgaris* L. for resistance to *Aphanomyces cochliodes* Drechs. and *Rhizoctonia solani* Kühn. Proc. ASSBT 11:656-660.

Schneider, C.L. and G.L. Hogaboam. 1983. Evaluation of sugar beet breeding lines in greenhouse tests for resistance to *Aphanomyces cochlioides*. J. ASSBT 22:101-107.

Schneider, C.L. and E.D. Whitney. 1986a. Black root. *In:* E.D.Whitney and J.E.Duffus (eds.). Compendium of Beet Diseases and Insects. Amer. Phytopath. Soc., St. Paul, MN, p. 18.

Schneider, C.L. and E.D. Whitney. 1986b. Fusarium Yellows. *In:* E.D. Whitney and J.E. Duffus (ed.). Compendium of Sugar Beet Diseases and Insects. Amer. Phytopath. Soc., St. Paul, MN, pp. 18-19.

Schneider, C.L. and E.D. Whitney. 1986c. Phoma root rot. *In:* E.D. Whitney and J. E. Duffus (eds.). Compendium of Sugar Beet Diseases and Insects. Amer. Phytopath. Soc, St. Paul, MN, p. 20.

Schneider, C.L., and E.D. Whitney. 1986d. Pythium root rot. *In:* E.D. Whitney, and J.E. Duffus (ed.) Compendium of beet diseases and insects. Am. Phytopath. Soc., St. Paul, MN, p.20.

Schneider, C.L., E.G. Ruppel, R.J. Hecker, and G.J. Hogaboam. 1982. Effect of soil deposition in crowns on development of Rhizoctonia root rot in sugar beet. Plant Dis. 66:408-410.

Schneider, F. 1937. Sur un croisement de la betterave a sucre avec *Beta procumbens*. Inst. Belge Amelior. Betterave, Publ. 5:544-545.

Schneider, K., R. Schafer-Pregl, D.C. Borchardt, and F. Salamini. 2002. Mapping QTLs for sucrose content, yield and quality in a sugar beet population fingerprinted by EST-related markers. Theor. Appl. Genet. 104:1107-1113.

Scholten, O.E. 1997. Characterisation and inheritance of resistance to BNYVV in *Beta*. PhD diss., Agric. Univ. Wageningen, Netherlands.

Scholten, O.E., T.S.M. De Bock, R.M. Klein-Lankhorst, and W. Lange. 1999. Inheritance of resistance to BNYVV in *Beta vulgaris* conferred by a second gene for resistance. Theor. Appl. Genet. 99:740-746.

Scholten, O.E., L.W. Panella, T.S.M. De Bock, and W. Lange. 2001. A greenhouse test for screening sugar beet (*Beta vulgaris*) for resistance to *Rhizoctonia solani*. Eur. J. Plant Path. 107:161-166.

Scholten, O.E., R.C. Jansen, L.P.C. Keizer, T.S.M. De Bock, and W. Lange. 1996. Major genes for resistance to BNYVV in *Beta vulgaris*. Euphytica 91:331-339.

Schwanitz, F. 1938. Die Herstellung polyploider Rassen bei *Beta* Rüben und Gemüsearten durch behandlung mit colchicin. Züchter 10:278-279.

Schwanitz, F. 1940. Beitrage zur Züchtung und Genetik selbst-fertiler Rüben (*Beta vulgaris* L.). Züchter 12:167-178.

Scott, R.K. and K.W. Jaggard. 1993. An analysis of the efficiency of the sugar beet crop in exploiting the environment. J. Sugar Beet Res. 30:37-56.

Scott, R.K., F. Harper, D.W. Wood, and K.W. Jaggard. 1974. Effect of seed size on growth, development and yield on monogerm sugar beet. J. Agric. Sci. Camb. 82:517-530.

Sedlmayr, T.E. 1960. Inheritance of speed of germination in sugar beet (*Beta vulgaris* L.). PhD diss., Michigan State Univ., East Lansing, MI.

Shane, W.W. and P.S. Teng. 1992. Impact of cercospora leaf spot on root weight, sugar yield and purity. Plant Dis. 8:812-820.

Shavrukov, Y., M. Kawakatsu, and M. Tanaka. 2000. Correlation between polyovules and number of pistil lobes (stigmatic rays) in sugar beet (*Beta vulgaris* L.). Euphytica 1:17-21.

Sheikholeslami, M., G.A. Hedjaroude, and M. Okhovat. 1998. Fungi causing postharvest decay of sugar beet root in Kermanshah, Iran. J. Plant Pathl. 34:84-91.

Shepherd, R.J., F.J. Hills, and D.H. Hall. 1964. Losses caused by beet mosaic virus in California grown sugar beet. J. ASSBT. 13:244-251.

Shimamoto, Y., Tanada T., and Abe J. 1990. Analysis for bolting of sugar beet by means of test crosses of biennial lines with the annual line. Proc. Jpn. Soc. Sugar Beet Tech. 32:134-137.

Simon, M. 1947. Developpement actuel et perspectives des mesures de lutte contre le nematode de la betterave. Inst. Belge Amelior. Betterave Publ. Tech. 15:77-91.

Simpson, G.C. and C. Dean. 2002. *Arabidopsis*, the Rosetta stone of flowering time? Science 296:285-289.

Skaracis ,G.N. and G.A. Smith. 1987. Ploidy level effect on resistance to cercospora leaf spot in sugar beet. Proc. 24th Meet. ASSBT p. 26.

Skaracis, G.N. and E. Biancardi. 2000. Breeding for cercospora resistance in sugar beet. *In: Cercospora beticola* Sacc.: Biology, Agronomic Influence and Control Measures in Sugar Beet. Advances in Sugar Beet Research. IIRB, Brussels, vol. 2, pp. 177-195.

Skoyen, I.O., R.T. Lewellen, and J.S. McFarlane. 1975. Effect of powdery mildew on sugarbeet production in the Salinas Valley of California. Plant Dis. Rept. 59:506-510.

Slavov, K. 1984. The effect of plant density on the overwintering of tranplantless directly seeded fields of sugar beets for seed production. Rasteniev'd-Nauki 21:43-48.

Smeekens, S. 2000. Sugar-induced signal transduction in plants. Ann. Rev. Plant Physiol. Plant Molec. Biol.: 51:49-81.

Smigocki, A., S.D. Ivic, D. Wilson, et al., 2003. Molecular approaches for control of the sugarbeet root maggot. Proc. ASSBT/IIRB 1:419-428.

Smit, A.L. 1983. Influence of external factors on growth and development of sugar beet (*Beta vulgaris* L.). Agricultural Research Reports 914. Pudoc, Wageningen, Netherlands.

Smith, C.H. 1952. Heritable differences in germination of sugar beet seed at low temperature. Proc. ASSBT 6:411-414.

Smith, C.M., Z.R. Khan, and M.D. Pathak. 1994. Techniques for Evaluating Insect Resistance in Crop Plants. CRC Press, Boca Raton, FL.

Smith, G.A. 1980. Sugar beet. *In:* W.R. Fehr and H.H. Hadley (ed.). Hybridization of Crop Plants. Amer. Soc. Agron., Madison, WI, pp. 601-616.

Smith, G.A. 1987. Sugar beet. *In:* W.R. Fehr (ed.) Principles of Cultivar Development, MacMillan Co., New York, NY, vol. 2, pp. 577-625.

Smith, G.A. and J.O. Gaskill. 1970. Inheritance of resistance to cercospora leaf spot in sugar beet. J. ASSBT 16:172-180.

Smith, G.A. and E.G. Ruppel. 1974. Heritability of resistance to cercospora leaf spot in sugar beet. Crop Sci. 53:695-696.

Smith, G.A. and E.G. Ruppel. 1986. Registration of tetraploid cercospora resistant sugar beet germplasm. Crop Sci. 2:392.

Smith, G.A. and S.S. Martin. 1989. Effect of selection for sugar beet purity components on quality and extraction. Crop Sci. 29:294-298.

Smith, G.A. and L.G. Campbell. 1996. Association between resistance to cercospora and yield in commercial sugar beet hybrid. Plant Breed. 115:28-32.

Smith, G.A., S.S. Martin, and K.A. Ash. 1977. Path coefficient analysis of sugar beet purity components. Crop Sci. 17:249-253.

Smith, G.A., R.J. Hecker, and S.S. Martin. 1979. Effects of polyploid level on the components of sucrose yield and quality in sugar beet. Crop Sci. 19:319-323.

Smith, G.A., R.J. Hecker, G.W. Maag, and D.M. Rasmuson. 1973. Combining ability and gene action estimates in an eight parent diallel cross of sugar beet. Crop Sci. 13:312-316.

Snyder, F.W. 1963. Selection for speed of germination in sugar beet. J. ASSBT. 12:617-622.

Snyder, F.W. and G.A. Smith. 1986. Heritability of taproot weight: leaf weight ratio in sugarbeet. J. ASSBT 23:142-147.

Snyder, G.W., J.C. Ingersoll, A.C. Smigocki, and L.D. Owens. 1999. Introduction of pathogen defense genes and a cytokinin biosynthesis gene into sugar beet (*Beta vulgaris* L.) by *Agrobacterium* or particle bombardment. Plant Cell Rept. 18:829-834.

Solel, Z. and G. Minz. 1971. Infection process of *Cercospora beticola* in sugar beet. Phytopathology 61:463-466.

Speckmann, G.J. and T.S.M. De Bock. 1982. The production of alien monosomic additions in *Beta vulgaris* as a source for the introgression of resistance to beet root nematode (*Heterodera schachtii*) from *Beta* species of the section *Patellares*. Euphytica 31:313-323.

Speckmann, G.J., T.S.M. De Bock, and J.H. De Jong. 1985. Monosomic additions with resistance to beet cyst nematode obtained from hybrids of *Beta vulgaris* and wild *Beta* species of the section *Patellares*. I. Morphology, transmission and level of resistance. Z. Pflanzenzucht. 95:74-83.

Spettoli, P., G. Cacco, and G. Ferrari. 1976. Comparative evaluation of enzyme multiplicity in a diploid, a triploid, and a tetraploid sugar beet variety. J. Sci. Food Agric. 27:341-344.

Srivastava, H.M. 1995. Breeding of sugar beet for *Sclerotium* and high temperature tolerance. Proc. IIRB 58:177-179.

Srivastava, H.M. 1996. Genetic diversity for high-temperature tolerance in sugar beet. *In:* L.Frese, L.Panella, H.M. Srivastava, and W. Lange (eds.) Report of 4th *Beta* Genetic Resources Workshop and World *Beta* Network Conf. Izmir, Turkey, 28 Feb.- 3 Mar. 1996. IPGRI, Rome, Italy, pp. 52-57.

Stajner, D., N. Mimica-Dukic, and O. Gasic. 1995. Adaptability to drought in sugar beet cultivars. Biol. Plant. 37:107-112.

Stander, J.R. and J.C. Theurer. 1970. Inheritance and linkage of A Virescens and A Chlorina in *Beta vulgaris* L. Crop Sci. 10:548-549.

Stanghellini, M.E., D.C. Sands, W.C. Kronland, and M.M. Mendonca. 1977. Serological and physiological differentiation among isolates of *Erwinia carotovora* from potato and sugar beet. Phytopathology 67:1178-1182.

Stanghellini, M.E., P. von Bretzel, M.W. Olsen, and W.C. Kronland. 1982. Root rot of sugar beets caused by *Pythium deliense*. Plant Dis. Rep. 66:857-858.

Stanghellini, M.E., L.J. Stowell, W.C. Kronland, and P. von Bretzel. 1983. Distribution of *Pythium aphanidermatum* in rhizosphere soil factors affecting expression of the absolute inoculum potential. Phytopathology 73:1463-1466.

Steele, A.E. 1984. Nematode parasites of sugar beet. *In:* W.R. Nickle (ed.). Plant and Insect Nematodes. Marcel Dekker, Inc., New York, NY, pp. 507-569.

Steele, A.E. 1986. Nematode parasites of sugar beet. *In:* E.D. Whitney and J.E. Duffus (eds.). Compendium of Beet Diseases and Insects. Amer. Phytopath. Soc., St. Paul, MN, pp. 33-36.

Stehlik. V. 1933. L'hérédité de la richesse en sucre et du poids chez les hybrides de la betterave sucrière et fourragère. CR IIRB 3:74-109.

Stevens, M. and A. Dewar. 1996. Pest and disease problems in Chile: What lessons can be learnt? Brit. Sugar Beet Rev. 2:34-38.

Stewart, D. 1931. Sugar-beet yellows caused by *Fusarium conglutinas* var. *betae*. Phytopathology 21:59-70.

Stewart, D. 1950. Sugar beet x *Beta procumbens*, the F1 and backcross generations. Proc. ASSBT 6:176-179

Stibbe, C. and B. Märländer. 2002. Field emergence dynamics significance to intraspecific competition and growth efficiency in sugar beet (*Beta vulgaris* L.). Eur. J. Agron. 3:161-171.

Stout, M. 1954. Some factors that affect the respiration rate of sugar beets. Proc. ASSBT 8:404-409.

Stump, W.L., G.D. Franc, S.D. Miller, and R.G. Wilson. 2002. Azoxystrobin and post-emergence herbicide combinations for Rhizoctonia and weed management in sugarbeet. J. Sugar Beet Res. 39:37-58.

Tacconi, R. and R. Pola. 1996. Resistance to *Heterodera schachtii* of oleiferous radish, white mustard and kenaf. L'Informatore Agrario 52:76-77.

Taguchi, K., N. Ogata, H. Takahashi, T. Kuranouchi, M. Kawakatsu, and M. Tanaka. 2000. Breeding Aphanomyces root rot resistance in sugar beet: I. On-site screening for resistance to Aphanomyces root rot in disease-infested field. Proc. Jpn. Soc. Sugar Beet Tech. 42:52-58.

Tamada, T., M. Saito, T. Kiguchi, and T. Kusume. 1990. Effect of isolates of BNYVV with different RNA components on the development of rhizomania symptoms. *In:* R. Koenig (ed.). Proc. Ist Symp. Working Group on Plant Viruses with Fungal Vectors, Braunschweig. Eugen Ulmer, Stuttgard, Germany, pp. 41-44.

Tamada, T., Y. Shirako, H. Abe, R. Saito, T. Kiguchi, and T. Harada. 1989. Production and pathogenicity of isolates of BNYVV with different numbers of RNA components. J. Gen. Virol. 70:3399-3409.

Tatlioglu T. 1974. Vererbung und Umweltvariabilität der Monogermie bei *Beta vulgaris* L. Z. Pflanzenzücht. 72:233-244.

TeKrony, D.M. and E.E. Hardin. 1969. The problem of underdeveloped seeds occurring in monogerm sugar beets. J. ASSBT 15:625-639.

Thakur, R.P. and A.N. Mukhopadhyay. 1972. Nitrogen fertilization of sugar beet in relation to sclerotium root rot caused by *Sclerotium rolfsii* Sacc. Ind. J. Agric. Sci. 42:614-618.

Theurer, J.C. 1968a. Inheritance of a lutescens mutant in sugar beets, *Beta vulgaris* L. Crop Sci. 8:422-423.

Theurer, J.C. 1968b. Linkage tests of mendelian male sterility and other genetic characters in sugar beets (*Beta vulgaris* L.). Crop Sci. 8:698-701.

Theurer, J.C. 1971. Inheritance studies of a pollen restorer from 'Ruby queen' table beet. J. ASSBT 16:354-358.

Theurer, J.C. 1981. Registration of eight germplasm lines of sugar beet. Crop Sci. 18:1011.

Theurer, J.C. 1993. Pre-breeding to change sugar beet root architecture. J. Sugar Beet Res. 30:221-239.

Theurer, J.C. and G.K. Ryser. 1969. Inheritance studies with a pollen fertility restorer sugarbeet inbred. J. ASSBT 15:538-545.

Theurer, J.C., R.E. Wyse, and D.L. Doney. 1978. Root storage respiration rate in a diallel cross of sugar beet. Crop Sci. 18:109-111.

Theurer, J.C., C.C. Blickenstaff, G.C. Mahrt, and D.L. Doney. 1982. Breeding for resistance to the sugar beet root maggot. Crop Sci. 22: 641-645.

Theurer, J.C., D.L. Dooney, G.A. Smith, et al., 1987. Potential ethanol production from sugar beet and fodder beet. Crop Sci. 5:1034-1040.

Thomas, B. 1991. Phytochrome and photoperiodic induction. Physiol. Plant. 81:571-577.

Thomas, T. and N. Clarke. 1995. Are sugar beet problems being watered down. Brit. Sugar Beet Rev. 3:8-11.

Thomashow, M.F. 1993. Genes induced during cold acclimation in higher plants. *In:* P.L. Stepenkus (ed.) Advances in Low-temperature Biology. JAI Press, London, UK, vol.2, pp. 183-210.

Thomson, S.V., D.C. Hildebrand, and M.M. Schroth. 1981a. Identification and nutritional differentiation of Erwinia sugar beet pathogen from members of *Erwinia carotovora* and *Erwinia chrysanthemi*. Phytopathology 71:1037-1042.

Thomson, S.V., F.J. Hills, E.D. Whitney, and M.N. Schroth. 1981b. Sugar and root yield of sugar beets as affected by bacterial vascular necrosis and rot, nitrogen fertilization, and plant spacing. Phytopathology 71:605-608.

Thomson, S.V., M.N. Schroth, F.J. Hills, E.D. Whitney, and D.C. Hildebrand. 1977. Bacterial vascular necrosis and rot of sugar beet, general description and etiology. Phytopathology 67:1183-1189.

Thurau, T., S. Kifle, C. Jung, and D. Cai. 2003. The promoter of the nematode resistance gene Hs1pro-1 activates a nematode-responsive and feeding site-specific gene expression in sugar beet (*Beta vulgaris* L.) and *Arabidopsis thaliana*. Plant Molec. Biol. 52:643-660.

Toporovskaya, Y.S. 1975. The control of rots of sugar beet. Zashchita Rastenii 75:24-25.

Townsend, C.O. 1908. Curly-top, a Disease of the Sugar Beet. USDA: Bureau of Plant Industry, Bull. 122.

Tsuchiya, T. 1986. Chromosome engineering in sugar beet breeding for cyst nematode resistance. Proc. Sugar Beet Res. Asoc., Japan 28:166-177.

Tuitert, G. 1994. Epidemiology of Rhizomania Disease in Sugar Beet. Institut Rationele Suikerproductie, Bergen op Zoom, Netherlands.

Tungland, B.C., R.E. Watkins, and P.V. Schmidt. 1998. Sugar beet storage (Section 5.4). *In:* P. W. van der Poel. H. Schiewek, and T.K. Schwartz (eds.). Sugar Technology. Verlag Dr. Albert Bartens, Berlin, Germany, pp. 267-289.

Ulrich, A. 1961. Variety climate interactions of sugar beet varieties in simulated climates. J. ASSBT 11:376-387.

Vaisman, N.Y. 1981. Pseudo-self-incompatibility of initial and inbred populations of sugarbeet at two geographical points. Siberian Vestn. Nauki 1:25-31.

Van Dijk, H., P. Boundry, H. McCombie, and P. Vernet P. 1997. Flowering time in wild beet (*Beta vulgaris* ssp. *maritima*) along a latitudinal cline. Acta Oecologica 18:47-60.

Van Euwijk, F.A. and L.C.P. Keizer. 1997. Results of IIRB field trials for rhizomania. Report of CPRO-DLO, Wageningen, Netherlands.

Van Geyt, J.P.C., W. Lange, M. Oleo, and T.S.M. De Bock. 1990. Natural variation within the genus *Beta* and its possible use for breeding sugarbeet: a review. Euphytica 49:57-76.

Van Roggen, P.M., B. Debenham, P. Hedden, A.L. Phillips, and S.G. Thomas. 1998. A model for control of bolting and flowering in sugar beet and the involvement of gibberellins. Flowering Newsl. 25:45-49.

Van Schreven, D.A. 1936. De vergelingsziekte bij de biet en haar oorzaak. Meddelanden Institutet foer Suikerbiet 6:1-36.

Vince-Prue, D. 1975. Photoperiodism in Plants. McGraw Hill, London, UK.

Visser, H.R. 1951. Anatomisch-physiologische rasverschillen bij suikebieten. Studiekring voor Plantenveredeling, pp. 425-433.

Von Bretzel, P., M.E. Stanghellini, and W.C. Kronland. 1988. Epidemiology of Pythium root rot of mature sugar beets. Plant Dis. 72:707-709.

Wallis, R.L. and J.O. Gaskill. 1963. Sugar beet root aphid resistance in sugar beet. J. ASSBT 12:571-572.

Wallis, R.L. and J.E. Turner. 1968. Resistance of sugar beets to sugar beet root aphids, *Pemphigus populivenae* Fitch. J. ASSBT 14:671-673.

Watanabe, H., H. Uchino, and M. Hayasaka. 2000. Development of artificial inoculation method of Aphanomyces root rot for sugar beet. Proc. Jpn. Soc. Sugar Beet Tech. 42:47-51.

Watson, D.J. and M.A. Watson. 1953. Comparative physiological studies on the growth of field crop: III. The effect of infection with beet yellows and beet mosaic viruses on the growth and yield of sugar beet root crop. Ann. Appl. Biol. 40:1-12.

Watson, M. and G.E. Russell. 1956. The value of glasshouse tests with seedling plants tolerant to beet yellows virus. Ann. Appl. Biol. 44:381-386.

Watson, J.F. and I.L. Goldman. 1997. Inheritance of a recessive gene conditioning blotchy root color patterning in *Beta vulgaris*. J. Hered. 88:540-543.

Wauters, A. and J. Keleman. 1998. Economic and agronomic interest of nematode-resistant sugar beet varieties: Belgian experience from 1995 to 1998. Parasitica 54:63-75.

Weiland, J.J. and M.H. Yu. 2003. A cleaved amplified polymorphic sequence (CAPS) marker associated with root-knot nematode resistance in sugar beet. Crop Sci. 43:1814-1818.

Weiland, J.J., G.A. Smith, and L. Panella. 1999. Greenhouse assay for the evaluation of sugarbeet resistance to Rhizoctonia root rot. J. Sugar Beet Res. 36:95.

Weltzien, H.C. 1963. *Erysiphe betae* (Vanha) comb. nov., the powdery mildew of beets. Phytopath. Z. 47:123-128.

Westerdijk, C.E. and J.J. Tick. 1991. Investigations into advancing the growth period of sugar beet. Jaarboek Proefstation voor de Akkerbouw en de Groenteteelt in de Vollegrond, Lelystad, pp. 45-47.

Wewers, J.D.A. and W. Heijbroek. 1994. Selection of cultivars and choice of pelleted seed. Maandblad-Suiker Unie 12:13-15.

Whitney, E.D. 1982. The susceptibility of fodder beet and wild species of *Beta* to an *Erwinia* sp. from sugar beet. Plant Dis. 66:664-665.

Whitney, E.D. 1986a. Correlation among greenhouse tests and between field and greenhouse evaluations for Beet Necrotic Yellow Vein Virus (BNYVV) resistance in *Beta maritima*. Phytopathology 76:1074

Whitney, E.D. 1986b. Diseases caused by bacteria and bacterialike organisms. *In:* Compendium of Beet Diseases and Insects. Amer. Phytopath. Soc., St. Paul, MN, pp. 23-26.

Whitney, E.D. 1989a. Identification, distribution and testing for resistance to rhizomania in *Beta maritima*. Plant Dis. 73:287-290.

Whitney, E.D. 1989b. *Beta maritima* as a source of powdery mildew resistance in sugar beet. Plant Dis. 73:487-489.

Whitney, E.D. and R.T. Lewellen. 1977. Bacterial vascular necrosis and rot of sugar beet: effect on cultivars and quality. Phytopathology 67:912-916.

Whitney, E.D. and R.T. Lewellen. 1978a. Bacterial vascular necrosis and rot of sugar beet: Genetic vulnerability and selecting for resistance. Phytopathology 68:657-661.

Whitney, E.D. and R.T. Lewellen. 1978b. Registration of two sugar beet parental lines. Crop Sci. 18:920.

Whitney, E.D. and R.T. Lewellen. 1985. Bacterial vascular necrosis and rot of sugar beet: effect of moisture, age of plants, injury, inoculation and genotype on susceptibility to infection by *Erwinia*. J. ASSBT 23:28-38.

Whitney, E.D. and J.E. Duffus. 1986. Compendium of beet diseases and insects. Amer. Phytopath. Soc., St. Paul, MN.

Whitney, E.D. and B.E. Mackey. 1989. Differences in aggressiveness of *Erwinia carotovora* subsp. *Betavasclorum* strains and their reaction to sugar beet cultivars. Plant Dis. 73:220-222.

Whitney, E.D., R.T. Lewellen, and I.O. Skoyen. 1983. Reactions of sugar beet to powdery mildew: genetic variation, association among testing procedures, and resistance breeding. Phytopathology 73:182-185.

Wilhite, S.E., T.C. Elden, V. Puizdar, S. Armstrong, and A.C. Williamson. 1998. Root-knot nematode resistance genes in tomato and their potential for future use. Annu. Rev. Phytopath. 36:277-293.

Williamson, V.M. 1998. Root-knot nematode resistance genes in tomato and their potential for future use. Ann. Rev. Phytopathol. 36:277-293.

Winner, C. and W.R. Schäufele. 1977. Orientierende Untersuchungen uber del Einfluss der durch *Polymyxa betae* verursachten Wurzelbärtigkeit auf die Qualität von Zuckerrüben. Zucker 9:459-463.

Winter, H. and S.C. Huber. 2000. Regulation of sucrose metabolism in higher plants: localization and regulation of activity of key enzymes. Crit. Rev. Plant Sci. 19:31-67.

Wisler, G.C. and J.E. Duffus. 2000. A century of plant virus management in the Salinas Valley of California 'East of Eden'. Virus Res. 71:161-169.

Wisler, G.C., R.T. Lewellen, J.L. Sears, H.Y. Hiu, and J.E. Duffus. 1999. Specificity of TAS-ELISA for BNYVV and its application for determining rhizomania resistance in field-grown sugar beet. Plant Dis. 83:864-870.

Wisler, G.C., R.T. Lewellen, J.L. Sears, J.W. Wasson, H.Y. Liu, and W.M. Wintermantel. 2003. Effects of BNYVV, BSBMV and their fungal vector, *Polymyxa betae* on seedlings emergence, plant growth and virus concentration in sugar beet. Proc. ASSBT/IIRB 1:57-58.

Wolzniak, C.A. 2001. Transgenic sugar beet: progress and development. Appl. Plant Biotech. 10:301-324.

Wood, R.R. 1952. Selection for cold tolerance and low temperature germination in sugar beet. J. ASSBT 6:407-411.

Wood, R.R. 1954. Breeding for improvement of processing characteristics of sugar beet varieties. Proc. ASSBT 8:125-133.

Wood, R.R., H.E. Brewbacker, and H.L. Busch. 1950. Cold resistance in sugar beets. J. ASSBT 6:117-121.

Wood, R.R., R.K. Oldemeyer, and H.L. Bush. 1956. Inheritance of raffinose production in the sugar beet. J. ASSBT 9:133-138.

Wood, R.R., H.L. Bush, and R.K. Oldemeyer. 1958. The sucrose-sodium relationship in selecting sugar beets. J. ASSBT 10:133-137.

Wubben, M.J.E., H. Su, S.R. Rodermel, and T.J. Baum. 2001. Susceptibility to the sugar beet cyst nematode is modulated by ethylene signal transduction in *Arabidopsis thaliana*. Molec. Plant-Microbe Interact. 14:1206-1212.

Wyse, R.E. 1979. Parameters controlling sucrose content and yield of sugar beet roots. J. ASSBT 20:368-385.

Wyse R.E. 1980. Partitioning within the taproot sink of sugarbeet: effect of photosynthate supply. Crop Sci. 20:256-258.

Wyse, R.E. and S.T. Dexter. 1971a. Effect of agronomic and storage practices on raffinose, reducing sugar, and amino acid content of sugar beet varieties. J. ASSBT 16:369-383.

Wyse, R.E. and S.T. Dexter. 1971b. Source of recoverable sugar losses in several sugar beet varieties during storage. J. ASSBT 16:390-398.

Wyse, R.E., J.C. Theurer, and D.L. Doney. 1978. Genetic variability in post-harvest respiration rates of sugar beet roots. Crop Sci. 18: 264-266.

Wyss, U. 1992. Observations on the feeding behavior of *Heterodera schachtii* throughout development, including events during moulting. Fund. Appl. Nematol. 15:75-89.

Yonts, C.D., R.G. Wilson, and J.A. Smith. 1999. Influence of planting date and stand on yield and quality of sugar beet. J. Sugar Beet Res. 36:1-13.

Yu, M.H. 1978. Meiotic behavior of a disomic nematode-resistant sugar beet. Crop Sci. 18:615-618.

Yu, M.H. 1981. Sugar beets homozygous for nematode resistance and transmission of resistance to their progenies. Crop Sci. 21:714-717.

Yu, M.H. 1982a. Inheritance of resistance to *Heterodera schachtii* and chromosome segregation in triploid sugar beets. Can. J. Genet. Cytol. 24:567-574.

Yu, M.H. 1982b. Interpretation of mechanism for nematode resistance in sugar beet. J. ASSBT 21:351-361.

Yu, M.H. 1983. Sugar beet germplasm resistant to sugar beet nematode. Crop Sci. 23:1021-1022.

Yu, M.H. 1984a. Resistance to *Heterodera schachtii* in *Patellares* section of the genus *Beta*. Euphytica 33:633-640.

Yu, M.H. 1984b. Transmission of nematode resistance in the pedigree of homozygous resistant sugar beet. Crop Sci. 24:88-91.

Yu, M.H. 1992. Growth and reproduction performance of ovule-induced sugar beet plants. SABRAO J. 24:47-55.

Yu, M.H. 1994. Biological methods of nematode control in sugar beet. *In:* Toward Enhanced and Sustainable Agricultural Productivity in the 2000s: Breeding Research and Biotechnology. Taichung DAIS & SABRAO, Taipei, Taiwan, vol. 3, pp. 913-919.

Yu, M.H. 1995a. Root-knot nematode development and root gall formation in sugar beet. J. Sugar Beet Res. 32:47-58.

Yu, M.H. 1995b. Identification of a *Beta maritima* source of resistance to root-knot nematode for sugar beet. Crop Sci. 35:1288-1290.

Yu, M.H. 2002. Registration of sugar beet germplasm M1-3 resistant to root-knot nematode. Crop Sci. 42:1756-1757.

Yu, M.H. 2003. Development of root-knot nematode-resistant sugar beet. Proc. ASSBT/IIRB 1:763-765.

Yu, M.H. and A.E. Steele. 1981. Host-parasite interaction of resistant sugar beet and *Heterodera schachtii*. J. Nematol. 13:206-212.

Yu, M.H. and K.C. Jones. 1983. Preliminary biochemical assay on resistance to *Heterodera schachtii* in sugar beet. Genetics 104:s73.

Yu, M.H. and P.A. Roberts. 2002. Selection of root-knot nematode resistant sugar beet from field plantings. Nematology 4:240.

Yu, M.H. and R.T. Lewellen. 2004. Registration of root-knot nematode-resistant sugar beet germplasm M6-2. Crop Sci. 44:1502-1503.

Yu, M.H., W. Heijbroek, and L.M. Pakish. 1999. The sea beet source of resistance to multiple species of root-knot nematode. Euphytica 108:151-155.

Yu, M.H., L.M. Pakish, and H. Zhou. 2001. An isozyme marker for resistance to root-knot nematode in sugar beet. Crop Sci. 41:1051-1053.

Yun, Y.M. 1986a. Sugar beet root maggot. *In:* E.D. Whitney and J.E. Duffus (eds.). Compendium of Beet Diseases and Insects. Amer. Phytopath. Soc., St. Paul, MN, pp. 40-41.

Yun, Y.M. 1986b. Sugar beet root aphid. *In:* E.D. Whitney and J.E. Duffus (eds.). Compendium of Beet Diseases and Insects. Amer. Phytopath. Soc., St. Paul, MN, p. 42.

Yun, Y.M. 1986c. Aphids. *In:* E.D. Whitney and J.E. Duffus (eds.). Compendium of Beet Diseases and Insects. Amer. Phytopath. Soc., St. Paul, MN, p. 48.

Zhao, B., I.J. Mackay, P.D.S. Caligari, and R. Mead. 1997. The efficiency of between and within full-sib family selection in a recurrent selection program in sugar beet (*Beta vulgaris* L). Euphytica 95:355-359.

Zielke, R.C. 1973. Yield, quality, and sucrose recovery from sugar beet root and crown. J. ASSBT 17:332-344.

Zielke, R.C. and F.W. Snyder. 1974. Impurities in sugar beet crown and root. J. ASSBT 18:54-60.

3

Conventional Breeding Methods

3.1 Selection Methods (M. De Biaggi and G.N. Skaracis)

Increases achieved in sugar beet production through breeding have been very impressive and, initially at least, occurred at a considerable rate (McFarlane, 1971). Approximately half of the progress is attributable to genetic improvements (Sneep and Hendriksen, 1979). In achieving these results, mass selection was applied, followed by several more complicated schemes based on progeny evaluation and combining ability assessment (Smith, 1987). Further advances in the last fifty years have made possible the employment of various selection approaches, originally developed for other crops.

3.1.1 Mass Selection

Mass selection in sugar beet includes the following phases (Bosemark, 1993): (i) a number of phenotypically desirable plants are selected from a genetically variable population; (ii) selected plants are intercrossed by open pollination; and (iii) seed is bulk harvested from all plants and selection is renewed. The objective of the process is an increased frequency of superior genotypes in the population (McFarlane, 1971). The contribution of this method in sugar beet breeding specifically relates to success in: (i) adaptation of populations to new agroclimatic conditions; (ii) improvement of morphological characteristics; (iii) development of bolting resistant material; and (iv) the selection for resistance to several diseases.

Given that the units of selection are individual plants, practical difficulties encountered in the use of mass selection in sugar beet basically are those related to environmentally caused variation and to variation arising from interplant genotypic competition, the latter sometimes accounting for more than half the total phenotypic variance (Lichter, 1972). To reduce unequal environmental influence on the plants, materials undergoing selection should be sown with the best possible precision in highly homogeneous fields. Obviously, when selecting for tolerance/ resistance to a certain disease, the presence of other pathogens is unacceptable. Also, a very useful approach is the stratified (or gridded) mass selection, i.e. partitioning of the entire field into smaller areas and selecting a fixed number of plants from each (Bosemark, 1993). Evaluation in the absence of interplant competition theoretically could be effected through honeycomb breeding designs (Fasoulas and Fasoula, 1995) but no results of relevant work in sugar beet have been reported so far.

Another difficulty, common in more complicated selection systems as well, is the presence of strong negative correlation among various characters on which composite traits, such as sugar yield, depend. Since the agronomic value of commercial varieties depends on the balance of a large number of useful characters expressed at their maximum possible level (Biancardi, 1999), it is very important that selection in improving one of them does not result in adverse effects on the others.

Successful application of mass selection in sugar beet breeding requires an adequately high level of heritability of the traits, and the ability accurately to evaluate the genotypic worth of a plant solely on the basis of its phenotype (Hecker, 1967). Consequently, mass selection is expected to be quite efficient for qualitative (monogenic or oligogenic) traits and considerably less efficient for quantitative (multigenic) traits. Although variations do exist, these considerations have generally been confirmed in practice.

Development of highly bolting resistant populations by mass selection was reported by Knapp (1958). As a result of a breeding program initiated as early as 1925, continuous progress by mass selection in severely diseased fields led to the development of a series of curly top resistant varieties widely used in the USA (Coons et al., 1955). The same selection method allowed for the improvement of resistance to virus yellows and to diseases caused by pathogens such as *Aphanomyces*, *Fusarium*, *Erisyphe* etc. Good results were also obtained in producing many *Rhizoctonia* resistant lines (Panella, 1995). Applying a strict selection intensity under severe disease conditions, plants with the rhizomania resistance gene were chosen and the very successful variety "Rizor" subsequently developed (Biancardi et al., 2002).

More recently, a smooth-root beet, useful for minimizing soil tare at harvest, was developed through a 3-cycle mass selection between a high sucrose family and a family characterized by smooth root (Saunders et al., 1999). Mesken (1989), after five cycles of mass selection, was able to improve root shape and reduce crown height both in diploid and tetraploid material.

Exploiting mass selection in tetraploid families, used to develop pollinators of triploid hybrid varieties, is more difficult than for diploid families, due to the complexity of tetrasomic inheritance. There have been some successful applications however, e.g. maintenance of an acceptable level of bolting resistance of pollinators used for the production of autumn sown varieties (De Biaggi, unpubl. data).

Mass selection can give satisfactory progress in improving traits controlled by genes with significant additive effects, as evidenced in the case of sucrose content and the content of various non-sugars (Smith et al., 1973). Root yield, controlled by genes with substantial nonadditive action, shows poor response to mass selection, although it is quite useful when dealing with nonadapted starting material of a breeding program (Bosemark, 1993).

In a typical mass selection scheme, sugar beet selection fields are established concomitant with those of the industrial crop and the plants called "mother beets" (see Fig. 5.1). At harvesting, care is taken to avoid damaging roots and plants grown on sites of obviously irregular inputs (e.g. fertilizer, water etc.) removed. After this first morphological selection, plants are defoliated and washed so that a more accurate selection can be effected. Roots with undesirable characteristics (e.g. excessive hypocotyl size, multiple crown, deep root grooves, rough or abnormally colored skin, excessive fanging etc.) are discarded. With completion of this stage, approximately 10% of the beets in the plot, those closest to the desirable ideotype, are left. After numbering and weighing, the roots are sampled using special apparatus. It is necessary that samples represent the chemical characters of the entire root (Biancardi et al., 1979). Individual analyses include sugar content, usually measured by polarimetry and the most important nonsugars (sodium, potassium, amino-nitrogen, reducing sugars, betaine etc.). Beets selected on the basis of these analyses are treated with fungicides and kept under temperature and light conditions appropriate to induce vernalization. The roots can be transplanted to polyethylene sacks, filled with a mixture of soil and peat and stored during winter in rooms with natural light. This technique is useful in reducing root damage and allows maximum flexibility in moving the plants pots can be moved right up to the time of flowering (Biancardi, unpubl. data). Thus following overwinter vernalization, in the following spring the selected roots are allowed to intercross by open pollination in an isolated seed plot and the improved population produced.

In the case of complicated characters such as resistance to cercospora l.s., environmental variation is too high to safely select genetically superior plants on the basis of phenotypic characteristics. This is a situation wherein micropropagation finds useful application. Indeed, evaluation of disease infection in rows of identical plants is more precise than in individual plants. Also, when selecting for rhizomania resistance, plants can be grown in the greenhouse and subjected at early stages of development, to quantitative ELISA tests for viral content (Giunchedi et al., 1985). The importance of these tests lies in the fact that they can distinguish between homozygous (*RzRz*) and heterozygous (*Rzrz*) resistant plants (Burky and Büttner, 1989). Mass selection based on this kind of analysis allows for efficient elimination of the recessive alleles and therefore a seed production of only true-breeding resistant plants (De Biaggi, unpubl. data).

3.1.2 Progeny Selection and Line Breeding

Selection based on the phenotypes of the plants themselves, even when heritability is high, is not a very efficient method for improvement. In the case of a monogenic, dominant trait for example, it is impossible to totally eliminate undesirable recessive alleles as they soon become stably hidden in a heterozygous state. Evaluation of a genotype based on the characteristics of its progeny provides for a much more efficient means for improvement. This system was introduced around the middle of the nineteenth century by Louis de Vilmorin and is credited with the impressive initial increase in sugar yield. Since then, selection by progeny testing and line breeding, with various versions, has been in common use.

Family selection includes two methods known as half-sib and full-sib progeny selection, depending on whether only one or both parents are known respectively. Both methods permit accumulation of favorable genes with additive (Smith et al., 1973) and dominant (Helmerick et al., 1965) effects and have been successfully employed for the development of improved multigerm diploid and tetraploid lines and populations. However, where heterotic (nonadditive) traits are concerned (e.g. yield), response to selection in already adapted material is quite inefficient. As with all selection schemes, efficiency primarily depends on whether the populations under improvement have a sufficient variation for the trait of interest.

LEGENDA

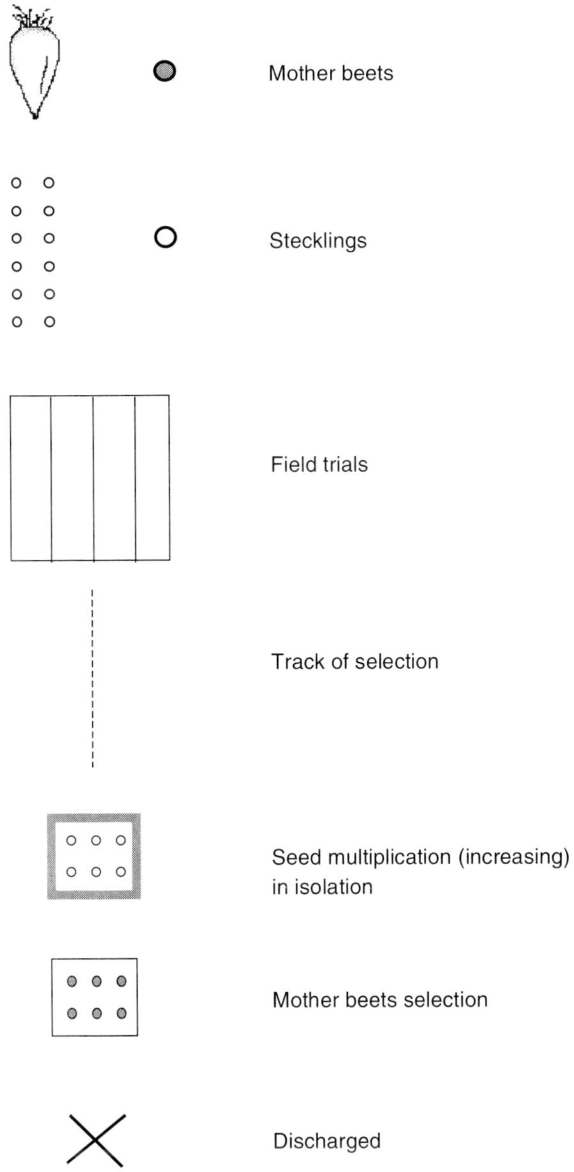

Fig. 3.0 Legenda for Figures 3.1 to 3.8.

3.1.2.1 Half-sib selection

Plants selected in the first year, in precisely the same manner as in mass selection, are intercrossed by open pollination under space isolation **(Fig. 3.1)** and harvested separately. Seed from a plant, a half-sib (HS) family, is a mixture of F_1 hybrids produced by the female gametes of the particular plant and by a random sample of the male gametes of the other plants in the crossing plot. Due to the possible presence of a variable degree of self-fertility, part of the seed on each plant can be the product of self-pollination. The seed from each of the HS families ideally should be sufficient for: i) replicated field trials to assess yield performance and other characters of interest (i.e. quality, disease and bolting resistance, etc.) on the basis of which the best families will be recombined; ii) production of stecklings (i.e. beets sown in August for flowering induction, see Figure 5.1) for subsequent synthesis of the improved population; and iii) the maintenance of enough reserves for further use.

Usually the seed quantities are insufficient for satisfying all these needs and hence the first year field testing is limited. In addition, when dealing with traits characterized by heterotic response, the value of an HS family mainly depends on its heterozygosity and the presence of favorable allelic interactions, both nontransmissible to the progeny. For these reasons, very often it is necessary to reproduce HS families selected in the initial limited trials, by sib-mating in separate isolations. Higher seed quantities and reduced heterozygosity in these reproductions allow for a more reliable evaluation in repeated field trials so that selection of superior families to constitute the new population is more efficient. Several modifications are applicable to the aforesaid scheme. For example, instead of stecklings, mother roots selected at harvest within each superior family can be used for next year's seed production.

In the case of diploid multigerm populations, another variation can be employed (**Fig. 3.2)**. Mother beets selected in the first year are halved, and one part is used to produce selfed seed, while the other part participates in an open pollination intercross. Seed is separately harvested on each plant. In practice, two distinct selection paths are followed, with field tests of S1 and corresponding HS families alternating in the years. Evaluation of HS families provides an indication for their general combining ability (GCA), while that of their respective S1s permits a rapid elimination of obviously inferior lines. In this way, a fairly efficient selection is achieved. The weak points of the method are: i) its practical application is complex; ii) loss of many lines due to the difficulty in producing selfed seed; and iii) relatively high costs considering that a single population is improved. After appropriate test-cross evaluations, the half-sib families obtained may be used directly as pollinators of commercial varieties.

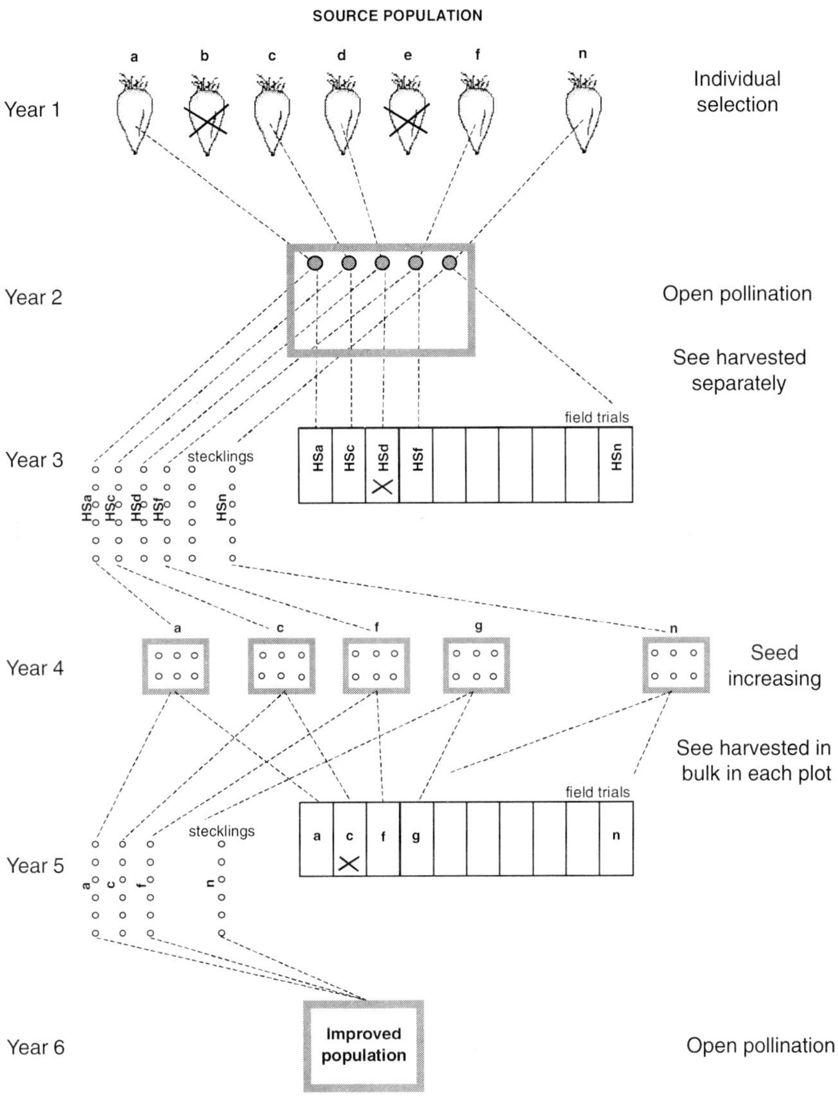

Fig. 3.1 Scheme of half-sib selection.

3.1.2.2 Full-sib selection

This method allows a more efficient selection because the parentage of every full-sib (FS) family, based on only two plants, is fully determined (Smith, 1987; Hecker, 1985). As with the scheme described above, the full-sib selection method is used for production of diploid and tetraploid pollinators of commercial varieties.

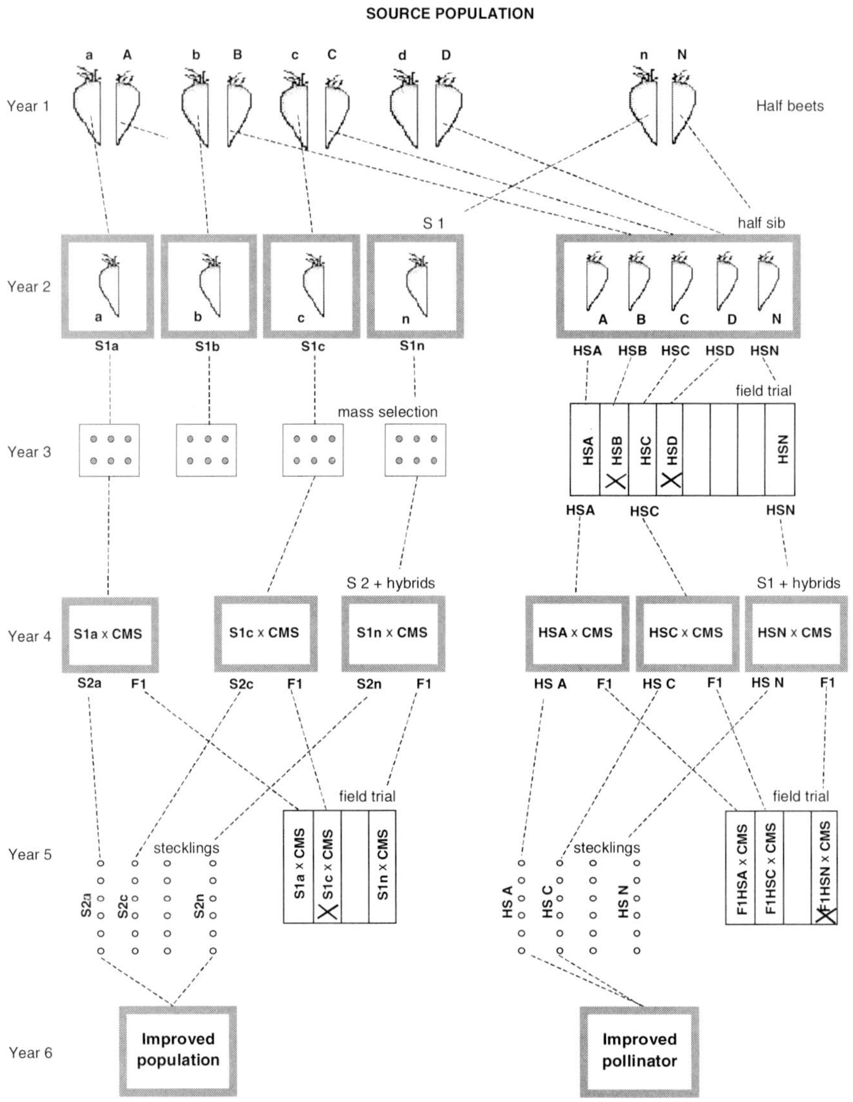

Fig. 3.2 Scheme of half-sib selection with halved mother beets.

Following mother beet selection, seed is produced in isolated paired crosses **(Fig. 3.3)**. All new FS families enter field trials in the third year, while in August the same families are sown in the nursery while results of the trials are awaited. Only the proven superior families are intercrossed in the following year to obtain an improved population with a superior range of favorable genetic combinations. As with the HS families, seed increase of

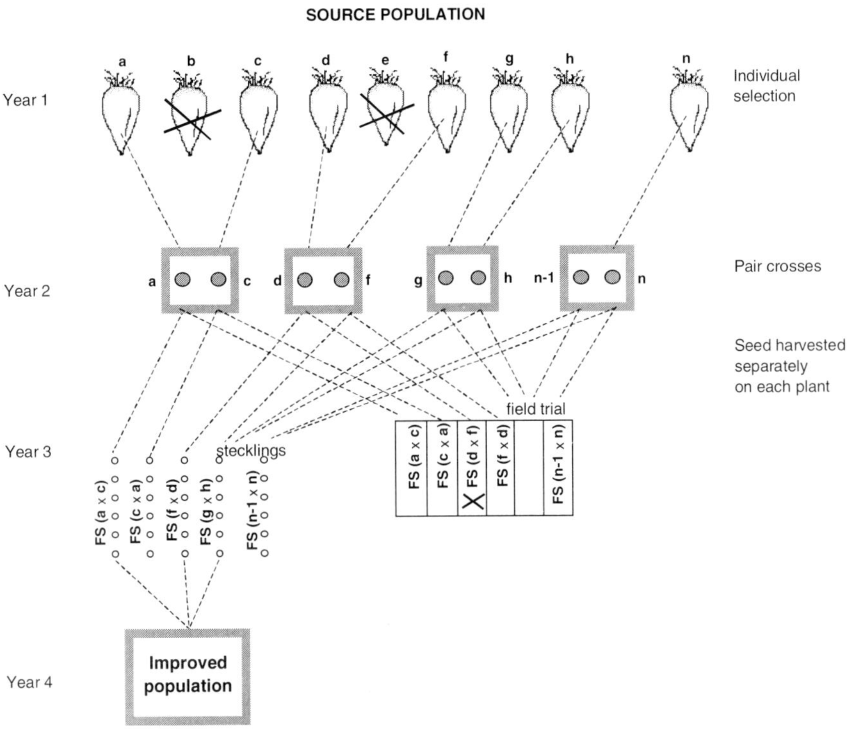

Fig. 3.3 Scheme of full-sib selection.

the initially produced FS families allows a more accurate assessment of their breeding value. The resultant seed however, being the product of full-sib mating, will already manifest an inbreeding effect that should be seriously considered. For this reason, unless inbred lines are targeted, no further cycles of selection are applied. Again, beets selected within each family can replace stecklings for seed production of superior families.

Helmerick et al., (1965) proposed a modification of the above method. The FS families produced in the second year are crossed in year three to a good combining ability CMS parent. In this manner, a series of monogerm hybrids is obtained to be tested in year four. Families that prove to be significantly superior are intercrossed in the fifth year to produce a population with improved GCA. Given that the application of the FS method, as already mentioned, actually is a strong inbreeding process, an advantage comes from the possibility of selecting the best families from field trials in the fourth year and subsequently using them as pollinators in commercial production of improved varieties. The low F_2 seed quantity of the FS families in the third year might necessitate another round of multiplication under isolation. In this way and although it is better to use FS-F_2 seed, the value of the FS-F_3 families obtained in the fourth year is verified through test with a commercial CMS in the fifth and subsequent multilocation field evaluation in the sixth year. The best families are then sown in the nursery along with a commercial CMS and used to produce seed of the improved variety. If all F_2 seed is used, new basic seed is produced from the F_3 seed reserves.

3.1.2.3 Progeny selection using clones

Both in half-sib and full-sib selection methods, superiority of the best progenies is due in part to a high degree of heterozygosity and partly to favorable but nonheritable allelic interactions. Micropropagation techniques (see Chapter 4) permit production of a desirable number of genetically identical individuals. Availability of many such plants instead of one, is very useful. Indeed, the genotypic worth of a plant is much more precisely evaluated when based on clones derived from it. At the same time, the relationship of homozygosity to heterozygosity and the level of favorable allelic interactions present in a genotype can be reproduced without resorting to inbreeding. It also allows for estimation of environmental (intraclone) variability and genotype $\times$ environment interaction if several locations are used for field trials. The advantages of this technique for family selection are quite obvious: gene frequencies can be maintained unaltered and the lines thus more efficiently evaluated for their combining ability and other useful characteristics. Generally, clones are used for the improvement of diploid and tetraploid multigerm pollinators. The technique is not so useful for O-type and corresponding CMS monogerm lines, whose production is based on continuous selfing and backcrossing respectively, resulting in very low within-line variability.

One of the schemes employed for population improvement using clones is shown in **Fig. 3.4**. The initial population should have a known pedigree, a sufficient variability for the trait to be improved, and a good agronomic value. It is possible to start with seeds randomly selected from the population, in a manner similar to the HS families. Each of 30 chosen seeds is used to produce cloned plants. In the second year, these cloned plants are crossed in isolation with the same CMS tester line and 30 new monogerm hybrids obtained. It is also necessary to harvest S1 seed from all families of clones. In the third year, the 30 hybrids for every line are tested in field trials and selected for superior sugar production. With a usual selection intensity (5-10%), the best genotypes are chosen. To further confirm the results, the tests could be repeated in the fourth year. However, to save time, it is preferable to use more test sites in the third year.

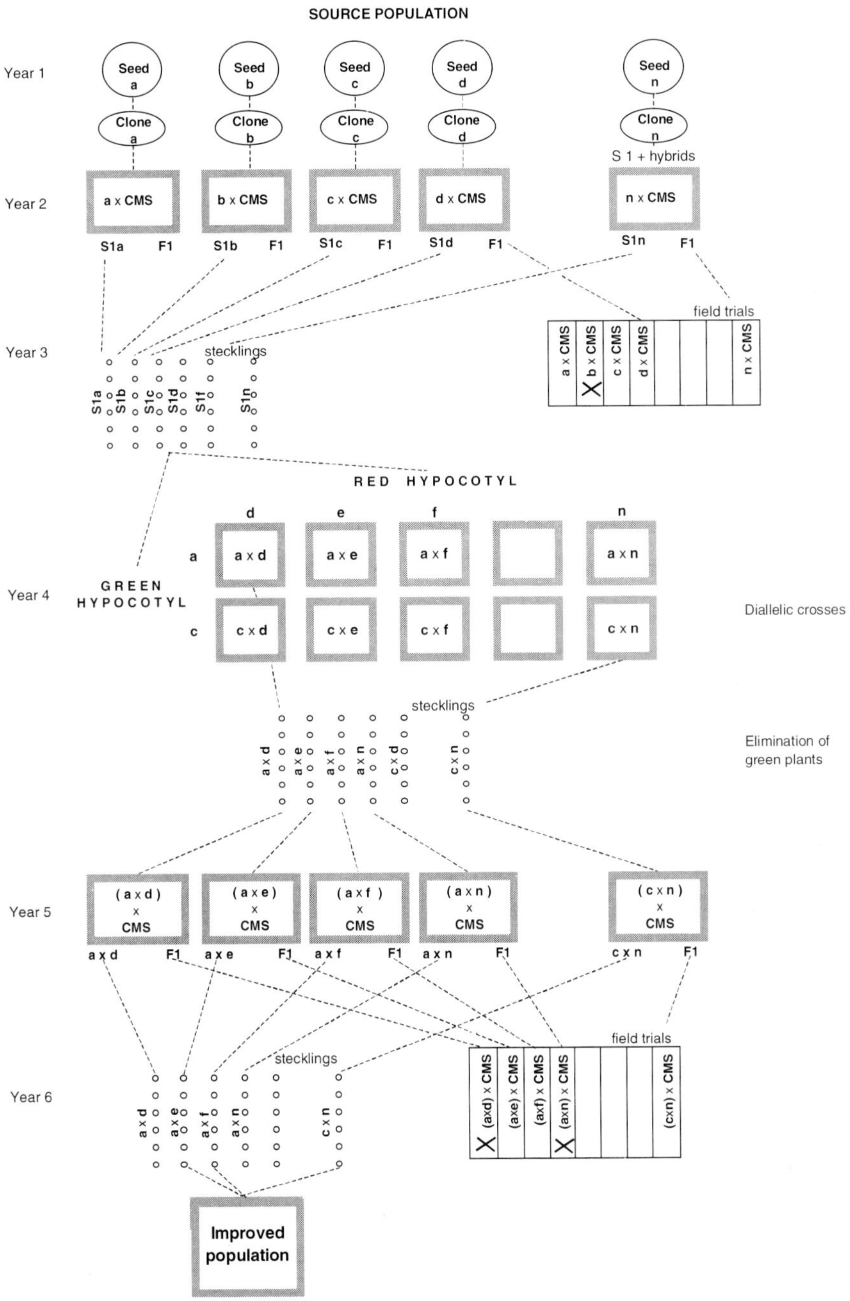

Fig. 3.4 Scheme of progeny selection using clones.

Direct use of the selected genotypes as pollinators of a new variety is not advisable. Indeed, it is necessary to carefully evaluate: (i) any decline in vigor of the pollinator line; (ii) the ability of the line to function effectively as a pollinator; and (iii) the sensitivity to certain climatic or disease conditions. It is more advantageous to effect diallel crosses between the superior clones in order to verify the level of incomplete heterosis obtainable in the F_1s (Smith et al., 1973). For organizing the partial diallel crosses, the clones are planted in a nursery in the third year and classified on the basis of their hypocotyl color. If S1 seed is used, individuals within each family are selected on hypocotyl color. Whether using plants raised as clones or from S1 seed, pairs of plants, one with a green hypocotyl, the other with red are established. Seed is harvested only on the plants with green hypocotyl. The F_1s are sown in the nursery in the fourth year and only plants with a red hypocotyl selected, as that confirms they are hybrid. These plants from each F_1 are crossed in the fifth year to one or more CMSs and the hybrids planted in year six in multilocation field trials. At the end of the year, the results of agronomic performance permit selection of the best combinations of pollinator lines to form the improved population or to be used to produce promising experimental hybrids. Cloned plants can be used in each step of the process instead of seed reproduction. Large quantities of seed of a proven superior pollinator can be obtained by crossing the cloned parental components.

3.1.3 Recurrent Selection

The most important trait for a sugar beet variety is high sugar production per unit area, i.e. the product of root yield and sucrose content. Also important are characteristics facilitating the factory process of sugar extraction, as well as genetic resistance to several biotic and abiotic factors. Mass and family selection methods, although initially producing favorable results, rapidly become of limited efficiency. Cyclical methods designed for a progressive accumulation of gene combinations with favorable additive and nonadditive effects are more likely to result in significant improvements in these traits.

Recurrent selection (RS) refers to a group of methods developed to accumulate superior allelic combinations, and thus to allow selection of lines with superior combining ability for use as parents of new varieties. The importance of assessing combining ability of the lines selected to produce hybrid varieties has long been known. It was observed in maize that the process of inbreeding and selection on the basis of inbred performance *per se* was of little use in the synthesis of superior hybrids. It is essential to test the combining ability by progeny assessment.

The method of RS used for maize has the following characteristics: (i) plants of a heterozygous variety are selfed and evaluated for one or more favorable characteristics; (ii) inferior plants are eliminated; (iii) superior plants are propagated through the selfed seed; (iv) all possible crosses between the progenies of the superior plants are effected; and (v) the population resulting from these crosses is used to initiate a new cycle (Allard 1960). Experience mainly gained in maize was subsequently transferred to other plant species.

Recurrent selection systems applicable to sugar beet breeding fall into four principal models (Bosemark, 1993):

— Simple recurrent selection (SRS), based solely on the phenotype or on evaluation of S1 progenies;

— Recurrent selection for general combining ability (RSGCA), with selection based on the value of test crosses to a heterozygous common tester;

— Recurrent selection for specific combining ability (RSSCA), where the tester lines, usually of a high homozygosity level, provide information on the specific combining ability of the families selected;

— Reciprocal recurrent selection (RRS), in which two populations are simultaneously improved, in the same manner as in GCA, one serving as tester to the other.

Discovery of the self-fertility (S^F) gene (Savitsky H., 1952) and of the gene (*a*) for nuclear or genetic male sterility (Owen, 1952), permitted development of a male-sterility facilitated recurrent selection (MSFRS) system, related directly to the four types of RS outlined earlier. Introduction of the male-sterility gene in monogerm self-fertile populations allows both selfing and hybridizing to be effected with no difficulty. Selfing a homozygous self-fertile, heterozygous male-fertile ($S^F S^F Aa$) line produces 25% male-sterile plants within its progeny that can be used for crossing purposes. In this manner, cyclical selection programs are greatly facilitated. An RS scheme was first presented by Gilmore (1964) and was subsequently improved by Doggett and Eberhart (1968) in sorghum.

A simplified approach for the development of monogerm O-type populations improved through MSFRS with characters transferred from diploid multigerm families, is shown in **Figure 3.5**. Male-sterile plants of the population are crossed in the first year to a series of multigerm, self-sterile, fertile families having both disease resistance and desirable agronomic traits. Rows of the population with sterile plants are alternated in the field with rows of the pollinator families so that no family dominates over the others, and genetic variation remains at the level desired. The resultant F_1

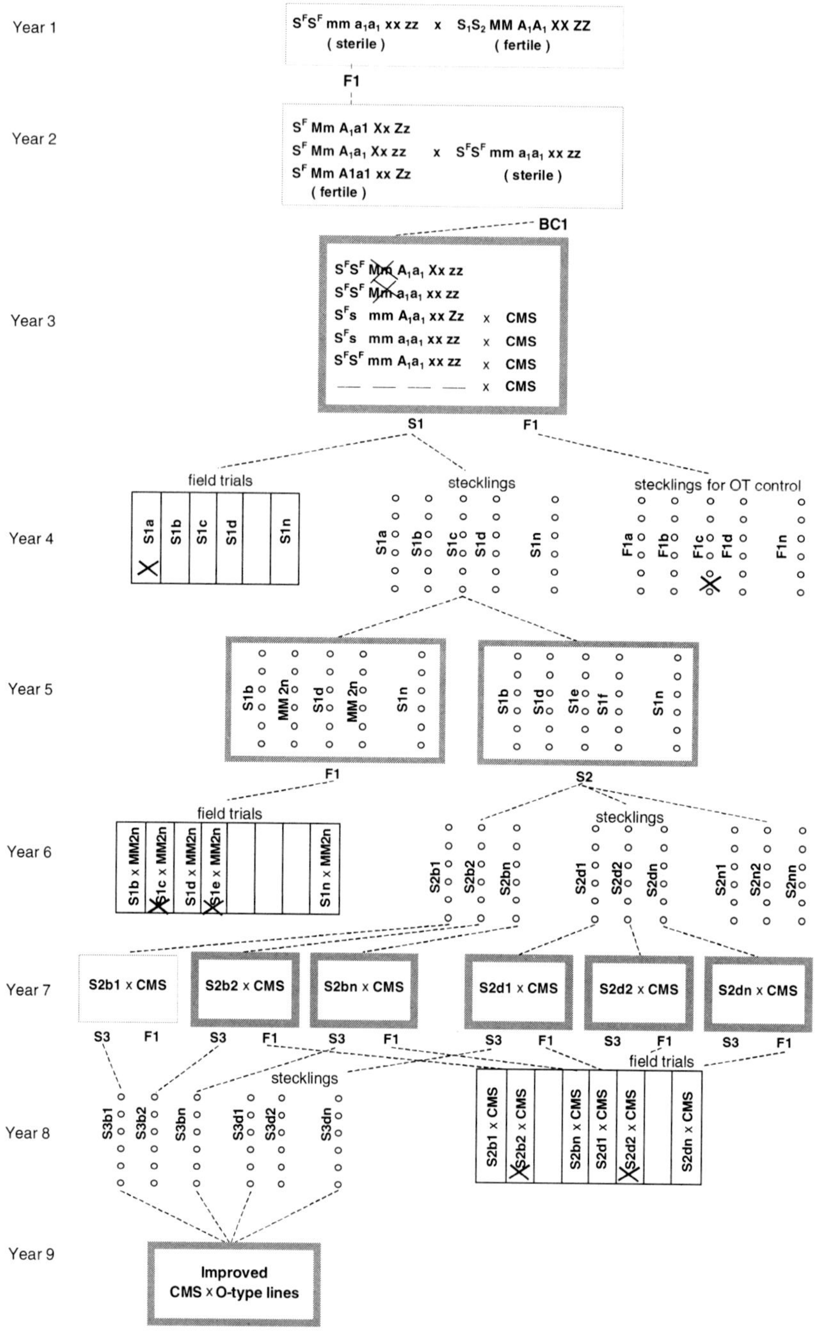

Fig. 3.5 Simplified scheme of recurrent selection.

hybrids are heterozygous self-fertile ($S^F s$), multigerm (*Mm*) and male-fertile (*Aa*) with various combinations at the loci of the two maintainer genes (*Xx* and *Zz*). To approach the genetic constitution of the recurrent parent, the F_1 plants are backcrossed, in the second year, to male-sterile plants of the original population. It should be noted that in the F_1 progenies, gene frequency for *x* and *z* might be higher than 50%, since recessive alleles are often present in the multigerm donor families. Stecklings of the new population, with only 25% of their germplasm derived from the initial multigerm families, are planted together with a monogerm CMS line. The field layout is such that each row of the population plants is flanked by rows of CMS plants. At the initiation of flowering, it is necessary to rogue the MSFRS population to leave only perfectly monogerm (*mm*) plants. Also, fertile (*Aa*) and sterile (*aa*) plants have to be individually tagged. Before the flowers open, a branch of every fertile plant is put in an isolation bag together with a branch from a nearby CMS plant. Selfed (S1) seed from every fertile plant as well as open pollinated seed (HS) from every sterile plant is harvested separately. F_1 hybrid seed, produced on the branch of CMS plant under isolation bags, is also harvested. Part of this hybrid seed is sown in the nursery.

Individual S1s with sufficient seed quantity are sown for agronomic trial in the fourth year. At the same time true O-type plants are determined by examining the sterility of the lines produced on the branches of CMS plants and sown in the nursery. Part of the remaining seed from S1 lines is sown in the nursery. Vernalized stecklings from the selected S1 lines, segregating in a 1*AA* : 2*Aa* : 1*aa* ratio, form in year 5 two isolation sites. In one of them, rows of the S1 lines are alternated with rows of a multigerm, fertile diploid family having a broad genetic base. Before flowering, fertile plants are removed, leaving only sterile S1 plants (about 25%) to be pollinated by the multigerm family. Upon seed harvesting, the progeny of these genetic male-sterile plants from each S1 line represent a series of hybrids with a heterozygous tester. In the second isolation site, where only S1 plants are participating, sterile plants are removed and S2 seed is harvested separately from at least, 10 plants of every S1 line. Since all plants are heterozygous for the dominant gene for self-fertility, no difficulties are encountered. In the sixth year, hybrids produced at the first isolation site are entered in agronomic trials to obtain the first indication of the potential GCA of the monogerm lines under development. Pending the results of the trials, all individual S2s harvested from the second isolation site are sown in the nursery. Based on the agronomic evaluation, S2s with superior GCA are selected.

A group of crosses for each of the selected S2 lines, is established under tent or space isolation in the next (seventh) year. Isolation is necessary because half of the S2 lines are segregating for the S^F gene, giving rise to some occasional presence of self-sterile plants.

In these crosses, the S2 lines function as O-type pollinators to an elite CMS inbred line, from which seed (F_1-CMS) is harvested. Concomitantly 5-10 new S3 lines are harvested separately on isolated branches of some randomly selected S2 plants and this seed used for further development following verification of the value for the 2-3 best original S2 lines. In the eighth year, the F_1-CMSs are compared in agronomic trials with the best single cross CMS used in commercial hybrid development. These trials serve to determine their value for bolting resistance, resistance to the most important diseases, seed-bearing performance, and any other useful trait. Based on such information, S3 lines of superior value are selected. In the next year, precommercial hybrid seed is produced using the best monogerm F_1-CMSs as seed bearers in crosses to already developed pollinators. In this way new monogerm hybrids are obtained to participate in variety registration trials.

The method was initially adapted to sugar beet breeding by Bosemark (1971) for the production of improved O-type lines. With the so called "Doggett" populations, several recurrent selection schemes of all the categories described above could be planned and applied for the improvement of both monogerm (O-type and respective CMS) and multigerm (pollinator) populations (Bosemark, 1971, 1993; Doney and Theurer, 1978; Hecker, 1978, 1985). MSFRS seems to have been the main selection scheme used by the European breeding firms in the 1980s and 90s. The method also served in the improvement of multigerm, diploid pollinator families. In this case though results were less pronounced because of the much higher effort dedicated to the development of tetraploid pollinators.

After years in use, recurrent selection has demonstrated limitations that cannot be underestimated. Apart from complexity in its application, the increasing number of traits to be improved necessitates construction of populations with very broad genetic variation. As is known, the combination of many desirable characters is a process of considerable difficulty.

3.1.4 Inbreeding and Heterosis

With the selection methods described, it is difficult to determine the frequency relations of homozygotes and heterozygotes in the progenies. This limitation can be overcome by the isolated intercrossing of a small number of genetically similar individuals by paired-plant crosses (sib-mating) or by selfing of single plants, the latter representing the highest level of inbreeding. Genotypes undergoing cycles of more of less intensive inbreeding are called "lines", while those reproduced by open pollination are generally called "families" (Allard, 1960).

Homozygosity of the lines under selection is the result of strict control of pollination. As the percentage of identical alleles at the various loci increases, expression of recessive alleles, including lethal and sublethal, also increases. The higher degree of inbreeding causes fixation of inbred lines within a few generations. There is a considerable differentiation between these lines and, in the absence of lethal genes, they reproduce in successive generations without substantial change in their genetic constitution.

Two hypotheses have been proposed to explain heterosis in crosses between inbred lines suffering from inbreeding depression. According to the dominance hypothesis, hybrid vigor is merely the result of accumulating dominant alleles at more loci. Maximum heterosis in this sense is obtained when at least one dominant allele is present at each locus. The overdominance hypothesis, on the other hand, assumes that there are loci at which a heterozygote is superior to either homozygote and heterosis is directly proportional to the amount of heterozygosity (Allard, 1960; Falconer, 1960). Heterotic response in sugar beet has been considered similar to that in maize (de Vilmorin, 1923). In a research conducted in 1932 (Stewart et al., 1940; Coons et al., 1955), 41 F_1 sugar beet hybrids between inbred lines were compared in field trials. All lines were multigerm and fertile; therefore the hybrid seed contained a variable percentage of selfed individuals. The agronomic evaluation indicated that the root yield of 31 of the 41 hybrids exceeded by 42.5% that of the parental lines. As in other cross-pollinated species, vigor is decreased when sugar beet undergoes inbreeding for a few generations. Obviously, this poses difficulties in the development of breeding programs. Upon selfing of individuals, a rapid increase of homozygous plants was observed, reaching about 90% at the third generation (Allard, 1960). The same percentage was reached after eight generations when sib-mating was followed. Inbreeding became progressively less efficient when other mating systems were employed. A similar situation holds true for tetraploid material. However, the approach to homozygosity is much slower; 15 generations of strict selfing are required to reach a 90% level of homozygosity.

Sugar beet is normally a self-sterile species so selfing presents several difficulties. It is possible however, to obtain small quantities of selfed seed on a few plants. According to a study of commercial material, 75% of the lines were lost due to their self-sterility (Oldemeyer et al., 1968). Better results were achieved when self-sterile plants were selfed at high altitudes or in locations with moderate temperatures during flowering (Oldemeyer and Bush, 1952). Pollination control for selfing or paired crossing was achieved by placing the plants under unique metallic frames covered before flowering with appropriate types of cloth (cotton or other) to prevent entry of undesired pollen. Production of inbred lines, as described above, was greatly facilitated by introducing the S^F gene for obligate self-fertility (Owen, 1942). The gene was used for the development of monogerm O-type lines derived from line SLC 101 (see Section 2.3). Examples of practical applications of inbreeding have been described by Coons et al., (1955).

A scheme employed to increase homozygosity in a multigerm diploid pollinator population, along with evaluation of combining ability, is illustrated in **Fig. 3.6**. In this case the trait for resistance to rhizomania is considered. The process is initiated by sowing in a homogeneously rhizomania-infected field, mother beets from a self-sterile family of good agronomic characteristics combined with a sufficient allelic frequency for the gene complex controlling rhizomania resistance.

The aim is to obtain multigerm lines with a high GCA accompanied by an elevated probability of the trait being in the homozygous dominant condition (*RzRz*), thereby minimizing the risk of transmission of recessive alleles in successive generations.

At the end of the growing period, only mother beets with minor disease symptoms are selected. Their number has to be sufficiently large to preclude reducing the genetic variability of the lines under selection. Every root, selected on the basis of criteria already described, is cut into four parts; this allows increased seed production upon selfing in the second year. The following year, all S1 lines obtained are tested under rhizomania infected and, if possible, disease free conditions. These tests enable elimination of lines with susceptibility to rhizomania, as well as lines with undesirable agronomic characteristics. Each of the selected S1 lines is crossed in the fourth year to a common monogerm, usually F1-CMS, female parent characterized by a desirable agronomic value coupled with rhizomania susceptibility. In this way the level of resistance of the heterozygous (*Rzrz*) hybrids can be determined. Some branches from plants of each S1 line are also used to produce S2 seed under small isolation bags. In the fifth year, the hybrids harvested on the CMS are tested both on rhizomania infected and rhizomania free fields and the S1 and derived S2 lines of poor productivity discarded.

The selected S2s are sown in the nursery and used to produce S3 lines the following year. These S3 lines are sown not only in the nursery, but also in the greenhouse to evaluate their level of resistance by the using ELISA techniques. Only lines manifesting complete resistance to the virus are selected and stecklings from the nursery subsequently crossed to CMS

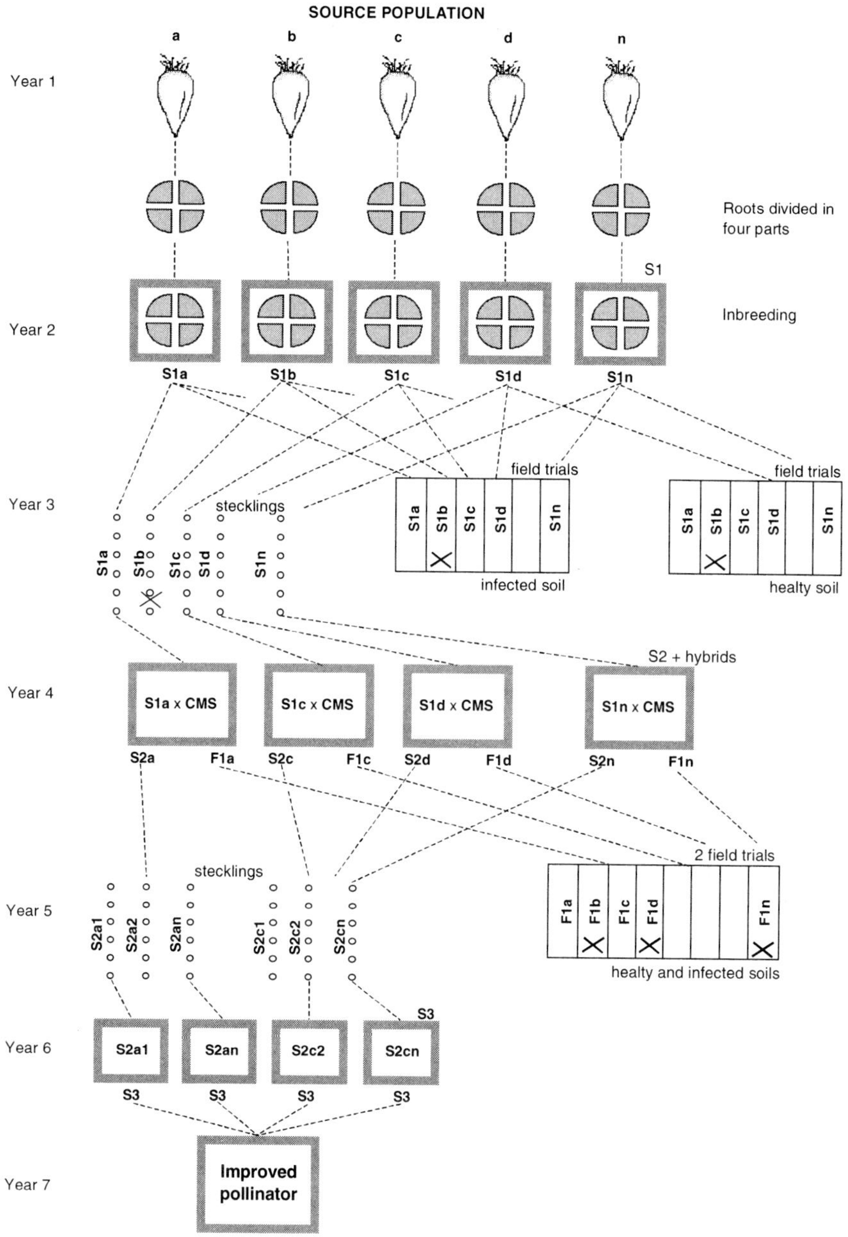

Fig. 3.6 Inbreeding with selection.

lines for the production of precommercial hybrids. Isolating some branches, as already mentioned, produces a new generation of S4 lines that can be used to increase gene fixation in the inbred lines if necessary. As homozygosity is approached however, multigerm sugar beet lines manifest considerable decline in vigor. Such an inbreeding effect is less pronounced in monogerm lines endowed with the self-fertility (S^F) gene.

3.1.5 Interspecific Hybridization

Hybridization with various species of the genus *Beta* has played an important role in sugar beet breeding. Indeed, interspecific crosses have enabled identification of several useful sources of resistance to diseases damaging the sugar beet crop. The first problem faced in these crosses was obtaining viable F_1 plants due to genetic incompatibilities, unsuccessful pollination or zygotic abortion which can occur at any phase following the first cell division (Allard, 1960). Beets of the *Beta* (syn. *Vulgares*) section have been successfully crossed only to species of the *Corollinae* and *Procumbentes* sections (Lange et al., 1999). Attempts with *Beta nana* failed (McFarlane, 1971). Obviously, some minor difficulty was encountered in crossing the species and subspecies of section *Beta.*

The sea beet (*Beta vulgaris* ssp. *maritima*) is the most important wild beet and the easiest to cross with cultivated beets, albeit considered their ancestor (Hjerdin *et al.*, 1994). The sea beet grows along the Mediterranean and North Atlantic coasts, from the British Isles to the Canary Islands (Doney *et al.,* 1990). According to Ulbrich (1934), the species is also commonly found in the belt that sweeps down from the Balkans, through Greece and Turkey as far as the Persian Gulf. The plant is also relatively common along the coasts of the Adriatic Sea. In such areas it can grow for 6-7 years and is therefore classified as a perennial, which flowers in its first year (Munerati *et al.,* 1913).

Resistances to cercospora leaf spot and rhizomania are the most important results obtained in transferring genes from different biotypes of sea beet to sugar beet. Crosses between sugar beet and sea beet are easy given the above-mentioned affinity between the subspecies. Hybridization experiments for transferring useful traits to cultivated varieties began toward the end of the nineteenth century in different countries, but only works carried out in Italy by Munerati have given significant results (Munerati *et al.,* 1913). A cycle of mass selections of *B.v.* ssp. *maritima* for resistance to cercospora leaf spot began on wild plants sown in cultivated soil, followed by inbreeding, with the main objective fixation of bienniality. Using the first predominantly biennial lines, he began a program of crosses of the selected sea beet with conventional sugar beet. This was followed by a number of backcrosses to eliminate the negative traits of the wild parent (fanged and fibrous root, tendency to bolting, etc.). Further selections improved the bolting resistance and after ten years led to the release of the line R581, considered the first substantial progress against the disease (Coons et al., 1955).

The family was distributed to public and private breeding stations. The American variety US201 is cited as one of the oldest, together with the Italian "Cesena R" and "Mezzano 71", Polish "Buszczynski CLR", French "Desprez RC2", and Dutch "Vanderhaven AC" (Bongiovanni et al., 1958). The increased effort of the breeding companies has produced an improvement in sugar yield and bolting resistance, whose poor showing a few years before were the main negative traits of the CLS resistant varieties. With recent breeding progress, sugar yield is today at levels similar to that of susceptible varieties.

The first assessments of commercial varieties in rhizomania diseased fields began in 1958 (Bongiovanni and Lanzoni, 1964), i.e. before the discovery of the disease's real causes (Canova, 1966). The results of field tests together with those obtained from 1966 onward from several seed companies (Gentili and Poggi, 1986) showed clearly that the "Alba P" multigerm diploid variety and several others of Italian origin were more productive in rhizomania-infected soils (Biancardi et al., 2002). The varieties in question also possessed good CLS resistance, so they must have been derived from crosses with the sea beet performed by Munerati (Skaracis and Biancardi, 2000). The Italian sea beet biotypes, from which CLS resistance was obtained, probably also provided the genes inducing resistance to rhizomania carried by variety "Alba P" (Lewellen and Biancardi, 1990; Biancardi et al., 2002). It has been ascertained that resistance of the "Alba type" is governed by different genes with additive effect (Lewellen and Biancardi, 1990). Other authors have confirmed the presence of rhizomania resistant genes in sea beet biotypes harvested in many countries (Whitney, 1989; Lewellen et al., 1987). From 1980 to 1985 variety 'Rizor' was bred at the SES-Italy breeding station. It contains a monogenic dominant gene for qualitative rhizomania resistance. It was much more productive than the resistant varieties cultivated at the time and derived from Alba genotypes (De Biaggi, 1987).

At the American company Holly Sugar, Erichsen discovered two monogerm lines, one O-type and its CMS, both of which had excellent rhizomania resistance. Analysis of the progenies revealed a dominant monogenic character in this case too (Lewellen et al., 1987). In 1986 lines with "Holly type" resistance were purchased by European seed producers and integrated into their genetic programs.

In the 1990s, many breeders used analyses based on recombinant DNA in order to locate resistance genes on the chromosomes. The factors

"Rizor type" and "Holly type" map in the same locus and should be considered components of the same gene (Barzen et al., 1997). The trait has been named *Rz* by Lewellen et al., (1987). Scholten et al., (1999) mapped another gene *Rz2*, derived from a biotype of sea beet, identified with the code WB42. The gene or genes that provide resistance to rhizomania (*Rz1*, *Rz2*) map in the same chromosomic fragment that was probably derived from sea beet (Barzen et al., 1997). Biancardi et al., (2002) confirm the hypothesis of the common origin both for the quantitative resistance of "Alba type" and the qualitative resistance of "Rizor type" and "Holly type" from the Munerati source.

In section *Corollinae*, the species display a variable chromosomic number, 2n = 18, 2n = 36, 2n = 54 (Scheibe, 1934; Walther, 1961, 1963). Attempts to cross these with sugar beet have so far produced poor practical results (Seitz, 1935, 1936, 1938; Stehlik, 1947). Some cases have been observed in which hybrids obtained using the *Beta lomatogona* have conferred some resistances to cold, drought, and soil salinity (Scheibe, 1934). Experiments with *Beta macrorhiza*, which Oldemeyer referred to in 1964, did not overcome the initial difficulties. The only successful hybridization involving species of the section *Corollinae* was carried out by Helen Savitsky in 1969, when she crossed sugar beet tetraploid genotypes with *Beta corolliflora,* that was also tetraploid and resistant to the beet curly top virus (BCTV). The experiences produced no practical developments, but the methodology used in the experiments proved valuable for later works (Coons, 1975).

Section *Procumbentes* consists of diploid species (2n = 18), such as *Beta procumbens* and *Beta webbiana*, while *Beta patellaris* is tetraploid (2n = 36). The species displays traits that are potentially useful for breeding, including resistance to cercospora leaf spot and curly top. Another interesting trait of the mentioned species is its resistance to cyst nematodes (*Heterodera schactii*). Many breeders have obtained hybrids using species of section *Procumbentes* and those of section *Beta* (Oldemeyer, 1954; Johnson and Wheatley, 1961). Thanks to previous works (Savitsky, 1960), in 1965 Savitsky and Price obtained a significant quantity of hybrids diploid, triploid, and tetraploid between species of section *Beta*, by using the painstaking technique of "bridging hosts". In summary, the lack of vitality found in interspecific crosses with section *Procumbentes* could be resolved by crosses between sea beet and *Beta procumbens*, which can produce viable F_1 hybrids. These hybrids can then be used in subsequent crosses with cultivated beet (Oldemeyer and Brewbacker, 1956).

Further experiments by Helen Savitsky in 1973 and 1975 transferred an alien chromosome carrying a gene for resistance to cyst nematode from

Beta procumbens to the sugar beet. Through repeated crossing, monosomic addition lines were obtained that transmitted the resistance to their progenies with a frequency of about 10%. This material was used by Heijbroek et al., (1988) to obtain a homozygotic diploid line which contained a fragment of alien chromosome carrying resistance to the cyst nematode. If the diploid material were used as the pollinator to produce resistant varieties, it very easily lost the genes for resistance. Therefore, every individual plant has to be tested using recombinant DNA techniques to ensure the gene is present. For the moment, this represents a significant obstacle in producing commercial resistant varieties. A second source of resistance to cyst nematode was found in a sea beet biotype collected in France (Hijner, 1951). The trait is recessive and multigenic (Heijbroek, 1977) and was used for the same purpose by Lange and De Bock (1994). As noted above, crosses between *Beta webbiana* and Swiss chard (*Beta vulgaris* ssp. *vulgaris* Garden Beet Group) resulted in sterile progeny (Savitsky and Gaskill, 1957).

3.1.6 Mutagenesis

Mutation, or rather the sudden appearance of new and heritable traits (Gaul, 1962), represents an important factor in the natural evolution of a species (Stadler, 1930; Elliott, 1961). The techniques used to increase the frequency of natural mutations have had various applications in plant breeding (Stubbe, 1966; Broertjes and van Harten, 1978). Mutagenesis has proved particularly effective in the production of genetic variability and in its recombination (Elliott, 1961). Use of ionizing and nonionizing radiation has led to several improvements, especially in weeds (*Gramineae*) and in crops reproduced vegetatively (Elliott, 1961; Broertjes and van Harten, 1978). Mutagenesis has not produced significant results in sugar beet, with the single exception of the synthesis of tetraploid genotypes obtained using chemical agents. According to Smith (1987), the lack of applications of mutagenesis in sugar beet is partially due to the wide availability of genes and traits present in the other species of genus *Beta*.

Raudsepp (1976) obtained beet with rounded roots after treatment with mutagenic chemicals. In 1980, Kinoshita took seeds of lines with normal cytoplasm and exposed them to gamma rays. He found a high frequency of CMS plants in the offspring. Forms of monogermity and pollen sterility were isolated by Golev (1975) after irradiating multigerm seeds. However, the modified trait was not transmitted to the progeny.

3.2 Production of Commercial Varieties (G.N. Skaracis and M. De Biaggi)

3.2.1 Introduction

All sugar beet varieties in cultivation prior to the 1950s were diploid, multigerm populations with considerable genetic variability. These varieties were improved initially through mass selection and basic seed was obtained through open pollination of selected mother roots. Subsequent improvements were achieved employing progeny selection methods. The number of families contributing to the production of basic seed progressively decreased, so that only high yielding plants with good GCA were included in the final population. Colchicine-mediated autotetraploidy permitted development of anisoploid varieties. Anisoploid varieties are produced by allowing a mixture of diploid and tetraploid parental populations to freely intermate (Schwanitz, 1938). Multigerm anisoploid varieties were quite popular in Europe until the early 1960s (Bosemark, 1993).

Lines selected in the USA, endowed with genes conditioning monogermity, self-fertility, and cytoplasmic male sterility, provided seed companies the basic germplasm for development of diploid or triploid, genetically monogerm sugar beet hybrid varieties. The resultant varieties were actually three-way top-cross hybrids, in which the diploid or tetraploid pollinator retained a broad genetic base, similar to that found in multigerm open-pollinated varieties. For more than a decade, hybrid varieties were produced on monogerm F_1-CMS seed parents while the pollinators were developed through mass selection and progeny testing. Further progress was achieved by exploiting the then recently discovered monogenic resistances to rhizomania and development of *in vitro* techniques for micropropagation. Beginning in the 1990s, diploid S1, S2, and S3 lines selected for the *Rz* allele conferring rhizomania resistance were used as parental components.

Development of triploid varieties frequently employed cloning of individual plants from tetraploid families or obtaining F_1s from pairs of such clones. The first seed increases were then directly used as pollinators of new triploid hybrids. The production of monogerm hybrid varieties in which the seed parents were F_1s, between CMSs and unrelated O-types with high combining ability, resulted in considerable improvements in agronomic performance. In addition, the breeders were interested in improving the F1-CMS's to increase the efficiency of commercial seed multiplication. In fact, the need to reduce seed production costs became as important as the need to improve traits related to field emergence and crop establishment.

Another breeding objective focused on the production of disease-resistant hybrids through the development of O-type and CMS lines carrying resistance genes. Normally, these monogerm F1-CMSs manifested low seed yield and quality characteristics when used as commercial seed parents. This problem has not been completely solved.

The majority of the varieties currently grown are monogerm diploid hybrids with single or multiple disease resistance. Diploid and triploid varieties without disease resistance are still grown in northern Europe, where rhizomania and cercospora leaf spot were not a problem until recently. Triploid varieties, resistant to both diseases, are sown in southeastern Europe, while diploid varieties with various disease resistances (ie. to cercospora, rhizomania, rhizoctonia, curly top etc.) are available in North America.

3.2.2 Open-pollinated, Synthetic, and Anisoploid Varieties

Flower morphology and self-sterility of sugar beet have long dictated the production of varieties without pollination control. Selected plants were freely intercrossed, producing open-pollinated (OP) populations. The seed on each plant was a bulk of hybrids resulting from random pollination of this plant by its more or less nearby neighbors. Pollination control in normal populations, i.e. without the use of a male-sterility or self-fertility mechanism, is only possible by selfing single plants in isolation chambers or under bags. Another method of controlling pollination in self-sterile genotypes is to place two plants in an isolation chamber: in this way, small amounts of hybrid seed can be produced. A variety is open pollinated when obtained through random pollination among progenies produced directly from selected mother roots.

Progress in progeny selection and line breeding techniques led to identification of families with good GCA and their use as parents in commercial seed production. Varieties whose seed was always obtained anew by intercrossing a few families were designated "first generation synthetics" (Bosemark, 1993). The main concern in the synthesis of multigerm varieties was the number and characteristics of the families contributing to the final seed production, not the method of random pollination. An appropriate array of different combinations permitted the seed companies to blend multiplied elite seed to obtain varieties with the specific yield types explicitly indicated on their packaging (Coons, 1936).

Although breeding methodologies employed by seed companies differed, production of open-pollinated and synthetic varieties followed certain general schemes. In open-pollinated multigerm varieties (**Fig. 3.7**), the process is initiated by selection of mother roots at the end of the campaign for desirable morphological, yield and disease resistance traits. The selfed seed produced on each plant in isolation is multiplied separately the following year and the progenies evaluated in agronomic trials. Meanwhile, plots of mother roots from the same genotypes are grown for individual selection within the best families and elite seed produced in the

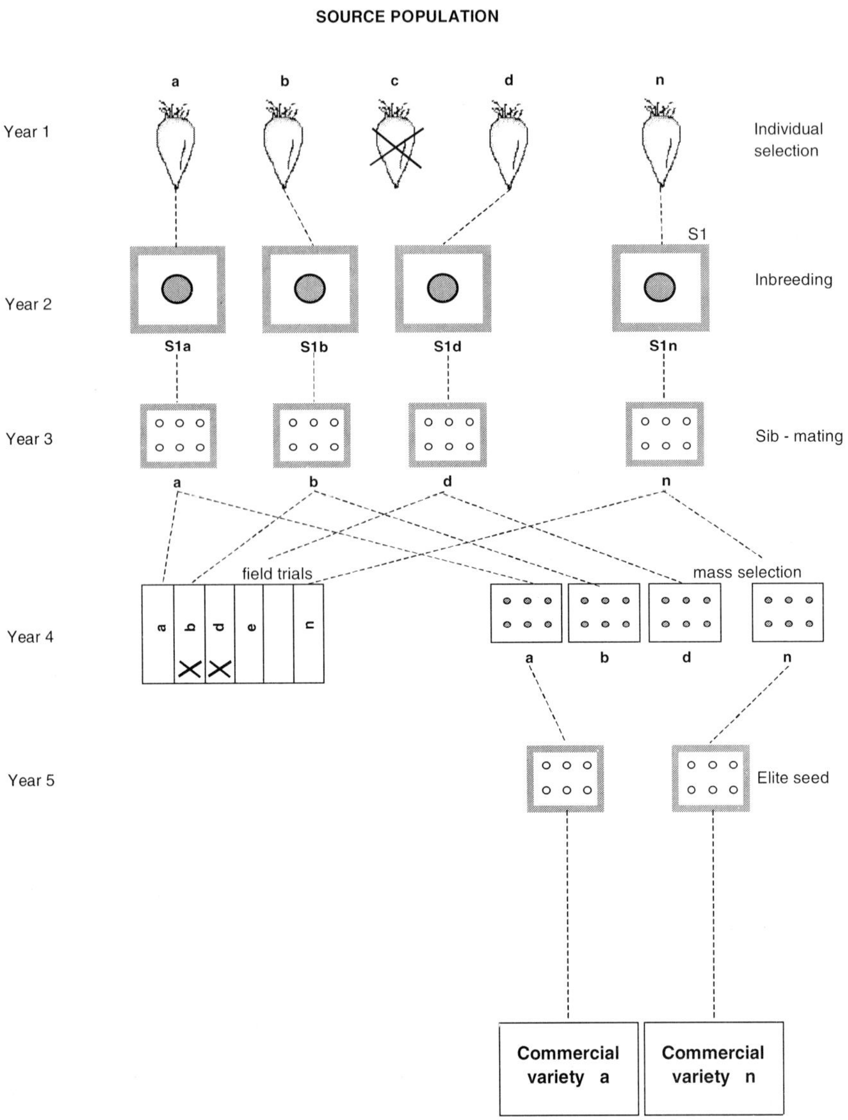

Fig. 3.7 Production of open-pollinated commercial varieties.

next year, using the selected mother beets. The different lots can undergo an additional field evaluation cycle before the best ones are mixed and sown in a nursery for commercial seed production.

According to Coons (1936), schemes to develop synthetic sugar beet varieties usually follow the same path until the fourth years field trials

(Fig. 3.8). A selected family is then both singly reproduced and crossed to other selected families for combining ability assessment in the next years, usually multilocation trials. Based on evaluations, the best blends of F_1 hybrids to form the synthetic are chosen and the variety produced using the appropriate mixture of reserved parental seed. In this way, the loss of heterosis usually observed in F_2 is precluded (Bosemark, 1993). Besides single-cross evaluations, top crosses or polycrosses can be used to test the GCA of the families selected.

Availability of tetraploid families has enabled production of anisoploid (sometimes improperly called polyploid) varieties, consisting of a mixture of diploid, triploid, and tetraploid plants, obtained by free intermating of tetraploid and diploid families. Flower development in tetraploid plants is generally slower than in diploid plants and this may lead to a lack of synchronization with consequential negative effects on pollination and seed ripening. Furthermore, diploid pollen is less abundant and not as effective as haploid pollen. Experience has led to the practice of mixing and transplanting stecklings of the tetraploid and diploid parental components in a 3 to 1 ratio. This approach has enabled production of commercial seed with a high percentage of triploids (around 50-60% on average), which are more productive in terms of sugar yield, the reason anisoploid varieties were introduced in the first place. Also, the amount and quality of the seed proved satisfactory. However, under specific circumstances and for some parental components, the optimal ratio may not be 3 to 1.

3.2.3 Monogerm Hybrid Varieties

The discovery of cytoplasmic male sterility in sugar beet (Owen, 1945) made truly hybrid seed a possibility and allowed breeders to capitalize on heterosis, similar to what had already been accomplished in maize. Genetic monogermity, developed only a few years later (Savitsky, 1950), provided a solution to serious concerns about labor costs for thinning. With the introgression of the monogerm character into cytoplasmically male-sterile (CMS) lines and their corresponding maintainer (O-type) lines, genetically monogerm hybrid varieties became a reality. To this end, selection work accomplished in multigerm material (synthetic and OP populations) was exploited in the development of pollinators for commercial hybrids. On the other hand, new breeding strategies and activities had to be implemented for the development and improvement of genotypes destined as components in the female lineage of these varieties.

3.2.3.1 Development of CMS seed parents

Synthesis of monogerm hybrid varieties requires the use of elite cytoplasmically male-sterile seed parents, whose development and improvement is of utmost importance in a sugar beet breeding program.

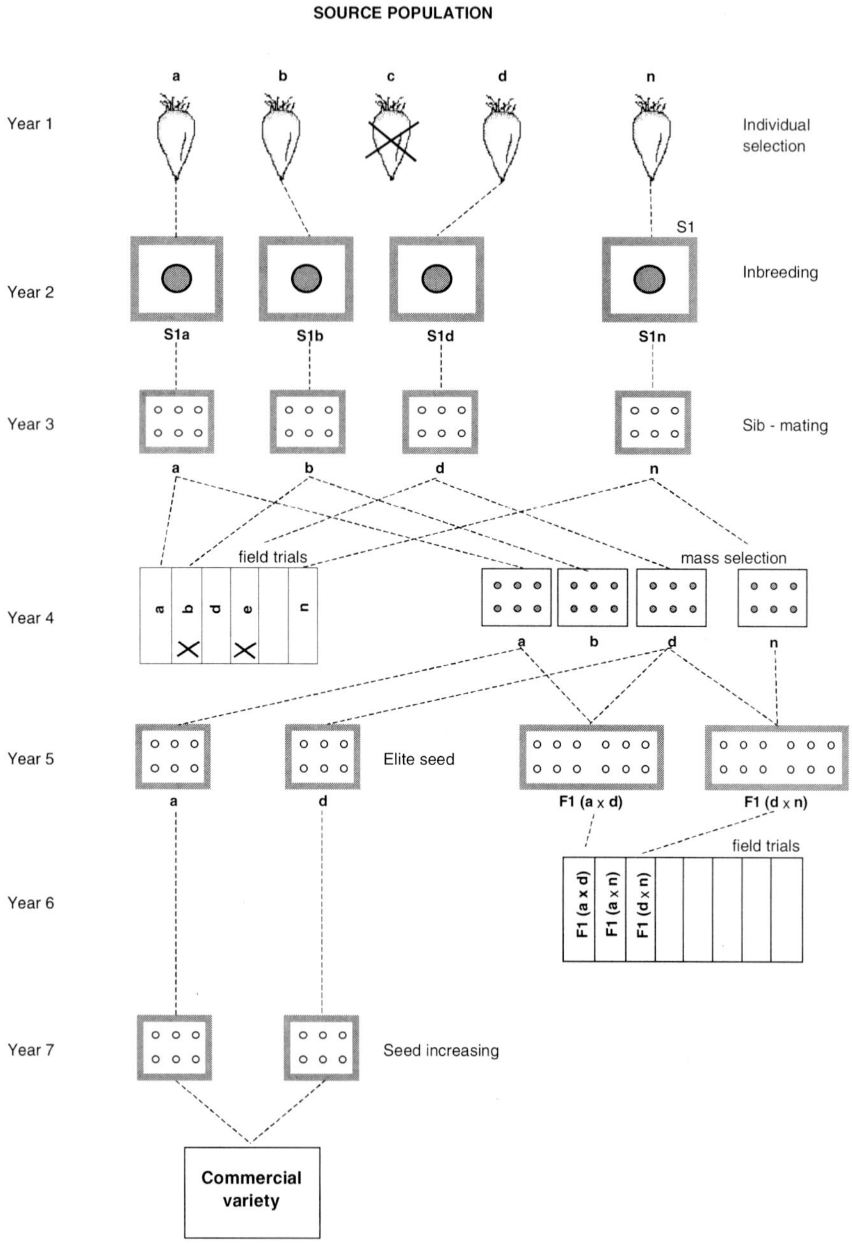

Fig. 3.8 Production of synthetic commercial varieties.

CMS lines are reproduced only when pollinated by their corresponding maintainer genotypes, thus protecting breeder rights to their commercial varieties. The operational sequence of relevant breeding activities usually begins with development of inbred maintainer lines and then the production of their male-sterile equivalents by repeated backcrossing to plants with the sterile cytoplasm. A concurrent operation of both stages is quite frequently practiced also.

Plants endowed with the O-type character occur at a very low frequency in sugar beet populations. It is this feature that, along with the need for monogermity and good combining ability, renders the entire task particularly laborious, time consuming and hence costly. Employing the O-type character and monogermity trait of line SLC 101, European and American seed companies were immediately involved in an intensive process of adapting O-type monogerm germplasm to specific needs. This effort in general has been successful, although some undesirable traits in the initial material proved quite difficult to remove. It should be mentioned in this respect that, apart from monogermity derived from the aforesaid source, it is relatively easy to find plants of some multigerm cercospora-resistant germplasm with bigerm or monogerm flowers (Biancardi, unpubl.). Normally, an adequate level of monogermity for further use is achieved with modest inbreeding of these individuals.

Identification of maintainer genotypes, a process referred to as O-type indexing, commences with strictly controlled crossing of a candidate monogerm individual to a CMS plant and the seed is separately harvested in order to obtain S1 reserve seed (on the pollinator plant) and test-cross seed (on the CMS plant) for male sterility assessment. If all flowering plants produced from the latter are sterile, the original CMS plant was pollinated by a true O-type individual, which harbored both pertinent nuclear genes in the required homozygous recessive condition. In cases in which not all, but a high proportion of the test-cross progenies are sterile, further evaluation of S1 plants with the same procedure will frequently produce useful O-types as well. To expedite the procedure, annual CMS plants are frequently used for the initial cross for O-type indexing. Following the identification of maintainer genotypes, 5-6 cycles of backcrossing to each cycle's CMS progenies are considered adequate for the production of a reasonably equivalent male-sterile line (**Fig. 3.9**). Throughout the backcrossing process, for obvious reasons taking place under isolation, flowering characteristics of the O-type and CMS progenies should be noted.

Self-sterile sugar beet genotypes undergoing indexing are difficult to maintain (Smith, 1987) unless vegetative propagation is employed. Given that such techniques (i.e. clonal propagation) were not always available,

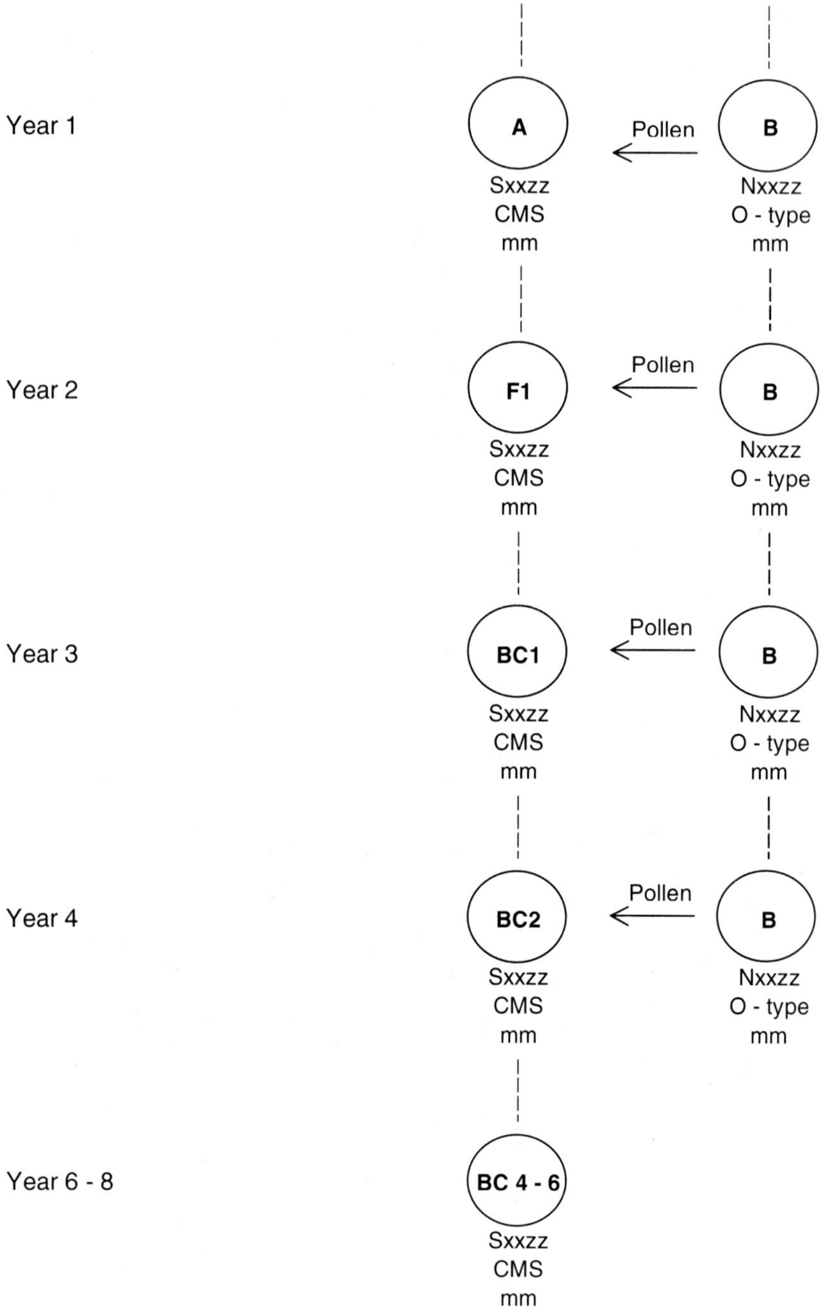

Fig. 3.9 Backcrossing of CMS lines.

self-sterile O-types have been used to a limited extent, primarily in the USA. Most of the breeding work has focused on self-fertile germplasm, carrying the S^F gene (Savitsky, 1954) thereby ensuring production of selfed seed and allowing for selection, development, and maintenance of inbred O-type lines. The disadvantages of self-fertile germplasm relate to the inability or extreme difficulty in making crosses for improvement purposes. However, this drawback may be overcome by the introduction of nuclear male sterility into self-fertile populations (Bosemark, 1971).

Prospective maintainer genotypes have to be evaluated for their combining ability. Once detected, O-types can be propagated by intermating S1 plants and the seed increase used for *per se* evaluation and crossing to elite CMS lines. Genotypes that perform well can immediately enter the mainstream breeding program, while having their male-sterile equivalents developed and used for the production of F_1 CMS female parents of commercial hybrids. With self-fertile germplasm however, a common practice is to complete 2 to 3 cycles of inbreeding with simultaneous selection for desirable seed traits and then determine the O-type combining ability. This can be efficiently estimated by crossing to an elite male-sterile tester for evaluation of GCA and then using a partial diallel crossing scheme with other O-types and their male-sterile equivalents for specific combining ability (SCA) evaluations (Bosemark, 1993). It is of interest to note that a number of male-sterile "synthetics" (advanced generations of a composite cross among the three parental components) have also been developed and exploited for various purposes, for example to satisfy specific disease resistance needs (Smith and Ruppel, 1979).

Whatever precedes the selection of male steriles as candidate female parents of commercial hybrids, their final use depends their performance in extensive multilocation and multiyear field comparison of their hybrids with a series of pollinators and current varieties.

3.2.3.2 Development of pollinators

Development and improvement of pollinators for the production of hybrid varieties is considerably simpler since selection takes place in multigerm fertile populations and there is no need for introgression of the monogerm or CMS trait. On the other hand, because seed companies usually rely on a small number of elite male-sterile lines to produce an array of varieties, pollinators have to be developed and replaced at a much faster rate. Therefore, breeders prefer to begin selection in populations endowed with the greatest possible genetic variation so that several unique pollinators may be obtained and hence a number of hybrids with the same male-sterile female are feasible.

Diploid pollinators are improved using the breeding schemes described above. Selection progress at the tetraploid level is significantly slower and more difficult due to the tetrasomic inheritance involved. It is therefore preferable to produce new tetraploid pollinators by doubling the chromosomes of diploid genotypes already selected for the desired traits. However, in many cases selection is initiated and practiced in tetraploid populations. The choice of a particular breeding method depends, among other factors, on the level of self-sterility of the materials to be improved. For example, if the frequency of self-sterility is too high, introducing the S^F gene may be beneficial. Breeding pollinators directly at the tetraploid level may be of interest when desiring to improve disease resistance controlled by genes with significant additive effects (such as resistance to cercospora leaf spot).

Recent commercial hybrids are produced on CMS lines enriched with resistance characters. Pollinators of these hybrids usually originate from broad genetic base populations such as multigerm synthetics or mixtures of lines with different degrees of inbreeding. Their selection is based primarily on combining ability in testcrosses to elite male-sterile females or to the intended commercial hybrid CMS parent itself. The resultant substantial genetic variability assures stable performance of the hybrids in diverse environments (Bosemark, 1993). This approach also allows for easier and faster modification of the hybrid's features in subsequent years, by altering the composition of the lines involved in synthesizing the pollinator. Thus, grower requests for minor modifications in the characteristics of a variety can be met efficiently and at a lower cost. Finally, some diploid rhizomania-resistant varieties have recently been produced using pollinators comprising a single inbred line or an F_1 between two inbred lines.

3.2.3.3 Synthesis of hybrid varieties

Through the years synthesis of hybrid sugar beet varieties has depended primarily on the use intended in a market with ever-changing demands, continuous progress in breeding methodology and related technology, time availability, and genetic resources. As in many other cross-pollinated crops, several types of hybrids have been produced and exploited, with varying degrees of success. Sugar beet hybrid varieties can be classified according to their ploidy level, the number and genetic base of constituent lines and the crossing modes involved. Breeders in the USA, where polyploidy was never attractive, concentrated almost exclusively on the development of diploid hybrids. Subsequently success with anisoploid varieties in Europe stimulated interest in the development of triploid hybrids and triploids were predominant until almost a decade ago. Since then, especially given the need for rhizomania-resistant varieties and incorporation of molecular techniques into mainstream breeding, seed companies have increasingly delivered diploid hybrids. With regard to the number of parental components involved, the overwhelming majority of sugar beet varieties have, for a long time, been three-way hybrids.

Three-way hybrids

Due to inbreeding depression, male-sterile lines are usually unsuitable as final seed parents. To overcome the seed yield and quality shortcomings of inbred male steriles, F_1s between a CMS line and an unrelated O-type line are commonly used as female parents for commercial hybrids **(Fig. 3.10).** Obviously, even a low frequency of fertile or bigerm plants cannot be tolerated in the F_1 seed parents, as they would cause problems in production and excessive losses in seed processing. Another item of paramount importance in producing a high yield of quality seed is the need for simultaneous flowering of the hybrids parents and this should be considered when determining which CMSs and O-types should be used in combination. As already mentioned, selection of these combinations is usually based on both a series of diallel crosses among the lines and on subsequent crosses of the better F_1s to a number of candidate multigerm pollinators. The best choice of parents to produce a three-way top-cross hybrid, $(A \times B) \times C$, can be predicted by evaluating the nonparental $(A \times C)$ and $(B \times C)$ single crosses (Skaracis and Smith, 1984). Pollinators of commercial three-way crosses usually possess a broad genetic base (open-pollinated or synthetic populations) and the resultant top-cross hybrid varieties have been the most common sugar beet varieties grown both in Europe and the USA for a long time.

Widely available adapted multigerm tetraploid populations used in the synthesis of anisoploid varieties were immediately engaged as pollinators in the development of triploid monogerm hybrids in Europe. Despite the higher cost of producing quality hybrid triploid seed, triploid hybrids enjoyed undisputed commercial success, in part because of better environmental stability, compared to diploid hybrids. In addition, bolting and cercospora resistance were enhanced by the higher ploidy (Nakamura et al., 1992). These, and other advantages have been attributed, at least in part, to increased heterozygosity and the possibility of accumulating more favorable genes controlling yield in polyploids (Bosemark, 1993). Breeders in the USA, preoccupied with improving disease resistance at the diploid level, were reluctant to develop triploid varieties and virtually all of their commercial hybrid varieties have been diploid. It was not until the 1990s that triploid hybrids, coming from Europe, gained widespread acceptance in northern sugar beet production areas, Minnesota and North Dakota.

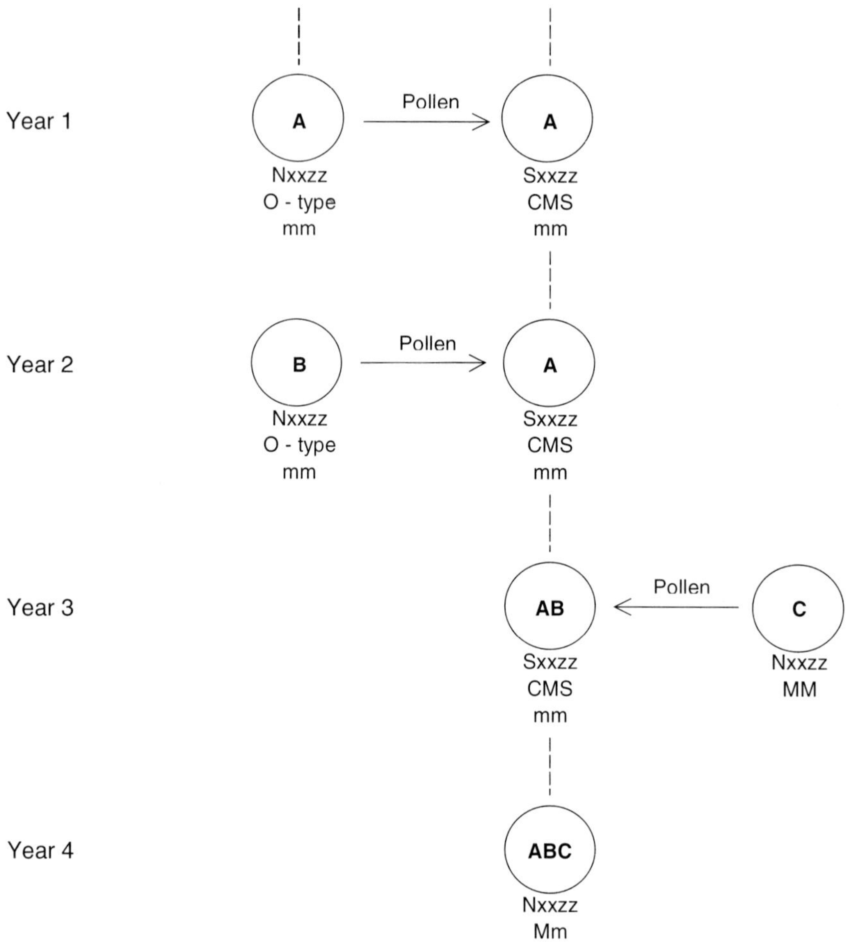

Fig. 3.10 Production of three-way hybrid varieties.

Triploid hybrids can be produced if one or the other parent is a tetraploid. However, due to difficulty in obtaining acceptable quality hybrid seed on tetraploid monogerm male steriles, all current commercial triploid sugar beet hybrid varieties are derived from broad based tetraploid pollinators and diploid male-sterile seed parents. Despite progress in developing monogerm O-types with improved germination and seed-plant traits (Bosemark, 1993), some advantages in avoiding contamination during pollination (Scott and Longden, 1970) and several indications for a probable superiority over their conventional counterparts (Fitzgerald, 1975; Lahousse, 1976), triploid hybrids produced on tetraploid CMSs (reverse triploids) have never attracted commercial interest.

Double-cross hybrids

As soon as good diploid, inbred, multigerm lines could be developed, they were exploited as pollinators for production of hybrid varieties. A possibility that attracted limited interest was use of advanced generations of F_1's between two such lines in self-incompatibility mediated crosses (Bosemark, 1993). Owen (1954) was the first to outline a scheme for production of double-cross hybrids that utilized both cytoplasmic and nuclear male sterility. In this system the female parent of the commercial hybrid was the usual F_1-CMS line and the pollinator an F_1 between two multigerm inbred lines, one of which was segregating for nuclear male-sterility (**Fig. 3.11**). To avoid the high cost for producing the pollinator (part of the plants of the segregating line were fertile and had to be removed), Theurer and Ryser (1969) proposed that the final F_1 pollinator be produced on a CMS line pollinated by a pollen fertility restorer multigerm inbred line. The problem with this method was the difficulty and uncertainty of fertility restoration-mediated pollination. There had been some commercial interest in double-cross sugar beet hybrid varieties in the past, mainly in areas of the USA with no significant disease problems (Hecker and Helmerik, 1985).

Single-cross hybrids

Genetic heterogeneity in three-way top-cross hybrids, although providing useful broad adaptability, presents several drawbacks. From a breeding

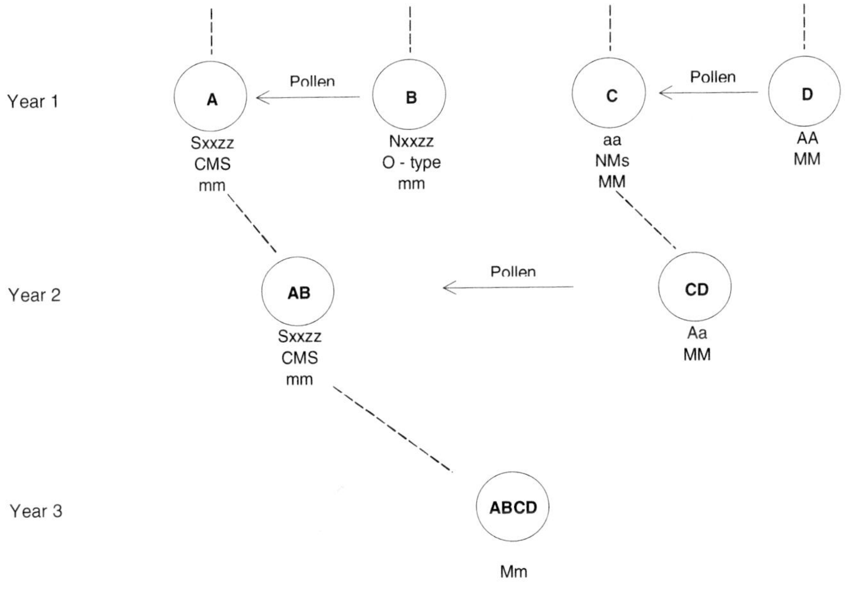

Fig. 3.11 Production of double-cross hybrid varieties.

viewpoint for example, altering three-way top-cross hybrids is slow compared to hybrids involving inbred pollinators which, as mentioned earlier, are easier to replace or to improve for a specific trait. Variation in root morphology adversely affects harvesting and processing of the crop. Single-cross hybrids between highly inbred lines (**Fig. 3.12**) could theoretically minimize such problems, provided that sugar yield and adaptability traits are not sacrificed, that seed yield and quality are acceptable and seed production is economically feasible. Single-cross hybrids were used to a very limited extent in regions of the USA where curly top resistance is required (Hecker and Helmerik, 1985). Despite a few promising reports (Le Cochec, 1982), commercial single-cross sugar beet hybrids seemed a rather remote possibility. However, as breeding methodology for the development of outstanding male-sterile and inbred multigerm pollinator lines has progressed and molecular technology is exploited, the situation seems to be changing and there is evidence that single-cross hybrids might replace the present types of hybrids in future (Bosemark, 1993).

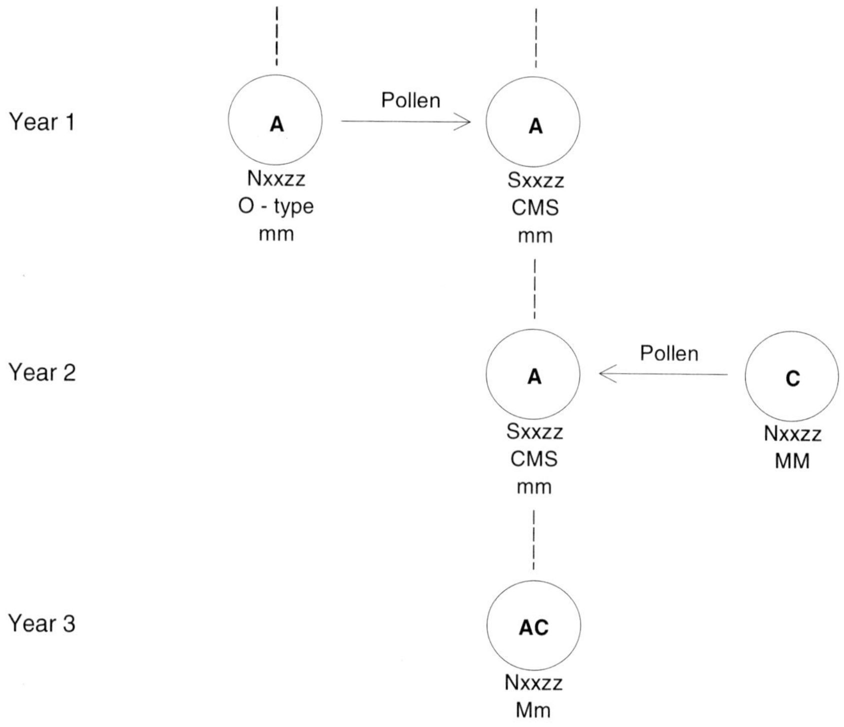

Fig. 3.12 Production of single cross hybrid varieties.

3.2.4 Maintenance of Parental Lines

Grower demand for consistent reliable varieties requires that the specific genetic constitution of the hybrid parental components remain unchanged. Given that these parents vary in type (diploid or tetraploid, inbred or more broad based families, clonal hybrids, etc.), it is necessary to choose the most appropriate process for maintaining allelic frequencies. It is very important to verify, both in the field and in the laboratory, that each new generation of basic seed retains all the specific traits desired (Smith, 1987).

An example can be drawn from the maintenance of diploid rhizomania-resistant varieties in which breeders must examine through modern molecular techniques, all precommercial seed production stages for homozygosity of the resistant parent and discard any heterozygous individuals (Horemans, 2000). If the pollinator of a variety is the F_1 or F_2 of parents stored in vitro as clones, all appropriate storage and regeneration techniques must be followed so that each new generation is indeed identical to the original parent.

Any reproduction of the basic seed of hybrids requires a representative sample from earlier seed lots and the most careful handling of the seed for establishment of the steckling nursery. It is also necessary to avoid contamination of the lines due to mixtures during seed processing. Limited seed quantities from a few sugar beet plants can be produced effectively under bags or small tents that prevent pollen contamination (Brewbaker, 1934). For a relatively large number of plants, greenhouses provide better protection (Biancardi and Mandolino, 1997). Production of large quantities of high quality basic seed is usually effected with space isolation in open fields (Kline *et al.*, 1938; Archimowitsch, 1949; Biancardi and Mandolino, 1997). However, distances normally considered satisfactory for isolation from other sugar beet (400-600 m) or related *Beta* pollen sources (1,000 m), may not prevent contamination of lines with high inbreeding depression. In this situation it is preferable to increase lines in a greenhouse or under other safer conditions.

Experience indicates that a critical factor for avoiding contamination by foreign pollen and obtaining high yields of quality raw seed is simultaneous flowering of the parental lines. In this respect tetraploid genotypes usually require additional attention (Scott and Longden, 1970).

3.2.5 Winter Varieties

Lack of sufficient rainfall in many parts of the world exposes the sugar beet crop to drought stress and reduces yields. If the winter temperature is not too low, sugar beet may be planted in the autumn or early winter and harvested in early summer (Jonsson, 1999). This is practiced in some areas of the Northern Hemisphere between the 40th parallel and the tropic of Cancer and includes Southern Spain, Southern Italy, Morocco, California, Iran, Syria, among other countries (Esteban Baselga, 1999). Autumn-winter sugar beet crops are relatively new, having been introduced in the 1950s.

During winter the plants are subjected to vernalization conditions, causing some bolting in normal genotypes in the late spring. Consequently the most important characteristic required in winter varieties is strong bolting resistance. Other traits to be considered in the selection process include resistance to low temperatures and low fibrosity of the root, in addition to high sugar yield, processing quality, and disease resistance.

A typical autumn variety for southern Europe is a triploid monogerm bolting-resistant hybrid. Diploid rhizomania-resistant varieties are necessary when rhizomania is present, as in Southern Italy and California, where curly top resistance is also necessary (McFarlane, 1971). These varieties are produced by the usual procedures of crossing tetraploid or diploid pollinators with monogerm F_1-CMS seed parents. Either or both parental components have to be characterized as having strong bolting resistance (Jonsson, 1999). The frequency of bolting usually becomes too high to recommend autumn planting north of 41 to 42° degrees latitude.

Attempts to obtain bolting-resistant lines were initiated in the early 1900s. Around 1925 in Hungary, Nemeth selected bolting resistant plants with a reasonably good sugar yield. The first experimental trials of autumn-sown sugar beet in southern Italy began in 1952 with multigerm varieties partially derived from Munerati's selection program (Biancardi, 1999). In California, breeding of autumn-winter varieties began with multigerm European genotypes and two to three cycles of mass selection on mother beets improved bolting resistance substantially. In this environment, the greatest progress was achieved with inbred lines developed through evaluation of progenies from single-plant selections. Some bolting resistant lines responded more to a long photoperiod while others responded to a long thermal induction period (McFarlane, 1971).

According to Bosemark (1993), several cycles of mass selection followed by progeny selection and line breeding are required for adaptation of sugar beet populations to winter growing conditions and further improvement is possible with selection of locations more favorable for bolting than areas of intended use.

3.3 Field Evaluation and Registration (E. Biancardi and M. De Biaggi)

3.3.1 Field Evaluation

The conclusive phase of a breeding effort is that of accessing the agronomic value of improved genotypes. It requires organization of field trials to forecast performance of varieties with acceptable precision, and to verify that the new genotypes are superior in some way to those already on the market (Simmonds, 1979).

Field trials are also essential throughout the breeding process. The precision required is generally proportional to the number of entries to be evaluated. Hence, in the pre-breeding phase, selection can be based on visual evaluations. As an example, the choice of resistant individuals in a diseased nursery can be based on single rows or plots (Petersen, 1994). In such screening nurseries, evaluation of the primary yield characters (sugar content and yield) is based on individuals (Cornish et al. 1990) or plots with few replications (Simmonds, 1979). In advanced stages of selection, a more comprehensive set of characters that includes processing quality must be evaluated (**Tab. 3.1**).

Uniformity and stability characteristics should be examined as they are necessary for success in diverse environments. Field evaluation of sugar beet requires multi location trials and the harvest of at least 50 kg of roots from each experimental plot. Consequently increased costs are incurred, especially when trial sites are far from the laboratory where the roots will be analyzed (Kristensen and Hill, 2002).

Certain principles must be recognized when sugar beet variety trials are organized. The trial site should be uniform and representative of the area. Information on culture inputs and practices during the previous 5 to 10 years assists in identifying probable pathological and nutritional problems or problems associated with crop rotations, all of which may affect homogeneity. Soilborne pathogens should be avoided, unless genetic resistance to a specific disease is to be evaluated. In this case, differential response of the varieties becomes clearer when the disease acts alone and not in association with other diseases and pests. Limiting the risks of complicating diseases requires longer than usual rotation cycles.

Cultural practices should be similar to those used in commercial sugar beet production in the area, as the new varieties will encounter these same conditions. In this regard important considerations are sowing and harvesting dates, fertilization, chemical treatments, and irrigation. Control or check varieties should include the most widely grown varieties in the area, to obtain a strong indication of the superiority, adaptability and productivity of the prospective variety compared to current commercial

Table 3.1 Analyses of production traits of sugar beet.

Plant stage	*Growing units*	*Sampling and analysis*	*Purpose*	*Organizations*
Mother beets	Rows; Large plots	Individual	Mass selection	Research institutes; Seed companies
Sugar beets	Replicated plots	Bulk (70-120 roots)	Evaluation of: Breeding lines – Selection	Research institutes; Seed companies
			Pre-commercial varieties – Registration	Official institutes
			Commercial varieties – Registration; Recommendation	Official institutes; Grower associates

varieties. There is a tendency to reduce the number of control varieties in the trials or to employ varieties with a limited use. This practice diminishes the validity of comparisons between the existing commercial varieties and the genotypes under evaluation.

Seed of control varieties to be included in the trials should be from commercially available lots and not from small lots of uncertain origin. This is an important prerequisite as small seed lots may have superior emergence potential, giving these varieties an advantage over the others in the trial.

Randomized block designs provide sufficient precision, provided that the trials are repeated for more than one year in several locations and the number of varieties per trial does not exceed 25-30 (Simmonds, 1979). For evaluation of more entries, more complex designs, such as the lattice, are recommended (Cochran and Cox, 1957). With lattice designs, replications are subdivided into smaller units or miniblocks. This reduces the error variance considerably compared to randomized blocks (Yndgaard, 2002; Yndgaard et al., 2002).

When several years and sites are involved, it is not necessary to obtain the highest possible precision in a single trial. For example, reducing the number of replications from 6 to 4 and adding additional trial sites may be a better use of resources. Before planting a trial, one must decide between traditional drilling followed by thinning and drilling to a stand. There have been two opposing lines of thought, both sustained by valid arguments (Morselli and Biancardi, 1996; Voss et al., 1996). In short, the thinning approach ensures homogeneous populations even under difficult emergence conditions. The plant to stand approach, on the other hand, compares varieties under more realistic conditions but may increase variability, although the latter should not influence performance data. Drilling to stand definitely involves more risks and consequently may increase costs. It is preferable however, because emergence in trial locations is generally high and stable over years. If seed germination continues to improve at the present rate, precision sowing will inevitably be adopted in all sugar beet field trials (Morselli and Biancardi, 1996; Fauchere, 2002; Wauters, 2002).

The importance of border rows in the plots is another controversial issue (Kerr, 2000). It is recognized that adjacent rows of different varieties might interfere with each other due to differences in foliage and root development (Büchse, 2002; Escriou, 2002). This effect seems quite low in normal field conditions.

It should also be acknowledged that use of border rows increases plot and trial size and therefore soil variability. Error variances without border rows may be lower than error variances with borders. Border rows also create obstacles to mechanical harvesting and hence their use does not seem justified in sugar beet variety trials.

In establishing variety trials one should be aware that rows compacted by implement traffic may perform differently than rows without compaction. This has to be taken into account when determining the area to be sampled and the equipment used for sowing and cultivation (Morselli and Biancardi, 1996; Guerif et al., 1996).

The number of harvest dates depends on the objectives of the experiment. In variety trials a single, usually medium-late harvest is preferred so that differences among varieties in productivity are maximized. Plants to be harvested at different times create problems in movement of the harvesters and may necessitate planting border rows to separate sampling areas. It is important that harvesting operations are mechanized not only for economic reasons, but also to avoid unrealistic estimates or even oversighting important qualities such as uniformity of root weight, shape and soil adherence. These characters have a significant effect on root yield and soil tare. So manual harvesting not only penalizes the more uniform varieties, but it also constitutes an unrealistic situation since all commercial production fields are harvested mechanically.

During field experimentation several observations must frequently be taken on characters such as emergence, susceptibility to diseases, wilting of foliage during stress periods, bolting tendency etc. Records on factors that affect the uniformity of a trial may explain eventual abnormal production data. Meteorological records from nearby stations may assist in interpreting results of multiyear trials.

The mean expected agronomic performance and processing value of a variety is estimated on the basis of multiyear and multilocation field trials including interactions between varieties, locations, and years.

A very important character, variety stability, is also evaluated by assessing the genotype x environment interaction (Perkins and Jinks, 1968). It has been noted for sugar beet that year effect and year x variety interaction is more important than location and location x variety effects (Marländer, 1991). It has likewise been estimated that variation due to year x variety interaction accounts for about 80% of the total variation. Because of this, in attempting to access environmental influence on the genotype, one has to give more consideration to years than to locations. Due to limitations in resources, this is not always possible. To compensate, it is important that variation accounted for by the location effect be increased by conducting experiments in as many unique environments as possible, provided they represent the region in which the varieties will be grown.

The agronomic evaluation of new materials become more precise if the comparisons to existing varieties increase. When performance data of a variety are compared to the average of all varieties in the same environment, a regression line is obtained (Simmonds, 1987). If the variety's performance coincides with the average of all varieties, the line passes through the origin ($a = 0$) and has a slope of 1 ($b = 1$). A variety is considered to have stable, average or less than average environmental impact on performance when $b \leq 1$. In the opposite case ($b > 1$), a variety can be characterized as less stable (unstable) or more responsive than average to environmental differences. Performance stability of a variety, most often related to a wide genetic base of both parental components, is definitely a desirable quality in commercial production **(Fig. 3.13)**. It should also be remembered that stable varieties are usually more productive in multiyear and multisite trials as well.

Pertinent to the above discussion, the following literature may be consulted: i) plot samples, Beiss and von Müller, 1974; ii) conservation of samples, Wauters, 2002; Burba et al., 1975; Oltmann et al., 1984; iii) chemical analyses and transformation of data, Oltmann et al., 1984; iv) experimental designs, Chocran and Cox, 1957; Clarke, 1980; v) statistical analysis, Petersen, 1994; Yndgaard, 2002; Kristensen and Hill, 2002; and vi) software and data processing, Schwarzbach, 1984; Utz, 1991; Kristensen and Hill, 2002.

3.3.2 Variety Registration and International Regulations

The ultimate goal of a breeding program is to have the new variety placed on the national registration list. The purpose of national lists is to certify the distinctiveness, uniformity and stability (DUS) of varieties. After this step, comes inclusion on the recommended list, which emphasizes relative productivity more than the national or official list. Companies can sell seed and acquire a market share only if their varieties are on both lists. While the ministries of agriculture are generally responsible for the national registration lists, professional organizations are charged with the development of recommended lists.

In the first part of the last century, the rights of seed companies were not sufficiently protected, although some rules had been established in USA, Germany, and the Netherlands (Hendriksen and Sneep, 1979). This situation led to the rapid development of hybridization methods in a number of crops. In fact reproduction of hybrid seed is protected *per se*, considering both the difficulties of multiplication without the parental lines and rapid decrease in productivity of generations beyond F_1. In 1961, the *Convention International pour la Protection des Obtentions Végétales* in Paris codified the protection of breeding rights in Western Europe. To obtain the right of

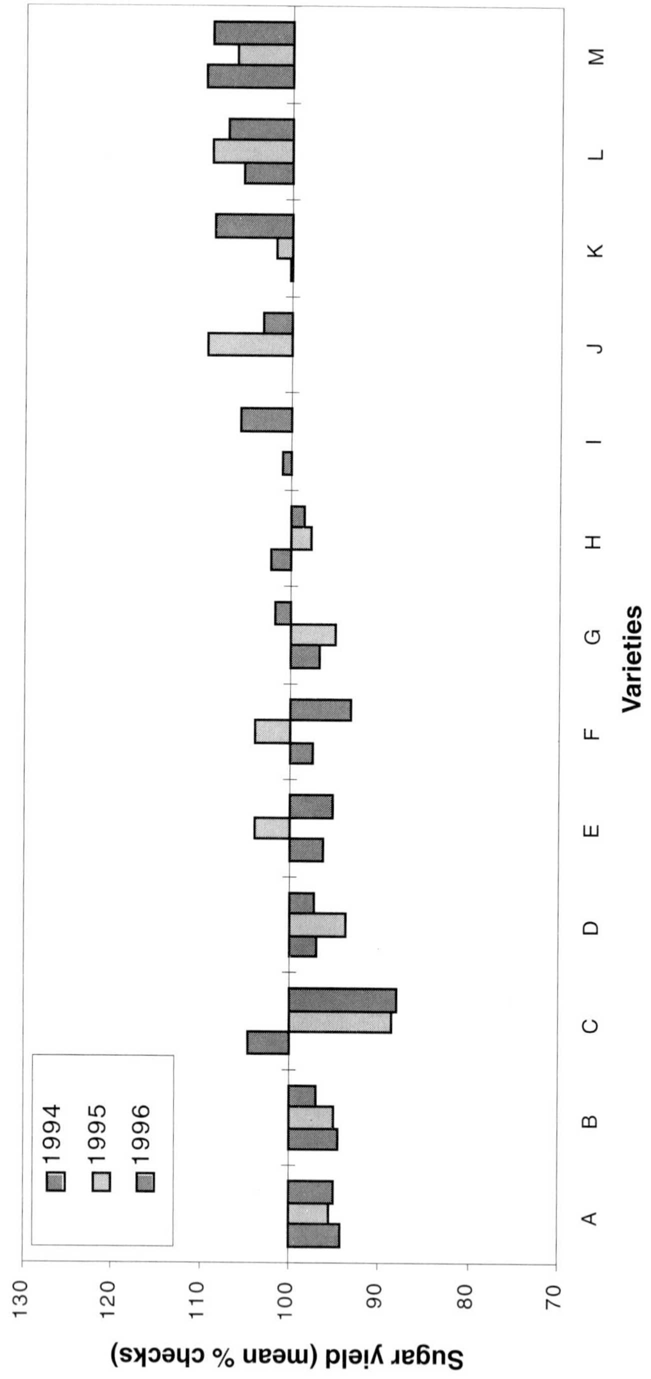

Fig. 3.13 Yield stability of varieties in Italy (From Boschetti et al., 1997).

protection, the varieties had to be new, distinct, homogeneous, and stable (Hendriksen and Sneep, 1979). Nowadays, in all countries with developed agriculture, official authorities must verify the characteristics of new varieties entering the national market. International Seed Testing Association (ISTA) methodologies are generally employed for controlling seed quality. ISTA procedures are periodically updated based on advancements in breeding and seed technology.

When a seed company wishes to have new varieties registered and recommended, it must follow standard procedures. European procedures are similar to rules used in other parts of the world. In countries with very different climatic areas, the seed companies must submit separate applications for each area. For example, in Russia there are six main production areas and four in the USA. European countries operate both the national and the recommended list, except for Switzerland and Finland, which have only a recommended list (Wevers, 2002). In Belgium, Ireland, the Netherlands, and the United Kingdom, the trials needed for the two lists are the same (Wevers, 2002).

Although the general rules are similar, each country has different registration procedures and criteria for acceptance on the recommended lists (Wevers, 2002). Seed companies have to indicate on the application form the characteristics of the varieties, such as disease resistances, production traits and so on. In many cases, new varieties are placed on recommended lists when three years of multisite trials indicate a higher sugar yield than the mean of the standard varieties. The companies can continue marketing as long as these varieties continue to perform well. In Europe, no transgenic sugar beet variety has been registered because of the existing restriction.

References

Allard, R.W. 1960. Principles of Plant Breeding. John Wiley & Sons, New York, NY.

Archimowitsch, A. 1949. Control of pollination in sugar beet. Bot. Rev. 15:613-628.

Barzen, E., R. Stahl, E. Fuchs, D.C. Borchardt, and F. Salamini. 1997. Development of coupling-repulsion-phase SCAR markers diagnostic for the sugar beet Rrl allele conferring resistance to rhizomania. Mol. Breed. 3:231-238.

Beiss, U., and A. von Müller. 1974. Beiträge zur Methodik der Ertrags und Qualitätbestimmung bei Zuckerrüben. Zucker 4:173-178.

Biancardi, E. 1999. Miglioramento genetico. *In:* B. Casarini, E. Biancardi, and P. Ranalli (eds.) La barbabietola da zucchero in ambiente mediterraneo. Edagricole, Bologna, Italy, pp. 57-145.

Biancardi E. and G. Mandolino. 1997. L'impollinazione e l'isolamento spaziale della barbabietola.Sementi Elette 5:9-13.

Biancardi, E., M. De Biaggi, L. Alessandrini, and G. Venturi. 1979. Confronto tra metodi di campionamento per la selezione tecnologica delle bietole madri. Sementi Elette 25:65-70.

Biancardi, E., R.T. Lewellen, M. De Biaggi, A.W. Ericsen and P. Stevanato. 2002. The origin of rhizomania resistance in sugar beet. Euphytica 127:383-397.

Bongiovanni, G.C., and L. Lanzoni. 1964. La rizomania della bietola. Progresso Agricolo 2:209-220.

Bongiovanni, G.C., G. Gallarate, and G. Piolanti. 1958. La barbabietola da zucchero. Edizioni Agricole, Bologna, Italy.

Bosemark, N.O. 1971. Use of mendelian male-sterility in recurrent selection and hybrid breeding in beets. Proc. Eucarpia-Fodder Crops Section, Lusignan, France Sept. 1970.

Bosemark, N.O. 1993. Genetics and breeding. *In:* D.A. Cooke and R.K. Scott (eds.) The Sugar Beet Crop. Chapmann & Hall, London, UK, pp. 67-119.

Brewbaker, H.E. 1934. Self-fertilization in sugar beet as influenced by type of isolator and other factors. J. Agric. Res. 4:323-337.

Broertjes, C. and A.M. van Harten. 1978. Application of Mutation Breeding Methods in the Improvement of Vegetatively Propagated Crops. Elsevier Sci. Publ. Co., Amsterdam, Netherlands.

Büchse, A. 2002. Effect of interplot interference on yield and performance in sugar beet variety trials. *In:* Variety trials in sugar beet - Methodology and design. *In* Advances in Sugar Beet Research. IIRB, Bruxelles, Belgium, vol. 4, pp. 73-84.

Burba, M., W. Haufe, and W. Krüger. 1975. Verhalten von qualitätbestimmenden Inhaltsstoffen der Zuckerrübe während der Herstellung und Lagerung von Gefrierbrei. Zucker 8:411-418.

Burky, K. and G. Buttner. 1989. Rizomaniatolerante Zuckerrübensorten. II. BNYVV-gehalt junger Testplanzen und Ertrag im Feld unter Rizomaniabefall. Zuckerindustrie 114:225-228.

Canova, A., 1966. Ricerche virologiche nella rizomania della bietola. Rivista Patologia Vegetale 2:3-41.

Chocran, W.G. and G.M. Cox. 1957. Experimental Designs. Wiley, Interscience, New York, NY.

Clarke, G.M. 1980. Statistics and Experimental Design. Edward Arnold Ltd., London, UK.

Coons, G.H. 1936. Improvement of the sugar beet. Yearbook of Agriculture USSR, Washington, DC, pp. 625-656.

Coons, G.H. 1975. Interspecific hybrids between *Beta vulgaris* L. and the wild species of *Beta*. J. ASSBT 18:281-306.

Coons, G.H., F.V. Owen, and D. Steward. 1955. Improvement of sugarbeet in the United States. Adv. Agron. 7:89-139.

Cornish, M.A., M.C. Smith, and I.J. Mackay. 1990. An evaluation of single plant randomised field trials of sugar beet (*Beta vulgaris* L.). Euphytica 45:1-7.

De Biaggi, M. 1987. Methodes de selection: un cas concret. Proc. IIRB 50:157-161.

De Vilmorin, J.L. 1923. L'hérédité chez la betterave cultivée, Genthiers-Villars, Paris.

Doggett, H. and S.A. Eberhart. 1968. Recurrent selection in *Sorghum*. Crop Sci. 8:119-121.

Doney, D.L. and J.C. Theurer. 1978. Reciprocal recurrent selection in sugar beet. Field Crops Res. 1:173-181.

Doney, D.L., E.D. Whitney, J. Terry, L. Frese, and P. Fitzgerad. 1990. The distribution and dispersal of *Beta vulgaris* L. ssp. *maritima* germplasm in England, Wales and Ireland. J. Sugar Beet Res. 27:29-37.

Elliott, F.C. 1961. Plant Breeding and Cytogenetics. McGraw-Hill, New York, NY.

Escriou, H. 2002. Study of interplot competition in sugar beet experiments at ITB. Variety trials in sugar beet—Methodology and design. *In:* Advances in Sugar Beet Research. IIRB, Bruxelles, Vol. 4, pp. 55-72.

Esteban Baselga, J.A. 1999. Caractérisation de la betterave de semis d'automne. Proc. IIRB 62:1-8.

Falconer, D.S. 1960. Introduction to Quantitative Genetics. Oliver & Boyd, Edinburgh, Scotland.

Fasoulas, A.C. and V.A. Fasoula. 1995. Honeycomb delection designs. Plant Breed. Rev. 13:87-139.

Fauchere, J. 2002. Influence de la qualitè des graines sur le varietee experimentees. Proc. IIRB 64:257-268.

Fitzgerald, P. 1975. Value of sugar beet triploids as affected by reciprocal crosses. Rep. IIRB Breeding and Genetics Study Group. Carlow, Ireland. IIRB, Bruxelles.

Gaul, H. 1962. Mutationen in der Pflanzenzüchtung. Zeitschrift Pflanzenzücht 50:194-307.

Gentili, P. and G. Poggi. 1986. Ritmo, esperienze italiane contro rizomania e cercospora. Maribo Italia, Bologna, Italy, Techn. Publ.

Gilmore, E.C. 1964. Suggested method of using reciprocal recurrent selection in some naturally self-pollinated species. Crop Sci. 4:323-325.

Giunchedi, L., M. De Biaggi, and C. Poggi Pollini. 1985. Evaluation of ELISA technique for the screening of rhizomania tolerant sugar beet genotypes. Proc. IIRB 48:385-390.

Golev, I.F. 1975. Results of selecting induced mutants in populations of multigerm sugar beet. Sakhara 5:91-100.

Guerif, J., G. Richard, C. Durr, H. Bolzard, and Y. Duvall. 1996. Compactage du sol et croissance racinaire de la betterave. Proc IIRB 59:199-212.

Hecker, R.J. 1967. Evaluation of three sugar beet breeding methods. J. ASSBT 14:305-318.

Hecker, R.J. 1978. Recurrent and reciprocal recurrent selection in sugar beet hybrids. Crop Sci. 18:805-809.

Hecker, R.J. 1985. Reciprocal recurrent selection for the development of improved sugarbeet hybrids. J. ASSBT 23:47-58.

Hecker, R.J. and R.M. Helmerick. 1985. Sugar beet breeding in United States. *In:* G.E. Russell (ed.) Progress in Plant Breeding. Butterworths, London, UK, vol. 1, pp. 37-71.

Heijbroek, W. 1977. Partial resistance of sugarbeet to beet cysts eelworm (*Heterodera schactii* Schm.). Euphytica 26: 257-262.

Heijbroek, W., A.J. Roelands, J.H. De Jong, C.G. van Hulst, A.H.L. Schoone, and R.G. Munning. 1988. Sugar beet homozygous for resistance to beet cyst nematode (*Heterodera schactii* Schm.) developed from monosomic additions of *Beta procumbens* to *Beta vulgaris*. Euphytica 38:121-131.

Helmerick, R.H., R.E. Finker, and C.W. Doxtator. 1965. Paired-plant crosses in sugar beets. J. ASSBT 13:548-554.

Hendriksen, A.J.T. and J. Sneep. 1979. Plant Breeding Perspectives. Ed. CAPD. Wageningen, Netherlands.

Hijner, J.A., 1951. De gevoeligheid van wilde bieten voor het bietecysteaaltje (*Heterodera schactii*). Meded. Inst. Rat. Suikerprod. 21:1-13.

Hjerdin, A., T. Sall, N.O. Nilsson, C.H. Bornman, and C. Hallden. 1994. Genetic variation among wild and cultivated beets of the section *Beta* as revealed by RFLP analysis. J. Sugar Beet Res. 31:59-67.

Horemans, S. 2000. Resistenza alla rizomania. Sementi Elette 3-4:13-14.

Johnson, R.T. and G.W. Wheatley. 1961. Studies on the backcross generations of interspecific hybrids between *Beta vulgaris* and *Beta webbiana*. J. ASSBT 11:429-435.

Jonsson, B.O. 1999. Development of a winter beet alternative for South Europe. Proc. IIRB 62:69-76.

Kerr, S. 2000. Variety interaction with sowing, soils and harvest. Brit. Sugar Beet Rev. 1:18-21.

Kinoshita, T. 1980. Induction of cytoplasmic male sterility by gamma ray and chemical mutagens in sugar beet. Gamma Field Symposia 1:27-48.

Kline, O.L., C.D. Tolle, and E.M. Nelson. 1938. Airplane collection of sugar beet pollen. Science 88:507-508.

Knapp, E. 1958. *Beta* Rüben. *In:* R. Roemer and W. Rudorf (eds.) Handbuch der Pflanzenzüchtung. Paul Parey, Berlin, vol. 3, pp. 196-284.

Kristensen, K. and J.Hill. 2002. Multi-environment variety trials: anslysis and prediction. Variety trials in sugar beet—Methodology and design. *In:* Advances in Sugar Beet Research. IIRB, Bruxelles, vol. 4, pp. 19-54.

Lahousse, R. 1976. Use of monogerm male-steriles and realization of triploid varieties of sugar beet. Rept. IIRB Breeding and Genetics Study Group, Bologna, Italy. IIRB, Bruxelles, Belgium.

Lange, W. and T.S.M. De Bock. 1994. Pre-breeding for nematode resistance in beet. J. Sugar Beet Res. 31:13-26

Lange, W., W.A. Brandenburg, and T.S.M., De Bock. 1999. Taxonomy and cultonomy of beet (*Beta vulgaris* L.). Bot. J. Linn. Soc. 130:81-96.

Le Cochec, F. 1982. Les varietees monogermesde betterave sucriere. Suggestions pour la selection d'un autre type de varietees: lae hybrides entre deux lignes fixees. Selectionneur Français 30:45-48.

Lewellen, R.T. and E. Biancardi. 1990. Breeding and performance of rhizomania resistant sugar beet. Proc. IIRB 53:69-87.

Lewellen, R.T., I.O. Skoyen, and A.W. Erichsen. 1987. Breeding sugarbeet for resistance to rhizomania: evaluation of host-plant reactions and selection for and inheritance of resistance. Proc. IIRB 50:139-156.

Lichter, R. 1972. A method for evaluating competitional variance in plant populations. Z. Pflanzenzüchtung 68:51-63.

Märländer, B. 1991. Einfluss der Bestandesdichte auf Ertrags- und Qualitätskriterien sowie über mögliche Ursachen der Konkurrenz in Zuckerrüben. J. Agron. Crop Sci. 164:120-130.

McFarlane, J.S. 1971. Variety development. *In:* R.T. Johnson, J.T. Alexander, G.E. Bush, and G.R. Hawkes (eds.) Advances in Sugar Beet Production. Iowa State Univ. Press, Ames IA, pp. 402-435.

Mesken, M. 1989. Response to divergent mass selection for top height in diploid and tetraploid sugarbeet. Prophyta 1:49-51.

Morselli, A. and E. Biancardi. 1996. Metodologia sperimentale in prove agronomiche di barbabietola da zucchero. Sementi Elette 3-4:53-56.

Munerati, O., T. Mezzadroli, and V. Zapparoli. 1913. Osservazioni sulla *Beta maritima* L. nel triennio 1910-1912. Stazioni Sper. Agrarie Ital. 6:347-371.

Nakamura, C., G.N. Skaracis, and I. Romagosa. 1992. Cytogenetics and breeding in sugar beet. *In:* T. Tsuchiya and P.K. Gupta (eds.) Chromosome Engineering in Plants: Genetics, Breeding, Evolution, Part B. Elsevier, Amsterdam, Netherlands, pp. 295-313.

Oldemeyer, R.K. 1954. Viable interspecific hybrids between wild species of the section *Vulgares* and species in the section *Patellares* in the genus *Beta*. Proc. ASSBT 8:153-156.

Oldemeyer, R.K. 1964. *Beta macrorhiza*. J. ASSBT 12:637-640.

Oldemeyer, R.K. and H.L. Bush. 1952. Influence of geographical locations and type of bag on selfed-seed setting for sugar beets. Proc. ASSBT 2:360-363.

Oldemeyer, R.K. and H.E. Brewbaker. 1956. Interspecific hybrids in the genus *Beta*. Proc. ASSBT 9:15-18.

Oldemeyer, R.K., W.H. Davis, H.L. Bush, and A.W. Erichsen. 1968. The evaluation of and the use of the top-cross test as a method of selecting inbred lines of sugarbeets for general combining ability. J. ASSBT 15:49-60.

Oltmann, W., M. Burba, and G. Bolz. 1984. Die Qualität der Zuckerrübe, Bedeutung, Beurteilungskriterien und züchterische Massnahmen zu ihre Verbesserung. Paul Parey, Berlin, Germany.

Owen, F. V. 1942. Inheritance of cross- and self-sterility and self-fertility in *Beta vulgaris*. J. Agr. Res. 64:679-698.

Owen, F.V. 1945. Cytoplasmically inherited male sterility in sugar beets. J. Agric. Res. 71:423-440.

Owen, F.V. 1952. Mendelian male sterility in sugar beet. Proc. ASSBT. 7:371-376.

Owen, F.V. 1954. Hybrid sugar beets made by utilizing both cytoplasmic and mendelian male sterility. Proc ASSBT 8:64.

Panella, L. 1995. Registration of four multigerm sugarbeet germplasm resistant to rhizoctonia root rot: FC716, FC717, FC718, and FC719. Crop Sci. 35:291-292.

Perkins, J.M. and J.L. Jinks. 1968. Envirnmental and genotype x environmental components of variability. Heredity 29:237-245.

Petersen, G. 1994. Agricultural Field Experiments. Marcel Dekker Inc., New York, NY.

Raudsepp, A.D. 1976. Forms of Sugar Beet Obtained by Treatment with Chemical Mutagens. Nauka, Moscow, pp. 236-241.

Saunders, J.W., J.M. McGrath, J.M. Halloin, and J.C. Theurer. 1999. Registration of SR94 sugar beet germplasm with smooth root. Crop Sci. 1:297.

Savitsky, H. 1952. Selective fertilisation studies and recovery of self-sterile from self-fertile races of monogerm sugar beets. Proc. ASSBT 7:339-343

Savitsky, H. 1954. Self-sterility and self-fertility in monogerm beets. Proc. ASSBT 8:29-33.

Savitsky, H. 1960. Viable diploid, triploid, and tetraploid hybrids between *Beta vulgaris* and species of the section *Patellares*. J. ASSBT 11:5-35.

Savitsky, H. 1969. Meiosis in hybrids between *Beta vulgaris* and *Beta corolliflora* and transmission of resistance to curly top virus. Can. J. Genet. Cytol. 11:49-67.

Savitsky, H. 1973. Meiosis in hybrids between *Beta vulgaris* L. and transmission of sugar beet nematode resistance. Genetics 74:241.

Savitsky, H. 1975. Hybridization between *Beta vulgaris* and *B. procumbens* and transmission of nematode (*Heterodera schactii*) resistance to sugar beet. Can. J. Genet. Cytol. 17:197-209.

Savitsky, H. and J.O. Gaskill. 1957. A cytological study of F_1 hybrids between Swiss chard and *Beta webbiana.* J. ASSBT 9:433-449.

Savitsky, H. and C. Price. 1965. Resistance to the sugar beet nematode *Heterodera schactii* in F_1 tetraploid hybrids between *Beta vulgaris* and *Beta patellaris.* J. ASSBT 13:370-373.

Savitsky, V.F. 1950. Monogerm sugar beets in the United States. Proc. ASSBT 6:156-159.

Scheibe, A. 1934. Über die Wildzuckerrüben Anatoliens, *Beta lomatogona* F.& M., *B. intermedia* Bunge und *B. trigyna* W.&K., Angew. Bot. 16:305-349.

Scholten, O.E., T.S.M. De Bock, R.M. Klein-Lankhorst, and W. Lange. 1999. Inheritance of resistance to beet necrotic yellow vein virus in *Beta vulgaris* conferred by a second gene for resistance. Theor. Appl. Genet. 99:740-746.

Schwanitz, F. 1938. Die Herstellung polypoider Rassen bei Beta-Rüben und Gemüsearten durch Behandlung mit Colchicin. Der Züchter 10:278-279.

Schwarzbach, E. 1984. A new approach in the evaluation of field trials. Vortr. Pflanzenzüchtung 6:249-259.

Scott, R.K. and P.C. Longden. 1970. Pollen release by diploid and tetraploid sugar beet plants. Ann. Appl. Biol. 66:129-135.

Seitz, F.W. 1935. Contribution à l'etude de la karyologie de l'hybride de *Beta trygina* avec la betterave sucrière. Publ. IRBAB 3:399-400.

Seitz, F.W. 1936. Ein Beitrag zur Zytologie eines seltenen Rübenbastardes und einer seiner Rückkreuzungen. Zeitschr. Wirtsch. 86:357-370.

Seitz, F.W. 1938. Ein Beitrag zur Artbastardierung von Zuckerrüben mit *Beta trygina*. Deutsche Zuckerind. 63:439-444.

Simmonds, N.W. 1976. Principles of Crop Improvement. Longman, Harlow, UK.

Skaracis, G.N. and G.A. Smith. 1984. Prediction of three-way top cross sugar beet hybrid performance. Crop Sci. 24:55-60.

Skaracis, G.N. and E. Biancardi. 2000. Breeding for cercospora resistance in sugar beet. *In: Cercospora beticola* Sacc., biology, agronomic influence and control measures in sugar beet. Advances in Sugar Beet Research. IIRB, Bruxelles, Belgium. vol. 4, pp. 177-195.

Smith, G.A. 1987. Sugar beet. *In:* W.R. Fehr (ed.) Principles of Cultivar Development. McMillan, New York, NY, pp. 577-625.

Smith, G.A. and E.G. Ruppel. 1979. Registration of F606 and F607 CMS sugar beet germplasm. Crop Sci. 19:300.

Smith, G.A., R.J. Hecker, G.W. Maag, and D.M. Rasmusson. 1973. Combining ability and gene action estimates in an eight parent diallel cross of sugar beet. Crop Sci. 13:312-316.

Sneep, J. and J. Hendriksen. 1979. Plant breeding perspectives. Pudoc, Wageningen, Netherlands, pp. 203-233.

Stadler, L.J. 1930. Some effect of x-rays in plant. J. Heredity 21:3-19.

Stehlik, V. 1947. Essais de croisment de betterave sucrière avec *Beta trygina*. IRBAB Publ. Tech. 15:103-108.

Stewart, D., C.A. Lavis, and G.H. Coons.1940. Hybrid vigor in sugar beet. J. Agric. Res. 60:715-738.

Stubbe, H. 1966. Induced Mutations and Their Utilization. Akademie Verlag, Berlin, Germany.

Theurer, J.C. and G.H. Ryser. 1969. Inheritance studies with a pollen fertility restorer sugar beet inbred. J. ASSBT 15:538-545.

Ulbrich, E., 1934. *Chenopodiacee. In:* A. Engler and H. Harms (ed.) Natürlichen Pflanzenfamilie. Verlag Wilhelm Hengelmann, Leipzig, Germany, vol. 16, pp. 379-584.

Utz, H.F. 1991. Plabstat 2F, a computer program for statistical analysis of plant breeding experiments. University of Stujest, Stujest, Gemany. User's Manual.

Voss, M.C., E. Ladewig, and U.E. Pfleiderer. 1996. Einfluss der Ablageentfernung sowie eines simulierten Feldaufgangs auf Ertrag und Qualität von Zuckerrüben—Zur Frage der Anlage von Sortenversuchen mit und ohne Vereinzelung. Zuckerindustrie. 121:586-589.

Walther, F. 1961. Eine neue cytologische Untersüchungen für Betarüben. Züchter 14:474-476.

Walther, F. 1963. Untersüchungen an Wildrüben in der Turkei—Cytologische Untersüchungen an Wildrüben im Gebiet von Eskisehir. Z. Pflanzenzucht. 49:173-180.

Wauters, A. 2002. Trial techniques: influence on experimental error. *In:* Variety trials in sugar beet—Methodology and design. Advances in Sugar Beet

Research. IIRB, Bruxelles, Belgium, vol. 4, pp. 87-93.

Wevers, D.A. 2002. Variety testing in some IIRB member countries. *In:* Variety trials in sugar beet—Methodology and design. Advances in Sugar Beet Research. IIRB, Bruxelles, Belgium, vol. 4, pp. 139-156.

Whitney, E.D. 1989. Identification, distribution and testing for resistance to rhizomania in *Beta maritima*. Plant Dis. 73:287-290.

Yndgaard, F. 2002. How to do sugar beet variety trials: single and multiple environment trials. *In:* Variety trials in sugar beet— Methodology and design. Advances in Sugar Beet Research, IIRB, Bruxelles, Belgium, vol. 4, pp. 3-12.

Yndgaard, F., B.V. Pedersen, and B.O. Jonsson. 2002. Analysis of sugar beet variety trials using mixed model equations I. Lattice and randomised complete block designs. J. Swedish Seed Assoc. 3:90-100.

4

Molecular Biology and Biotechnology

The impressive knowledge acquired in the area of plant molecular biology and the exciting developments in its practical applications have revolutionized the manner in which plant breeding programs are designed and conducted. As in other crops, the sugar beet breeder has begun to better comprehend the genetic mechanisms underlying the traits to be improved, given provision with appropriate biotechnological tools for a considerably more efficient manipulation of genetic material. It is expected that a rational employment of all related novel techniques will increasingly play an important role throughout the mainstream breeding operations and thereby most significantly enhance their productiveness. This chapter outlines the scientific information available and describes basic current and possible applications in the genetic improvement of sugar beet. Topics include progress in genomic research and its perspectives, various molecular marker-based methodological approaches as well as biotechnological applications of *in vitro* culture and genetic engineering.

4.1 Genomics in Genetic Improvement (M. McGrath)

4.1.1 Introduction

Genomics is a relatively recent idiom and initially described the activities associated with mapping, sequencing, and analyzing genomes, in particular the three billion nucleotides of the human genome (Hieter and Boguski, 1997). The genome is an established concept, defined as the basic (monoploid) set of chromosomes in an organism, and hence represents the sum total of all genes in that organism. In the broadest sense, genomics pertains to the study and action of the genome, much as genetics pertains to the study and action of genes. This chapter deals with concepts of genomics as applied to sugar beet improvement where possible and land plants in general, which result from technological advances in applying the tools of molecular biology developed over the past 50 years. Walbot and Goldberg (1979) provide a comprehensive pregenomics era overview of genome organization in plants.

Genomics in practice has been facilitated by instrumentation for large-scale nucleotide sequencing and genetic analysis in a parallel fashion (Venter et al., 2003), resulting in data unit cost savings due to economies of scale. Improvements are likely to continue in future and should result in further data unit cost savings as well as reduced capital equipment costs. Miniaturization of nucleotide sequencing and fragment analysis instruments is desired to reduce time and costs to the point that they become an integral aspect to practical plant breeding programs. Applications in breeding are many, such as evaluating allelic diversity at specific trait loci and phenotyping expressed genes in different environments at the molecular level, but the groundwork of building the biological information required for application to sugar beet is in an early stage of development.

4.1.2 Genome Organization of the Nucleus (Structural Genomics)

Genomes in genus *Beta* are based on x = n = 9 chromosomes. Specific numbering of individual chromosomes, defined in genetic linkage maps, has not been fully standardized to date. However, work by Schondelmaier and Jung (1997) should be referenced since their work has integrated previous cytogenetic information. It is presumed that individual chromosomes are homologous in genus *Beta,* but this remains to be conclusively demonstrated. Chromosome rearrangements are common in plants and often occur as a cause or consequence of speciation (Levin, 2001).

DNA content (C-value) of *Beta vulgaris* has been reported as 714 to 758 million base pairs per haploid genome, with variation reported among subspecies (Arumuganathan and Earle, 1991; Bennett and Smith, 1976). Variation in genome size within subspecies has not been investigated. The nine chromosomes of sugar beet are morphologically similar at mitotic metaphase, except that centromeres are either metacentric or submetacentric and a terminal constriction, or satellite, occurs on chromosome I. This constriction carries the major cluster of 18S-5.8S-25S ribosomal RNA genes, with about 2,300 tandem repeated copies (Ingle et al., 1975) at this site and about 20 copies at another locus (Schmidt et al., 1994). The 5S ribosomal RNA genes of *Beta vulgaris* have been cytologically and genetically located to an interstitial site near the centromere on chromosome IV (Schmidt et al., 1994; Schondelmaier et al., 1997).

Highly repetitive DNA sequences comprise 60% or more of the beet genome (Flavell et al., 1974). Excluding the 18S-5.8S-25S ribosomal RNA repeats, the highly repetitive fraction of the genome consists of numerous families of short (140 to 160 nt) repeating units present at high copy numbers (10^5 – 10^6 copies per genome) (Schmidt and Heslop-Harrison, 1996) and various classes of transposable elements (Schmidt et al., 1995; Staginnus et al., 2001). Each chromosome in sugar beet has a characteristic pattern of repeat-sequence distribution, supporting the notion that sugar beet is a true diploid with little or no duplication of its primary chromosome set in its recent evolution (Hallden et al., 1998; Schondelmaier et al., 1996). Diversification of highly repetitive sequences among *Beta* genomes, particularly within different sections, has proven an advantage in characterizing interspecific hybrids in *Beta* (Desel et al., 2002). Schmidt and Heslop-Harrison (1998) presented a conceptual framework for chromosome structure, largely based on their work with beets (**Plate 4.1**). The central premise is that each chromosome has a defined structure that includes a telomere at each chromosome terminus, a single functional centromere at a single chromosome location (Gindullis et al., 2001), families of highly repetitive sequences adjacent to centromeres and telomeres, and regions of single and low copy sequences interspersed with highly and moderately repetitive sequences that occupy interstitial regions. Single and low copy sequence regions typically correspond to regions carrying protein-encoding genes, although not all genes have been identified nor all their locations in the genome determined.

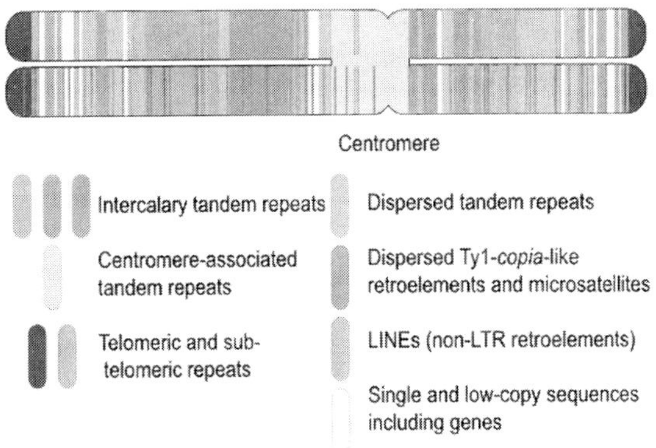

Plate 4.1 A Conceptual Framework for Chromosome Structure (Permission to reprint given by Elsevier Science Ltd).

Many molecular markers used for genome mapping in sugar beet and related species probably reside in highly repetitive regions, if for no other reason than that the majority of the genome comprises repetitive sequence families. Such single locus amplification probably results from minor sequence variants in individual members of repeat sequence families. An interesting observation has been that many markers are clustered at a single or a few loci on each sugar beet chromosome. Interpretation of marker clustering is that recombination has not occurred between markers defining these loci, and thus areas of limited or no recombination characterize regions in the sugar beet genome. This interpretation has serious consequences for breeding since recombination between agronomic genes to form enhanced combinations of alleles is the foundation of plant improvement. Low frequencies of recombination would require larger population sizes to effectively transfer traits via backcrossing for instance, and require larger population sizes or more generations of intercrossing to effectively recombine parental combinations in population improvement strategies. Areas of reduced recombination may not be a serious impediment however, if reduced recombination were strictly limited to areas of heterochromatin surrounding the centromeres and telomeres, as is likely in most eukaryotes (Schwarzacher, 2003).

4.1.3 Genomes of Chloroplasts and Mitochondria

Chloroplasts and mitochondria each contain DNA that codes for organelle specific functions and originated from ancient capture of bacteria by primitive eukaryotic cells (chloroplasts from cyanobacteria and mitochondria from the α-purple bacteria group), with subsequent loss or lateral gene transfer to the nucleus (Lewin, 1997). Importation of nuclear-encoded proteins into organelles compensates for these lost functions, and allows coordination between organelle and nuclear genome functions and gene expression. Although the selective advantage of organelles retaining vestiges of their progenitor genomes is mysterious (Allen, 2003), each organelle has a central role in energy metabolism in the plant: chloroplasts in harvesting light energy and mitochondria in respiration. Both organelles are typically inherited uniparentally through the cytoplasm of the female. In beets, these cytoplasmic genes play an essential role in the expression of cytoplasmic male sterility in hybrid seed production (Bosemark, 1993); however fertility restoration genes reside in the nucleus (Hjerdin-Panagopoulos et al., 2002). Nuclear-cytoplasmic interactions are beginning to be addressed at the molecular level (Schnable and Wise, 1998), and when defective may contribute to reduced vigor, viability, and fertility (Levin, 2003).

Chloroplast genomes have been sequenced from 19 organisms (De las Rivas et al., 2002), among which the spinach genome, the sequenced taxon most closely related to beet, is also most similar to *Arabidopsis thaliana* (mouse-ear cress). The spinach chloroplast genome is circular, slightly larger than 150 kb in size, and encodes more than 100 genes whose functions have mostly been described (Schmitz-Linneweber et al., 2001; Leister, 2003). Chloroplast gene content and identity vary slightly across plant taxa, however 13 chloroplast cytotypes have been described in sea beets (*B. vulgaris* ssp. *maritima*) (Desplanque et al., 2000). In other wild species as well as sugar beet, chloroplast genes have been found in the mitochondrial DNA, in which at least one is transcribed (Kubo et al., 1995).

The sugar beet mitochondrial genome has been sequenced (Kubo et al., 2000) and is one of the two complete higher plant mitochondrial genomes available, the other being *Arabidopsis* (Unseld et al., 1997). The beet mitochondria genome, as represented by a single male-fertile genotype, is more than twice as large as the chloroplast genome (368 kb), but only encodes 59 recognizable genes, two more than *Arabidopsis*, and duplicated sequences, introns, unidentified open reading frames, and foreign sequences imported from the chloroplast and nucleus comprise much of the mitochondrial genome. The conformation of higher plant mitochondrial genomes is not clear, but evidence suggests they differ from the circular chloroplast genomes in that circular (master circles and subgenomic circles), linear, and branched DNA molecules are found (Mackenzie and McIntosh, 1999; Backert et al., 1997; Knoop and Brennicke, 2002). These features may be partially responsible for the extensive recombination and size variation of mitochondrial genomes in plants.

Twenty-three mitochondrial cytotypes have been described in beets, and nonrandom associations between chloroplast and mitochondrial cytotypes may indicate common cytoplasmic ancestry in some populations of sea beets (Desplanque et al., 2000 and references therein). However, only two of these cytoplasms have been economically important in the deployment of CMS for hybrid seed production—the sterile (Svulg) and the fertile normal (Nvulg) cytotypes. These two mitochondrial genomes differ by at least 15 structural rearrangements including 35 kb of DNA inserted in five regions of the sterile cytoplasm (Kubo et al., 1999). These complex rearrangements have precluded precise identification of the gene(s) contributing to pollen sterility; however six mitochondrial genes that map in or near the rearrangements showed altered gene expression patterns between fertile and sterile lines, and only one of these (*coxI*; cytochrome c oxidase subunit I) was tentatively implicated in restoration of fertility (Kubo et al., 1999).

4.1.4 Genes (Functional Genomics)

In most cases, agronomic traits in sugar beet can be assumed to be controlled by genes whose product is a catalytic or structural RNA or protein. Relatively few of these RNA and protein molecules have been characterized functionally in beets; yet it is precisely the biochemical mechanisms underlying agronomic traits that should be targeted for genetic enhancement of sugar beet. Functional genomics is essentially the global analysis of gene expression, generally contrasted between two or more physiological states of development or environmental response (Pruitt et al., 2003). While RNA *per se* can assume catalytic activities, functional genomics as currently practiced primarily analyzes those protein encoding messenger RNA molecules which can be isolated relatively easily by virtue of a polyadenylated tail on their transcripts (e.g. poly-A+ RNA or mRNA). Analyses of these transcripts by nucleotide sequencing (or a host of other methods such as microarray analyses, differential display of transcripts, etc.) reveal information about the identity and abundance of a specific transcript or set of transcripts, diversity of transcripts present in specific cell types and tissues, and biochemical complexity of the physiological process. Potentially novel or unexpected developmental or environmental responses may be evident from such analyses. All such deduced responses are, by definition, properties of the cells used in the analyses, and tissues may be composed of complex cell types in typical plant analyses.

Gene expression is the realization of genetic potential. Global gene expression analyses were not available for plant breeding prior to large-scale sequencing projects. Gene expression results in a phenotype, which itself is the interaction of genotype and environment. While the genotype is accessible through inheritance and selection over generations in different environments, phenotypic expression of specific traits is often limited spatially or temporally, and for traits such as yield these effects accumulate throughout the growing season. Beet quality traits, for instance, are measured at the end of the growing season after the plant has experienced numerous changes in temperature, humidity, light intensity, soil moisture, and other environmental perturbations, each presumably having a measurable impact on plant growth and perhaps on final yield and sucrose content. Phenotypic responses to abiotic (and biotic) stresses may be individually predictable but undoubtedly are complex in their cumulative effects. Global gene expression analyses can begin to address this complexity by determining for example, which gene products are regulated under each type of stress. Each regulated gene product would have some probability for involvement in response to stress and would represent a potential target for breeding and selection. Perhaps more importantly these types of analyses can narrow the potential targets for breeding and selection to a few of the more important targets, which could then be incorporated into larger breeding objectives.

There are a number of inherent assumptions of functional genomics that may limit its direct application to breeding *per se,* but these are currently outweighed by benefits building a catalog of expressed genes in sugar beet. One assumption is that where gene functions have been identified in one species, identical or very similar nucleotide sequences in another species should encode the same or similar proteins, which are functionally equivalent. This form-follows-function argument is a reasonable first approximation but needs further rigorous confirmation at the biochemical level (Hofer and Ellis, 2002). A second assumption is that the presence of a transcript in an experimental analysis suggests a role of that product in the tissue under analysis. Gene expression is not necessarily precise and large differences may be encountered for a variety of reasons in replicated analyses. Large differences in abundance of specific transcripts under well-defined conditions would at least suggest some biological relevance. A third assumption is that identification coordinately expressed genes will ultimately allow genetic manipulation of specific biochemical pathways. This may be generally true but may be tempered by biochemical compensation among affected but unrelated pathways. Many abiotic stress responsive pathways for instance, may share elements with other pathways even though the specific stimuli may be quite unrelated (Knight and Knight, 2001).

A prerequisite to global gene expression profiling is knowledge of nucleotide sequences of expressed genes. The typical approach to gaining this information has been to sequence cDNA (complimentary, or copy, DNA reverse transcribed from mRNA) clones, which by definition are derived from expressed genes present in the source plant organs. These sequences are then compared for similarity to the ever-increasing number of nucleotide sequences held in data bases, and sequences with high similarity to genes with known function are used to assign putative gene functions. As many as 40% of the gene products identified to date have no known biochemical function. A catalog of expressed genes is often generated by mass sequencing of cDNA libraries, with each cDNA clone sequenced once, resulting in a collection of Expressed Sequence Tags (ESTs). ESTs by definition are preliminary and unsubstantiated indications of actual nucleotide sequence and often do not span the entire length of a gene, or may otherwise contain errors introduced during cDNA library construction (e.g. chimeric sequences derived from more than one gene product) or through the sequencing reactions themselves (e.g. misreadings due to polymerase failure).

Care is usually exercised in attributing specific functions to individual ESTs in the absence of confirming information. Land plants are thought to be monophyletic (Qiu and Palmer, 1999) and their genes derived from a common ancestor. Depending on the level of similarity between any two sequences, a putative protein functional class can be assigned for many ESTs (Burks, 1999). Most of the functional classes belong to various biochemical pathways, so a complete set of nucleotide sequences for an organism can define the types of biochemical reactions that might occur in that organism. Different cell types have characteristic patterns of biochemical processes, and global analyses of gene expression, such as an EST survey, can quickly indicate the biochemical status of the cells and tissues sampled. Differential expression of genes, especially among genes of a common pathway, provides a measure of the importance of any particular biochemical process on the organism's growth and development. This information can be used directly for selection and breeding during the more critical development phases and stress responses encountered during beet growth.

Progress in gaining nucleotide sequence from beet and other plants, has been remarkable in the past five years. The first complete nucleotide sequence of a plant genome, *Arabidopsis thaliana*, was completed in 2000 (The *Arabidopsis* Genome Initiative, 2000), providing an unparalleled opportunity to access plant genes. Gross characterization of *Arabidopsis* gene content revealed some striking features, which have general relevance for plant improvement. First, *Arabidopsis* has ca. 25,500 genes, although the gene discovery process is not complete (Allen, 2002). Approximately 70% of the genes could be assigned to a functional category, such as metabolism, cell growth and division, plant defense, energy production, etc., about 30% showed sequence similarity to expressed genes (i.e., ESTs) for which no function could be reliably categorized. In relation to other fully sequenced multicellular eukaryotic genomes of fruit fly (*Drosophila melanogaster*) and nematode (*Caenorhabditis elegans*), *Arabidopsis* shared most similarity in genes with basic metabolic functions and least similarity in genes that sense and respond to environmental and developmental signals. Development and responses to the environment involve changes in gene transcription, and *Arabidopsis* showed a greater number, diversity, and uniqueness of transcription factors and related proteins than previously surmised, supporting a role for these types of processes in the diversification of plants and perhaps varieties as well. Genomic characterization of rice (Yu et al., 2001; Goff et al., 2002), tomato (van der Hoeven et al., 2002), and other crop plants supports these generalizations obtained from *Arabidopsis*, except that gene number has been estimated between 30,000 and 50,000. *Beta vulgaris* would thus be expected to fall within this range for total gene number (Herwig et al., 2002), although significant differences in gene regulation, gene copy number, and presence or absence of specific gene classes could be expected from differences in beet form and function relative to other species in other plant families.

In *Arabidopsis* and other organisms, the number of protein families (i.e., those with similar biochemical activities or protein structures) was estimated at 11,000 protein families. Redundancy of genetic information is a hallmark of plant genomes, and duplicate genes are frequent. In *Arabidopsis*, the level of gene duplication appears to be about 60%, and in other diploid species such as rice this figure may approach 80%. Although these estimates are based on nucleotide sequence comparisons, different members of a gene family often have different roles in development and in response to environmental changes, but their function and expression may range from full complementation to completely different functions (Hofer and Ellis, 2002). There appears to be little or no regularity in the organization of duplicated chromosome regions in beets (Hallden et al., 1998; Schondelmaier et al., 1996), which suggests that the evolution of *Beta vulgaris* has not involved polyploidization or large-scale genome rearrangement in its recent past. However, sugar beet appears to have a substantial number of duplicated gene sequences. In preliminary experiments using 17 ESTs as probes against a BAC (Bacterial Artificial Chromosome) library with five sugar beet genome equivalents, on average 13 BAC clones per probe were identified (2.6 genes per probe per genome equivalent), with a range of 1 to 39 BAC clones per probe (McGrath, 2003).

One task for future sugar beet research is determination of which specific genes are of interest in germplasm improvement. A list of genes expressed in sugar beet is one of the earliest objectives to be accomplished. It is perhaps cost prohibitive to sequence the entire sugar beet genome at present. EST projects are more affordable and a great deal of progress on this objective has occurred recently. Insufficient time has elapsed to have fully explored these new resources and rapid progress can be expected. As of July 1, 2003, nearly 20,000 *Beta vulgaris* nucleotide sequences had been deposited in the National Center for Biotechnology Information (i.e. GenBank, www.ncbi.nlm.nih.gov). The majority of these are ESTs from mRNAs expressed in seedlings, roots, leaves, flowering inflorescences, and various pathogen-infected materials. The GABI set is unique in that clones were preselected prior to sequencing to remove a large proportion of redundant transcripts and thus represents a "unigene" set of over 10,000 unique expressed gene sequences covering the important developmental stages of beet growth (Herwig et al., 2002). Sugar beet researchers now have direct access to over one third of the expected

Beta vulgaris expressed genes. Distribution of these ESTs into predicted protein functional classes has been done for stressed seedling ESTs (**Plate 4.2**, McGrath et al., 2001).

How these resources will ultimately benefit sugar beet breeding and genetics is difficult to predict, but their importance should not be underestimated. For the first time, breeders have direct access to testing specific gene functions, such as those genes differentially expressed in root tissues (Bellin et al., 2002; Kloos et al., 2002), not just the correlation of phenotype with genotype. Allelic variation at agronomic loci will be able to be precisely measured and specifically incorporated where desired. Gene interactions will be accessible to mechanistic analyses, and quantitative genetic data will be defined in terms of biochemical mechanisms. Basically, the internal workings of the beet plant will be transparent, and thus allow better, more efficient, and more rationale breeding objectives to be posed, with results more precisely measured and predictable. It may be possible to select plants via molecular gene expression patterns in the field, which could be a substantial advantage in choosing parents for further breeding. A great deal of work will be required to realize these goals and some of the early applications of genomic technologies toward these ends, specifically for sugar beet, are described below.

Genetic mapping in sugar beet is reviewed in section 4.2.2. One of the first applications on obtaining ESTs will be to map these genes onto sugar

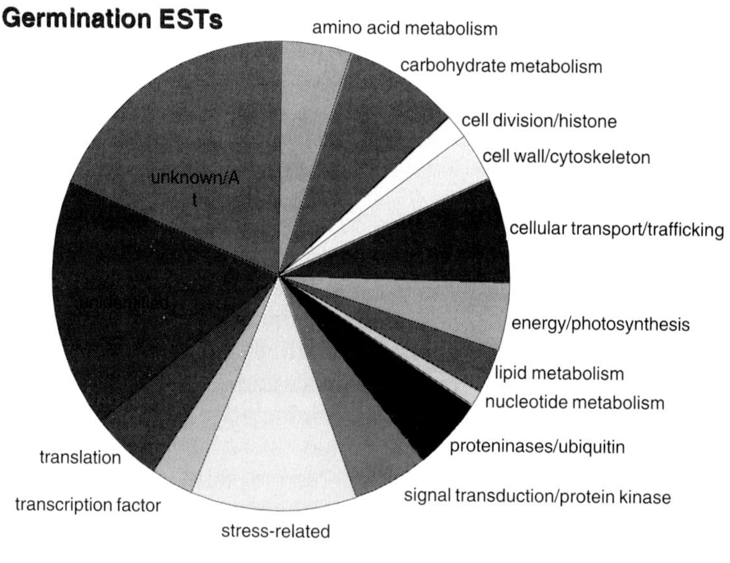

Plate 4.2 EST Pie Chart from PAG (from McGrath et al., 2001).

beet linkage groups and chromosomes, with applications possible for comparative mapping of crop genomes (Fulton et al., 2002). Of great interest are expressed genes that cosegregate with Quantitative Trait Loci (QTLs). One approach, the candidate gene approach (Tabor et al., 2002), supposes that certain proteins exert a major influence on expression of a trait, and that by mapping these candidates, some will map at the same genetic position as measured QTLs. Schneider et al. (1999) developed such a list of candidate genes involved in the accumulation of sucrose in sugar beets, isolated 54 separate gene products, and mapped 42 of these to the nine linkage groups of beet. Further work with these and additional ESTs, coupled with QTL analyses for a number of nondisease agronomic traits (e.g. sugar yield, beet yield, sucrose content, and impurity levels) uncovered many potentially useful associations of QTLs and specific genes (Schneider et al., 2002). This approach allows for further testing of association between QTLs and specific genes in other populations, and provides independent insight since QTL approaches alone have inherent limitations from being subject to potentially large environmental variation. Of particular interest is the recent determination of an entire class of disease resistance analogue genes (Meyers et al., 2003) in beets (Hunger et al., 2003) that will allow cosegregation analyses with disease resistance QTLs.

Another approach to gaining insight into the specific genes that control agronomic traits is to survey gene expression under the specific conditions in which a trait is expressed. This approach requires stringent control of the environment, such that condition can be repeated over many experiments and in particular across at least two genotypes with extreme contrasts for the response. Such an approach has been used to examine seedling vigor and has resulted in identification of at least two biochemical pathways leading to enhancement of seedling vigor (de los Reyes and McGrath, 2003). Using a simple *in vitro* solution germination regime that correlated reasonably well with field emergence (McGrath et al., 2000), differential gene expression analyses of mRNA profiles revealed a number of transcripts differentially regulated between extremes of high and low seedling vigor germplasm. Some of these transcripts were identified as members of a signal transduction pathway that were specifically expressed in the high vigor germplasm but not the low, identifying a genetic target for vigor enhancement, and another pathway for enhanced energy utilization through mobilization of lipid reserves during heterotrophic growth mediated by the glyoxylate pathway (de los Reyes, unpubl. data). Such promising candidates would not have necessarily been identified through a candidate gene approach; but their involvement in actual field emergence and suitability for gene-specific marker-based selection remain to be tested. It should be noted that development of suitable test environments, such as the *in vitro* germination assays, could find use as surrogate selection criteria, provided strong association with agronomic performance is assured.

A further application of gene nucleotide sequence data is in the current development of genetic markers based on single nucleotide polymorphisms (SNPs). SNP markers have a number of advantages over current technologies of hybridization- or PCR-based marker approaches. SNPs are perhaps the most frequent type of DNA polymorphism (Schneider et al., 2001; Rafalski, 2002), and considering that expressed genes are constrained from evolutionary changes by the requirement for gene function, uncovering genetic markers with sufficient polymorphism within genes could prove difficult. This does not appear to be a problem in beets. Schneider at al. (2001) examined and sequenced 37 genes developed from ESTs in two inbred sugar beet lines, and found 1 SNP per 283 bp within coding regions, and 1 SNP per 130 bp if introns and 5' and 3' flanking sequences were also considered. A brief comparison of 43 putative orthologous EST pairs in GenBank also showed that 74% of pairs differed in sequence with all showing more than 1 putative SNP present (Shaw and McGrath, unpubl. data). In 400 specific regions defined by ESTs, Schmidt et al., (2003) also showed that 75% of sequences derived from 16 divergent *Beta vulgaris* germplasm sources show SNPs, with an average of 4.6 SNPs per 200–600 bp. Although there are a number of competing technologies for SNP detection, the most difficult aspect in SNP development is the nucleotide sequence base from which comparisons can be drawn. SNPs are generally co-dominant; thus all products of recombination can be scored in mapping populations.

4.1.5 Other Genomic Considerations for Beet Improvement

Sugar beet breeding often follows a population improvement paradigm due to self-incompatibility in *Beta vulgaris*, and small flowers with windblown pollen that makes pollination control difficult, although this is slowly changing with deployment of a dominant self-fertility gene and development of doubled haploids. Cultivated sugar beet originated relatively recently from a small founding population, and allelic diversity in sugar beet is substantial but relatively narrow compared with allelic diversity in wild beets (e.g. *Beta vulgaris* spp. *maritima*). No convincing estimates of the difference in allelic diversity between cultivated and wild germplasm exist, but cultivated germplasm may only represent 10–25% of the alleles in *Beta vulgaris* spp. *maritima*, and perhaps less if all *Beta* species are considered. Much of this diversity is held in trust in germplasm banks worldwide. Expanding the diversity of cultivated germplasm is a priority and such wild species have contributed new sources of disease resistance, and promise to contribute additional useful genes or alleles in future.

Accessing genes and alleles in wild and unadapted germplasm is a long-term endeavor. The ultimate use of genomic technologies may be to bring this germplasm into use more efficiently. One possibility is the targeted evaluation of wild alleles at determined agronomic loci, followed by specific substitution for novel alleles by traditional backcross methods. Alternatively, multigenerational self-pollination enforced with use of the self-fertility gene between hybrids of sugar beet and a range of diverse germplasm types may reveal novel gene interactions, or uncover cryptic genes for agronomic and horticultural deployment. In either case, knowing the specific genes and alleles essential for beet productivity is essential.

Genomics may also suggest means for achieving alternative uses of sugar beet beyond sucrose, molasses, and fodder. Genetic transformation is one means of introducing novel variation for end use; deploying this technology is premature however, in light of the lack of fundamental knowledge about the number, identity, and diversity of genes and alleles already present in beets. One could imagine for instance, that a simple addition of one gene might resurrect a dormant pathway rendered inactive by gratuitous mutation at one step of a biochemical pathway, or perhaps nudging metabolic production toward a desired endpoint by allelic substitution. Analyses of the expressed gene portion of the genome will allow rationale targeting of such processes to effect accumulation of other useful industrial precursor chemicals besides sucrose for instance, or introduce biochemical modifications of sucrose for suitability as a precursor of polymers and other consumer products. Sucrose is the most abundant, chemically pure, renewable chemical resource, and advanced applications are prevented by unwanted side reactions due to the number of reactive chemical sites on the native sucrose molecule.

In conclusion, sugar beet genomics is an extension of traditional breeding and modern genetic methods. Its fundamental utility lies in the ability to define specific elements of the genome, first in terms of nucleotide sequence and later in terms of specific function. Such knowledge is essential for manipulating the beet genome for increased production efficiency and alternative uses. Such knowledge has already suggested specific targets for genetic improvement and will define genetic limitations inherent in the germplasm. Many long-standing production problems related to variety will be accessible through genomic analyses, and both pedestrian and surprising biochemical mechanisms to deal with these problems will be suggested from their results. Sugar beet breeders will need to incorporate this new knowledge into their breeding programs and should be intimately involved in generating and interpreting genomic information from their specific programs. Extensive similarities among plant genomes will expand the range of information about the inheritance of agronomic traits in different species, some of which will be directly applicable to beet improvement.

4.2 Molecular Breeding (G.N. Skaracis)

4.2.1 Introduction

Molecular breeding, broadly speaking, includes a series of methods based on the development and exploitation of appropriate molecular markers in the study and management of genetic material. Apart from enhancing our understanding of underlying genetics, use of all relevant technology aims at drastically improving overall efficiency of mainstream breeding programs. Basic molecular breeding applications refer to: i) construction of genetic linkage maps; ii) fingerprinting genotypes; and iii) marker-assisted selection through tagging of genes involved in the expression of agronomically important traits.

The term molecular marker describes polymorphism revealed at the molecular level as opposed to morphological genetic markers based on phenotypic differences. These latter were historically the earliest to be employed and their typical application in sugar beet relates to use of red hypocotyl coloration for ensuring selection of hybrid progenies in crosses between self-compatible or partially self-incompatible parents. The limited availability of morphological markers however, as well as the dependence of their expression on the environment and genic interactions, has considerably restricted their use for selection purposes.

Molecular markers can be grouped in two main categories: biochemical markers (storage proteins and isozymes) and DNA-based markers. Isozymes (i.e., different molecular forms of the same enzyme) have been successfully used for a number of applications in crop plants such as characterization and identification of genotypes and population genetic studies. An advantage of this type of markers is their ability to distinguish between heterozygotes and homozygotes while their shortcomings refer to the insufficient number, dependency on the environment, and developmental stage of the plant. Gene interactions might also complicate analysis and interpretation of the data obtained. DNA-based markers allow for detection of sequence variation directly at the DNA level. These markers are phenotypically neutral, environmentally independent, and not influenced by the developmental status of the plant. At the same time the number of this kind of markers is practically unlimited. DNA markers are grouped in two basic categories: i) markers based on hybridization approaches and ii) markers based on DNA amplification methods. DNA markers such as RFLPs, RAPDs, AFLPs, and SSRs have been most commonly used to date.

— Restriction Fragment Length Polymorphisms (RFLPs) were the first markers used in several crops (Tanksley et al., 1989), including sugar beet. Based on hybridization to specific probes, exploitation of variations in DNA restriction fragment length has the advantages of a strong discriminating power and a solid reproducibility. On the other hand, this methodology is costly, time consuming, and requires a large amount of pure DNA for analysis.

— Random Amplified Polymorphic DNA (RAPD) methodology is based on PCR amplification of anonymous stretches of DNA with use of short arbitrary primers (Williams et al., 1990). Its advantages are speed, simplicity, and relatively low cost, while no previous knowledge of genomic sequences is required and, like all PCR-based markers, automation of the process is possible. The disadvantages of this type of markers include reproducibility problems and their inability to discriminate homozygous and heterozygous genotypes.

— Amplified Fragment Length Polymorphisms (AFLPs) are derived from selective PCR amplification of restriction fragments upon digestion of total genomic DNA (Vos et al., 1995). The main advantage of this approach refers to the large number of loci analyzed in a single assay whereas the method is also highly sensitive and reproducible. On the hand, these markers are costly, technically demanding and dominant, although the latter can be overcome on the basis of gel scanning and use of special software.

— Microsatellites or Simple Sequence Repeats (SSRs) consist of DNA segments containing one or a few tandemly repeated nucleotides (Morgante and Olivieri, 1993). Polymorphism is revealed by differences in length (i.e., number of repeats) of the PCR amplification product using specific primers flanking the microsatellite. Although these codominant markers require considerable investment in sequence information, they are highly polymorphic (many alleles at a microsatellite locus) and distributed more or less throughout the plant genome. Once developed, simplicity of SSRs renders them quite attractive for breeding purposes.

Plant breeders' choice for type of markers to be employed depends on the particular goal as well as simplicity of the procedure involved and the possibility of automation, and usually a combination is desirable. In addition to the aforementioned marker systems, further promising techniques have recently been developed based on expressed sequence tags (ESTs) and single nucleotide polymorphisms (SNPs), as described in Section 4.1.

4.2.2 Genetic Mapping

Construction of genetic maps is accomplished by orderly placement of several markers according to relative genetic distances among them and their assignment to respective linkage groups based on all pairwise recombination values. Development of saturated linkage maps is considered a very important step for all studies related to gene function, regulation, and expression as well as for gene cloning purposes. In terms of practical breeding applications, they certainly become a prerequisite for any useful exploitation of molecular markers in marker-assisted selection, in particular as far as quantitatively controlled agronomically important traits are concerned.

Molecular linkage maps initially reported in sugar beet were based only on isozymes assigned to three (Wagner et al., 1992) and four (Abe et al., 1993) linkage groups. Morphological characters such as annuality, red hypocotyl-color, pollen fertility restoration, monogermity, and fasciation were also included in these first maps.

A first linkage map of sugar beet genome, including RFLP markers was reported by Pillen et al., (1992) and, upon its refinement, an extended linkage map was provided by the same research group (Pillen et al., 1993). The map included a total of 177 markers (168 RFLP markers, 7 isozymes and 2 morphological traits), distributed over nine linkage groups corresponding to the haploid chromosome number of sugar beet and covering a total length of 1057.3 cM with an average spacing of 6.0 cM. The morphological traits mapped were the hypocotyl color gene and the restorer gene (*X*). Later the map was further extended with more RFLP markers as well as the bolting gene and genes for nematode resistance derived from wild species (Schondelmaier et al., 1995). Immediately following Pillen's first report, a linkage map including 92 RFLP loci and loci controlling rhizomania resistance (*Rr1*), monogermity and hypocotyl color was reported by a different group, covering a total length of 540 cM (Barzen et al., 1992). A high density map, based solely on 413 RFLP markers was subsequently reported covering 621 cM with an average distance of 1.5 cM between markers, and these markers were profoundly clustered over the nine linkage groups (Hallden et al., 1996).

Use of RAPD markers in sugar beet mapping was first reported by Uphoff and Wricke (1995). This map had a total length of 738 cM and included 87 RAPD along with 1 RFLP marker, 5 isozymes and genes for resistance against the nematode, the restorer locus for male sterility, and the genes for annuality and hypocotyl color. Employing a considerable number of both types of DNA markers, Barzen et al., (1995) presented an extended map of the sugar beet genome containing 248 RFLP and 50 RAPD loci, the loci for rhizomania resistance (*Rr1*), hypocotyl color, and the locus controlling the monogerm character, to the nine linkage groups. Total coverage of the genome obtained was 815 cM.

Development of a RAPD-based map in sugar beet and its comparison with an RFLP map generated on the same mapping population was reported by Nilsson et al., (1997). The total map coverage of 508 cM included 160 RAPD and 248 RFLP markers characteristically clustered in the center of each of the nine linkage groups. Addition of the RAPD markers increased the total map length from 440 to 508 cM and led to minor changes in loci order.

AFLP markers have also been used for establishing linkage maps in combination with RFLP markers. Schondelmaier et al., (1996) reported integration of 120 AFLP markers in an RFLP-based previously developed map. AFLP loci were in general evenly distributed over the nine linkage groups. The integration caused a slight extension in total map length from 544 to 557 cM. A map exclusively constructed by the AFLP loci resulted in slight changes of marker order. It was suggested that AFLPs might be useful in QTL-analysis provided that map development includes one or two codominant markers (i.e., SSR, STS, RFLP) per linkage group. Another map, based on a combination of AFLP and RFLP markers, was reported comprising nine linkage groups and 891 loci. Loci detected by the various marker types presented a different distribution, but all marker classes were generally clustered in the same regions (Barnes et al., 1996). Recently, SSR markers have been mapped and assigned to each of the nine linkage groups with the aid of RFLP probes so selected as to give relatively full coverage of the sugar beet linkage groups (Rae et al., 2000).

Relatively dense maps described above have mainly been used for identification of major genes for characters such as fertility restoration, bolting, and resistance to rhizomania and nematodes. However, the association of markers with quantitative trait loci (QTL) controlling most of the agronomically important traits is highly desirable as it could provide for a more efficient indirect selection. QTL analysis and mapping in sugar beet has been reported for resistance to cercospora leaf spot (Schafer-Pregl et al., 1999; Setiawan et al., 2000); yield and yield components (Weber et al., 2000; Schneider et al., 2002).

The first genetic map incorporating not only anonymous markers, but also functional genes of sugar beet was presented by Schneider et al., (2002). Seventy-five ESTs related to sugar metabolism and transport, as well as 117 known markers (AFLP and RFLP) produced a map with a total length of 446 cM.

As mentioned earlier, successful exploitation of genetic maps depends on their highest possible completeness and saturation. The degree of possible saturation is related to the polymorphism present in the mapping population. Given that several maps exist for sugar beet, constructing a combined map could theoretically increase density and would allow for estimation of distances between markers that are not simultaneously polymorphic in the segregating populations. A minimum of two common polymorphic markers is required for any linkage group while combination is meaningful only if common markers are similarly located in the initial maps (Schumacher et al., 1997).

The possibility of effectively combining different sugar beet genetic maps from three populations (two of the same and one of different structure) was investigated by Schumacher et al., (1997). Three combined maps, based on maps from two populations each, were constructed. Maps from populations with the same structure could be integrated into a single map, a result consistent with findings of Hallden et al., (1996) who also combined maps from two populations of the same structure. On the contrary, the order of common markers proved different between maps stemming from populations of different structures, thus rendering the respective combined maps insufficient. It was suggested that the combined map could only be used as a tool for finding markers linked to traits of interest and not for genomic structure analysis (Schumacher et al., 1997). Difficulties encountered when comparing information from genetic maps of diverse origin were also revealed in studies by Weber et al., (1999) and Schafer-Pregl et al., (1999). It is evident that a uniform linkage group nomenclature and a common set of publicly available markers would definitely insure consistency and allow for successful combination of data originating from various laboratories.

A desirable feature following construction of molecular maps is the ability to associate linkage groups established with previous findings based on classical approaches. An important development in this direction has been the RFLP marker-mediated assignment of the nine linkage groups to each of the sugar beet chromosomes of the Butterfass (1964) complete trisomic series (Schondelmaier and Jung, 1997). The authors proposed that the nine chromosomes should be named according to the numbering system of the aforementioned trisomic series.

4.2.3 Marker-assisted Selection

Plant breeders have long been searching for genetic markers as a valuable tool to assist in their selection endeavors. Such indirect selection is mostly desired when dealing with traits whose phenotypic expression is evaluated with considerable difficulty and cost (i.e., large environmental dependency, resistance to biotic and abiotic agents, etc.). Molecular markers acting as genetic tags to the genes of interest can be scored easily and at an early developmental plant stage. In this manner, selection based on them (marker-assisted selection, MAS) becomes an efficient and reliable method to complement classical breeding procedures. In particular, important applications of marker-assisted selection include: i) hastening progress in backcrossing programs by selecting for a higher proportion of recurrent parent genome at early generations; ii) pyramiding different disease resistance genes, but with similar phenotypic expression, in various patterns in order to achieve a more robust and stable field performance; and iii) monitoring incorporation in elite lines of transgenes having major effects on traits such as herbicide and insect resistance as well as desirable characters from wild relatives.

A most important requirement for successful MAS in plant breeding is the availability of molecular markers tightly linked with trait genes (a distance of less than 1 cM is considered sufficient). A basic approach for the establishment of an association between a molecular marker and a trait is based on the study of their correlated inheritance in appropriate segregating generations (i.e., F_2 or BC1) from crosses between parents differing in the expression of the character and that are also polymorphic for the particular marker. Desirable characters can also be associated to molecular markers by several other methods such as: i) use of backcross-produced near-isogenic lines (NILs), essentially differing only for the presence of the target gene and for a specific marker (Martin et al., 1991); and ii) bulked segregant analysis (BSA) whereby two pooled DNA samples from a segregating population of a single cross are constituted on the basis of trait expression and screened for differences in the presence of the marker. Association is finally established upon linkage confirmation at the single plant level (Michelmore et al., 1991) and use of numerical and structural chromosomic aberrations such as monosomic addition lines, translocations etc. (Jung et al., 1992). Due to a significantly more complicated situation in quantitatively controlled characters (many interacting genes, several chromosomal regions involved, environmental evaluation required, etc.), increasing MAS efficiency for these traits is being sought through specific field design and statistical approaches to mapping of the respective quantitative trait loci (QTLs).

4.2.3.1 Rhizomania resistance

Identification of a DNA marker linked to the major gene *(Rr1)* conferring "Holly-type" rhizomania resistance was first reported by Barzen et al., (1992). The gene was mapped on linkage group IV in association with

RFLP marker located at 6.7 cM from the *Rr1* locus. Subsequently, two RAPD markers were identified as tightly linked (1.4 cM) to the *Rr1* allele (Barzen et al., 1997). These markers were converted to SCARs (sequence-characterized amplified regions) and their combined use in a single PCR assay could allow for the selection of homozygous-resistant individuals in segregating populations.

RAPD markers associated to the same type of rhizomania resistance were also studied later by a different group (Merdinoglu et al., 1995; Pelsy and Merdinoglu, 1996). A QTL for resistance accounting for 67.4% of phenotype variation was identified and located in a 4.6 cM interval between two markers; it was suggested that the identified QTL corresponded to the single major gene controlling the "Holly-type" resistance. However, it was concluded that a partial rather than complete dominance was involved and hence the resistant allele was designated *Rz1*.

Inheritance of resistance to rhizomania and identification of molecular markers linked to the trait were studied in various sources of resistance at CPRO-DLO. Plant material included two *Beta vulgaris* ssp. *vulgaris* accessions (Holly-1-4 from USA and R128 from Turkey) and two *Beta vulgaris* ssp. *maritima* accessions (R104 from Italy and WB42 from Denmark). Upon completion of the study it was concluded that resistance in Holly–1-4, R104, and probably R128, are situated at the same locus (*Rz1)* whereas resistance in WB42 lies at a different, closely linked locus *(Rz2)* with a distance of about 20 cM between *Rz2* and *Rz1*. Specific RAPD markers linked to both *Rz1* and *Rz2* loci were identified and converted into STS markers to be used for marker-assisted selection in breeding programs. But given the tight linkage between gene loci, combining the two genes in the same plant material would require markers flanking resistance at very short distances (Scholten et al., 1996, 1997, 1999).

Development of molecular markers linked to the "Holly" gene resistance to rhizomania has also been reported by Giorio et al., (1997), Francis et al., (1998), and Francis and Asher (2000).

4.2.3.2 Bolting trait

The first molecular marker tightly linked to the bolting gene *(B)* was an isozyme marker and the authors suggested that the isozyme locus (Idh_1) could be used as a marker for tagging the dominant *B* allele in segregating populations (Abe et al., 1993). In an earlier study in the same research group, five independently inherited isozymic loci were used as chromosome markers in an analysis of the annual and early flowering habit of *B. macrocarpa*. Two of these markers were linked to the 2-3 pairs of genes controlling the annual habit, and one of belonging to the same linkage as the sugar beet bolting gene (Abe et al., 1987).

Using information from an RFLP based map (Pillen et al., 1992), two RFLP markers flanking the *B* locus were identified, with respective distances of 3.8 and 5.2 recombination units (Boudry et al., 1994).

In a recent study (El-Mezawy et al., 2002) 17 AFLP markers loci were mapped at a distance of less than 3.2 cM around the bolting gene. Four closely linked markers were identified at distances ranging from 0.14 to 0.5 cM for the *B* locus. Perfect association with a gene was reported for two of these markers upon verification with a set of 88 sugar beet plants differing in bolting behavior. The usefulness of such markers, if proven in other genetic backgrounds as well, could very efficiently be used in breeding, seed lot control, and wild beet population studies.

4.2.3.3 Fertility restoration

Fertility restorer loci *X* and *Z* were located on sugar beet chromosomes III and IV according to chromosome nomenclature suggested by Schondelmaier and Jung (1997). For the restorer gene *X*, localized terminally on linkage group III, an RFLP marker at a distance of 9.6 cM was identified (Pillen et al., 1993); the exact location of restorer gene *Z* is still not known.

More recently, development of RAPD and AFLP markers associated to the restorer loci were reported. Of 812 RAPD markers studied, flanking markers have been observed at about 8 ± 3 and 16 ± 4 cM from the *X* gene. From a total of 1,936 AFLPs tested, 8 markers were found linked to the *X* locus which was restricted within 3.4 ± 1.5 cM by these markers and two of them were verified as cosegregating with the *X* gene (Hagihara et al., 1999).

Evenly distributed RFLP markers were used for identification of three major QTL regions associated with fertility restoration. Two QTL regions, 15 cM apart, mapped on chromosome IV, explained 79% of the variation observed. A third QTL region (5 cM) detected in a different population was assigned to the linkage group corresponding to chromosome III; it explained 72% of phenotypic variation. Placement of this QTL region corresponds to the earlier localization of the *X* gene, while that of the first two QTLs with the previous assignment of the *Z* gene. Nevertheless, as the authors state, it is still not clear which of the two restorer genes detected by QTL analysis, or do both correspond to restorer gene *Z* (Hjerdin-Panagopoulos et al., 2002).

A male fertility restorer locus associated with a new cytoplasmic male-sterility source (H-type) present in wild beets, was studied by BSA analysis and 9 RAPD markers linked to this restorer locus identified. RAPD and RFLP-marker based mapping of the gene region was compared to available sugar beet maps and the restorer gene (*R1H*) assigned to the same linkage group with the *Z* gene. Although the two genes appeared distant, it was postulated that a single locus repressing both the old and this new cytoplasmic sterility might be involved (Laporte et al., 1998).

4.2.3.4 Cercospora resistance

Being of a quantitative nature, resistance to cercospora l.s. could be traced by molecular markers tagging QTL regions affecting the trait and a few studies have reported identification and localization of such regions in sugar beet.

Based on 221 AFLP and 46 RFLP markers scored in a segregation for cercospora resistance population, five QTLs accounting for 7-18% of the phenotypic variation each, were detected by a composite interval mapping procedure. These QTLs were located on linkage groups 1, 2, 9 and two of them on linkage group 3 (Nilsson et al., 1999).

Schafer-Pregl et al., (1999) reported on identification of QTLs affecting cercospora resistance in an F_2 and a test-cross population using composite interval mapping. QTLs were analyzed under natural and artificial inoculation and repeated at different plant stages. Three major QTLs were detected on linkage groups 2, 6 and 9 according to data for both populations and under both inoculation conditions. Indications were also obtained for the presence of three more QTLs on linkage groups 2, 4 and 5 obtained in the F_2 population only, probably due to recessiveness of the alleles involved. The authors stressed that these QTLs appeared quite stable throughout plant development stages and the mode of infestation.

QTLs have also been studied in an F_2 population using a leaf disk approach that permits single plant evaluations (Koch and Jung, 2000). A total of 304 markers (RFLPs and AFLPs) were mapped on all sugar beet chromosomes and eight significant QTLs explaining 94.1% of the phenotypic variance detected. Four significant QTLs with alleles from the resistant parent were mapped, one on chromosome I (13.8% of the explained variance), two on chromosome III (11.4% each), and one on chromosome IV (20.5%). A similar study was almost simultaneously reported by Setiawan et al., (2000) with a different population, derived from the same parental material. Field and leaf disk tests were used to evaluate resistance level. Based on leaf disk test data, four QTLs were located on chromosomes III, IV, VII, and IX. The authors suggested use of QTLs located on chromosomes IV and IX for marker-assisted selection as well as for cloning purposes.

4.2.3.5 Powdery mildew

Considerable variation for resistance to powdery mildew (*Erysiphe betae*), within the germplasm of sugar beet has been found and appears of oligogenic nature. Several sources of resistance have also been identified in wild *Beta* accessions, some of which seem monogenic. Identification of several QTLs for oligogenic resistance and preliminary mapping of monogenic resistance to a single locus were recently reported (Janssen et al., 2003).

4.2.3.6 Sugar yield-related traits

Marker analysis for corrected sugar yield, sugar content, sodium, potassium, and amino-nitrogen was applied in two segregating populations (one segregating for rhizomania resistance and the other for cercospora resistance) tested in a series of environments in an attempt to identify QTLs for these characters (Weber et al., 2000). A number of QTLs were defined but they generally mapped in different chromosomal locations in the two populations and only a few were stably expressed in the same population across environments. It was concluded that selection on a QTL basis would require a more precise QTL estimation and that predictions are feasible only for similar environments.

Expressed sequence tags (ESTs) associated with functions in carbohydrate and nitrogen metabolism were used as molecular markers to map QTLs for 7 traits related to sugar production, including corrected sugar yield, beet yield, ion balance and content of sugar, amino nitrogen, potassium and sodium. Phenotypic evaluation of test-cross progeny was performed at six different locations. A total of 21 significant QTLs were detected for all traits analyzed across environments and linked ESTs were identified. Four of these QTLs (located on chromosome IV for root and corrected sugar yield, on chromosome IX for sucrose content and on chromosome II for potassium) were found environmentally stable. Validation experiments at the physiological expression level for these candidate genes are planned (Schneider et al., 2002).

4.2.3.7 Nematode resistance

Resistance genes to the beet cyst nematode (*Heterodera schachtii*) are found only in wild species of section *Pateralles*. At least three different resistance genes have been distinguished, i.e. *Hs1* in the homoeologous chromosomes I of each species, *Hs2* in the homoeologous chromosomes VII of *B. procumbens* and *B. webbiana,* and *Hs3* in chromosome VIII of *B. webbiana* (Kleine et al., 1998).

Isozyme markers were initially used to characterize all nine different component chromosomes in *B. procumbens* and *B. webbiana* monosomic additions and one such marker identified chromosome 7, containing a gene for resistance to the beet cyst nematode (Reamon-Ramos and Wricke, 1992). Genome specific DNA-based probes were used for identification of wild chromosome segments harboring the gene *Hs1* for resistance allowing for selection of plants in the segregating progenies of monosomic addition lines (Schmidt et al., 1990; Jung and Herrmann, 1991; Jung et al., 1992; Salentijn et al., 1992). RAPD markers were also identified linked to resistant genes *Hs*$^{pro\text{-}1}$ (Hallden et al., 1997) and *Hs*$^{pat\text{-}1}$ In the latter case, two of six markers developed were closely linked to the *Hs* $^{pat\text{-}1}$ and one was converted to STS markers usable for gene cloning (Salentijn et al., 1995). With the use of genome-specific satellite markers and chromosomal breakpoint analyses, the *Hs1* $^{pro\text{-}1}$ locus was successfully cloned (Cai et al., 1997) for further use in genetic engineering approaches.

Simply inherited resistance to root-knot nematode (*Meloidogyne* spp.) has been identified (Yu, 1995) from two rare accessions of the sea beet (*Beta vulgaris* ssp. *maritima*). Phosphoglucomutase (PGM) was found to be a potentially useful isozyme marker of resistance to *Meloidogyne* spp. from one (Mi-1) *Beta* source (Yu, et al., 2001); however PGM was absent in genotypes deriving from the other (M66) source. At least five DNA markers were found associated to the resistance from this source and a cleaved amplified polymorphic sequence (CAPS) marker cosegregating with resistance to root-knot nematode was produced (Weiland and Yu, 2001, 2003).

As evidenced in both cyst and root-knot nematodes, successful tagging of genes for resistance derived from different sources would greatly facilitate their simultaneous introgression into the genetic background of one recipient genotype through marker-assisted backcrossing.

Published results of research on the development of molecular markers reveal that most of the work is still at the stage of identifying markers tightly linked to genomic regions of interest. With a few probable exceptions, practical large-scale marker-assisted applications are still limited in sugar beet breeding. The amount of ongoing research work however, combined with the adoption of newer and rapidly advancing technologies (for example identification and use of single nucleotide polymorphisms, microarray-based high-throughput detection of polymorphisms, etc.) is expected to drastically change the situation in the near future.

4.2.4 DNA Fingerprinting

DNA fingerprinting or genotyping comprises a well-established marker-based technology with a variety of applications in the fields of research, plant breeding, and identification of commercially distributed plant or seed material. Applications in research include analysis of phylogenetic relationships, and monitoring of gene flow and genetic variation in populations. In plant breeding, genotyping serves for characterization and handling of germplasm, assessment of fidelity in germplasm multiplication, assigning parents to heterotic groups, management of exotic genetic material in prebreeding programs, etc. Finally, market-related applications of genotyping are addressed to diagnostics as far as seed purity control and variety identification are concerned.

4.2.4.1 Genetic diversity in cultivated sugar beet and related species

A number of molecular approaches have been used to investigate genetic variation in sugar beet and related species. Restriction fragment length polymorphism, analyzed by cDNA probing, revealed the clear difference of *Beta nana* from both cultivated sugar beet and *B. vulgaris* ssp. *maritima* accessions. Less pronounced variation has been detected between cultivated beets and *B. vulgaris* ssp. *maritima* accessions as well as within closely related genetic material of sugar beet (Nagamine et al., 1989). Both cDNA and genomic clones, derived from a sugar beet variety and *B.v.* ssp. *maritima*, were used to assess the degree of genetic variation between and within different species from the *Procumbentes* and *Vulgares* sections. The two sections were found clearly divergent while *B. procumbens* and *B. webbiana* were the most closely related within their section. Considerable variation was found among sugar beet varieties. Genomic probes generated greater variability in beet than cDNA probes (Mita et al., 1991).

Variation in nuclear ribosomal DNA (rDNA) was used to characterize and differentiate species of three *Beta* sections *(Beta, Corollinae* and *Procumbentes*); the *Procumbentes* section appeared to be the most distant. Cultivated and wild beet groups were clearly differentiated by a unique molecular marker. *Corollinae* could be divided into two subsections, one including *B. lomatogona* and *B. intermedia* and the other *B. corolliflora, B. trigyna,* and *B. macrorhiza* species (Santoni and Berville, 1992).

RFLP markers based on repetitive DNA sequences (sugar beet minisatellites) have been used to fingerprint accessions from all sections of genus *Beta*. Sugar beet varieties showed a low level of genetic diversity and displayed high similarity only to wild *Beta vulgaris* ssp. *maritima* accessions. The *Beta* section, except for *B. macrocarpa* showing a closer similarity to *Corollinae*, was clearly distinct from the other sections. Wild species in *Procumbentes* showed the lowest similarity to all other species (Jung et al., 1993). Satellite DNA from *Beta nana* revealed the considerable distance of *Procumbentes* from other sections of genus *Beta*. Moreover, this repetitive DNA proved a useful tool for detection of genetic variation in closely related *B. vulgaris* varieties, accessions, and species (Kubis et al., 1997). RFLP markers applied on many plant pooled samples were used to estimate genetic variation in a large number of monogerm and multigerm sugar beet breeding lines as well as wild beet in the *Beta* section. Variation among breeding lines was lower than that found in wild populations. A higher among-line variation compared to within-line variation was observed in the monogerm group (Kraft et al., 1997).

RAPD was employed in the study of variation and relationships in annual wild beet in the *Beta* section (Shen et al., 1996). Variation detected in the *adanensis* or *macrocarpa* groups was considerably lower than that in the *maritima* group. The latter was found closer to the *adanensis* group while *macrocarpa* was the most distinct group, in good agreement with earlier isozyme-based analysis (Abe and Tsuda, 1987). The same type of markers were successful in investigating the origin of *B. vulgaris* ssp. *maritima* populations at the Baltic sea coast and clearly distinguishing wild and cultivated beets (Driessen et al., 2001). RAPD markers were also analyzed in order to assess variation among USDA sugar beet releases and to monitor breeding-related genetic diversity changes over a fifty-year period (McGrath et al., 1999).

4.2.4.2 Characterization of germplasm and varieties

As in all crops, a very important aspect of sugar beet breeding refers to the availability of diagnostic tools to accurately characterize lines and varieties as well as to effectively examine the fidelity of seed reproductions across years and environments. A clear characterization of a commercial variety and of its parental components is also highly desirable for registration purposes and for safeguarding breeder rights. Traditionally, characterization and identification of lines and varieties has been based solely on distinctness, uniformity, and stability criteria as evaluated by morphological descriptors, which are not always an efficient means for discrimination.

With the development of molecular marker technologies, several attempts at identifying sugar beet varieties have been reported. Some encouraging indications were provided, with strong intracultivar variation inherent in sugar beet clearly demonstrated at the molecular level (Nagamine et al., 1989; Mita et al., 1991). Further, Jung et al., (1993) reported identification of 10 sugar beet varieties with use of two minisatellite RFLP markers. RAPD databased multivariate analysis pointed to the need for a large number of plants per variety for their probable discrimination (Ford-Lloyd et al., 1993). Evaluation of distinctness, uniformity, and stability of sugar beet varieties based on AFLP fingerprinting was recently reported by De Riek et al., (2001). Data from several varieties along with their consecutive seed deliveries were analyzed by three different approaches (cluster analysis based on marker-frequency data, analysis of molecular variance, and individual plant-based assignment tests searching for the closest relationships). It was concluded that a practical use of molecular markers might be successful for distinction purposes and that they could be useful for the prescreening of new applicants for registration of varieties and the grouping of similar varieties in appropriate field trials.

In contrast to the considerable difficulties encountered in the characterization of heterozygous sugar beet material, unequivocal identification of the completely homozygous double-haploid (DH) breeding lines appears to be achievable, as reported earlier using a simple repetitive DNA probe (Schmidt et al., 1993).

Various researchers have reported use of molecular markers in the study of somaclonal variation in sugar beet. Isozymes and RFLP markers were used to analyze sugar beet, as well as fodder and table beet plants regenerated from callus obtained from various explants. Both types of markers allowed for identification of somaclonal variants and the frequency of variant alleles present in any one of them was estimated between 0.05% and 0.1% (Sabir et al., 1992). The same type of DNA markers were also used to determine the extent of somaclonal variation occurring at the molecular level in somaclones regenerated from UV-treated calli and an average mutation frequency per allele was estimated to be around 0.03% (Levall et al., 1994). Using RAPD markers, somaclonal variant plants regenerated from leaf explants of a single sugar beet genotype were identified with an average somaclonal polymorphism frequency similar to those reported with isozymes and RFLP markers (Munthali et al., 1996). Changes induced during tissue culture in sugar beet have also been detected at the mitochondrial level employing both restriction fragment length and randomly amplified DNA polymorphism (Dikalova et al., 1993; Jazdzewska et al., 2000; Sadoch et al., 2000).

4.3 *In Vitro* Culture Technique (G.N. Skaracis)

4.3.1 Introduction

In vitro culture techniques, more often referred to simply as plant cell or tissue culture, comprise the oldest set of biotechnological tools successfully exploited by the plant breeder. Indeed, these techniques are rooted in the past with a series of coincidences, careful observations, and scientific interventions gradually having shaped their modern use in plant breeding. The key principle, based on which all relevant applications evolved, was definitely that of cell totipotency as introduced by Haberlandt (1902). As the term implies, somatic plant cells have the potential to regenerate entire plants if cultured *in vitro* under appropriate nutritional, hormonal, and environmental conditions.

An extensive series of basic studies conducted over the years has assessed the impact of nutrient media, plant hormones, cell growth conditions, types of explants, genotype dependency, etc. on tissue culture and morphogenesis. These studies, thoroughly reviewed by Krikorian and Berquam (1969), Murashige (1974), and Thorpe (2000), were of paramount importance in establishing the usefulness of *in vitro* culture technology in crop improvement. The main applications of these techniques include: (i) micropropagation and germplasm preservation; (ii) virus-free plant production (shoot-apex culture); (iii) haploid-dihaploid plant development (androgenesis and/or gynogenesis); (iv) somatic hybridization (protoplast isolation and fusion); (v) *in vitro* fertilization and embryo rescue; (vi) somaclonal variation and selection; (vii) plant genetic transformation; and (viii) secondary metabolite production (cells-bioreactors).

Although must of the applications listed above can contribute to sugar beet breeding, practically the most important are those concerning micropropagation, development of haploids, exploitation of somaclonal variation, and regeneration as a prerequisite for gene transfer.

4.3.2 Micropropagation

Micropropagation or clonal propagation refers to *in vitro* multiplication of a desirable donor genotype. This kind of propagation in sugar beet can also be achieved, as demonstrated below, through direct regeneration on mature tissues such as leaves, petioles, cotyledons, etc. However, due to the need for maintaining genetic stability and producing large numbers of true plant copies, it is essential that the multiplication process be based on tissues with preorganized meristems rather than *de novo* meristematic development. Consequently, from among the various candidate types of explants, the most widely used in routine sugar beet clonal propagation are axillary buds, shoot tips and early stage inflorescence tips (for example: Margara, 1977; Hussey and Hepher, 1978; Atanassov, 1980; Coumans-Gilles et al., 1981; Saunders, 1982; Miedema, 1982a, 1982b; Mezei et al., 1990; Goska and Szota, 1992; Zhong et al,. 1993).

Shoot formation takes place when the explants are cultured on defined nutrient media, such as MS (Murashige and Skoog, 1962) and derivatives therefrom (i.e., B5, Gamborg et al., 1968; LS, Linsmaier and Skoog, 1965; PG_{0B} , De Greef, 1978; and others), typically containing inorganic salts, a carbon and energy source, vitamins, and plant growth regulators. It is the presence of relatively elevated levels of cytokinins that triggers initiation of shoots and encourages subsequent axillary branching. The process can be continuous provided culture conditions remain favorable and multiplied material is regularly transferred to new media. In practice, protocols used for shoot multiplication differ slightly from those for shoot initiation. When a desirable number of shoots is obtained, these are induced to rooting by being transferred to a new culture media where the hormonal balance has been altered in favor of auxins. The stimulus to rhizogenesis however, seems to have been predetermined because even in the absence of auxins, or in a minimal presence of cytokinins, adequate rooting can usually occur (Ranalli and Mandolino, 1999). Apart from the synthesis of nutrient media and type of explant used, all relevant studies on the efficiency of sugar beet micropropagation point to the significance of the donor plant's genotype. At the practical level however, protocols developed in various laboratories are easily adapted to ensure efficient routine clonal propagation of all desired genotypes.

The potential offered by micropropagation for obtaining a large number of copies from a selected plant, at a relatively low cost and in short time, is of obvious usefulness in sugar beet breeding since:

- It is the most efficient way of preserving self-sterile genotypes while evaluating their breeding value on the basis of progeny testing. In this way, all relevant selection schemes based on progeny evaluation can be greatly enhanced.
- Selecting on a single plant base for characters suffering considerable environmental influence very often proves unsuccessful. Efficiency can be considerably improved by using a number of its clonal copies instead.
- A synthetic variety (or a population of specific use) is more reliably reconstructed through the years when its components are preserved and reproduced by micropropagation.
- Newly discovered or aquired cytoplasmically male-sterile genotypes can be preserved before integration in the main stream breeding with the appropriate maintainers.
- Wild germplasm that is difficult to germinate and reproduce can be preserved and exchanged between breeders in the form of clonal propagules.

As far as preserving valuable genotypes in particular is concerned, besides the usual practice of storing *in vitro* plants at 4-5°C, long-term maintenance can be optimized by cryopreservation techniques such as slow freezing (Braun, 1988), encapsulation-dehydration (Vandenbussche and De Proft, 1996), and vitrification (Vandenbussche et al., 2000)

4.3.3 Plant Regeneration

The ability to reliably regenerate entire plants by the *in vitro* culture of various cells and tissue explants is of paramount importance for most biotechnological applications in plant breeding and definitely comprises a key feature for any successful transformation endeavor. Regeneration of whole plants can proceed both by direct and indirect modes. Direct regeneration occurs on the cultured explant tissue itself, whereas indirect regeneration requires an intermediate callus phase. Regardless of the mode involved, there are two alternative morphogenetic pathways by which regeneration can be accomplished: adventitious organogenesis and somatic embryogenesis. The organogenetic process is initiated with alterations in a single or small group of cells followed by divisions resulting in a plastic globular mass. This meristemoid structure gives rise to a unipolar shoot or root primordium, the choice being directly determined by the hormonal balance (auxin/cytokinin ratio) of the culture media. Thus, a complete plantlet is produced in two stages, i.e., shoot development and root formation. Somatic embryogenesis on the contrary, more often associated with suspension cultures, is a process whereby single cells are induced with embryogenic competence and give rise to bipolar structures with a root/shoot axis, resulting thus in a one-stage complete plantlet development. Both direct and indirect regeneration, by either adventitious organogenesis or somatic emryogenesis have been intensively studied and achieved in sugar beet.

4.3.3.1 Callus-mediated morphogenesis

Successful sugar beet callus production, proliferation, and eventual regeneration has been obtained since the beginning of the 1970s by the use of different explants such as leaves, petioles, hypocotyls, cotyledons, axillary buds, inflorescences, root tips, seedling tissues, embryos, protoplast and cell suspensions. White friable callus composed of cells with thin walls, dense cytoplasm, and prominent nuclei, produced from blackened tissues is usually regenerative. Contrarily, compact callus developed from green tissues consists of thick-walled cells and presents no regenerative capacity. The most frequent morphogenetic path reported has been that of adventitious shoot development (Margara, 1970; Butenko et al., 1972; Hooker and Nabors, 1977; De Greef and Jacobs, 1979; Saunders, 1982; van Geyt and Jacobs, 1985; Saunders and Shin, 1986; Tetu et al., 1987; Freytag et al., 1988; Doley and Saunders, 1989; Ritchie et al., 1989; Yu, 1989; Krens et al., 1990; Abe et al., 1991; Owens and Eberts, 1992; Shimamoto and Mikami, 1993; Jacq et al., 1993; Zhong et al., 1993; Lenzner et al., 1995; Roussy et al., 1996). Indirect somatic embryogenesis has also been obtained by various researchers (Atanassov, 1976; Tetu et al., 1987; Freytag et al., 1988; Kubalakova, 1990; Tsai and Saunders, 1995; Hall et al., 1996a,b; Saunders and Tsai, 1999; Moghaddam et al., 2000).

4.3.3.2 Direct morphogenesis

Regeneration without intervention of a callus phase offers the advantages of better genetic stability of the regenerants and potentially improved simplicity, while it also presents the most efficient approach when employing biolistic transformation techniques. Direct shoot regeneration in sugar beet has been obtained from apices, petioles, leaf axils and leaf fragments, cotyledonary nodes, and thin cell layer explants (Rogozinska and Goska, 1978; Saunders and Mahoney, 1982; Detrez et al., 1988; Freytag et al., 1988; Ritchie et al., 1989; Sand, 1995; Toldi et al., 1996; Grieve et al., 1997; Saunders et al., 2001; Zhang et al., 2001). Direct somatic emryogenesis has been achieved from immature zygotic embryo explants (Tenning et al., 1992), zygotic cotyledons of mature embryos (Kulshreshtha and Coutts, 1997), and intact seedlings (Zhang et al., 2001). Further developments for production of this type of embryo in suspension cultures, should definitely provide considerable advantages over the classical micropropagation approach: (i) low-cost scaled-up batch culture in bioreactors is feasible; (ii) production of whole plants is more efficient than in the two-step organogenetic process; (iii) multiplication rates would be very high and the embryos could be encapsulated and treated as artificial seeds (Tsai and Saunders, 1999).

4.3.3.3 Regeneration efficiency

Most of the relevant studies have been concerned with improvement of regeneration efficiency through possible optimization of the various parameters involved. Given the numerous combinations of nutrient media, exogenous growth regulators, type and age of explants, condition of the donor plant, culturing conditions etc. employed in the different studies, referring to each would be quite confusing, especially since many contradictions become obvious. Some generalizations (also see Atanassov, 1986; Ranalli and Mandolino, 1999) can be deduced though:

- All morphogenetic processes are largely genotype dependent (with the probable exception of stomatal guard cell protoplasts) and also depend on the type and concentration of exogenous hormones, presence of endogenous hormones, as well as on various conditions during the *in vitro* culture period (i.e., temperature, light, tissue support matrix, nutrients)
- Callus induction and proliferation is usually enhanced when cotyledons, hypocotyls and new leaves (all explanted from young rather than older plants) are used. Although genotypic dependency causes considerable inconsistency, auxins are generally important for callus formation and results are often better when a cytokinin is also present at a lower concentration (e.g. auxin

to cytokinin ratio much higher than one). Good results are obtained however, even with a reversed hormonal ratio, while fully habituated (hormone autonomous) regenerative calli have also been produced.

- Given the presence of endogenous auxins, regeneration usually requires a case-by-case specific exogenous hormonal balance while combinations of other relevant parameters are easier to manipulate.
- Despite the significant advances in all related issues, regeneration in sugar beet still cannot be considered a totally routine process.

4.3.4 Protoplast Culture

Protoplasts are plant cells deprived of their rigid cell walls as a consequence of enzymatic digestion in cellulase and pectinase. These fragile spherical cells, surrounded by the plasma membrane, have the ability to survive under appropriate osmotic potential conditions, to regain their cell wall apparatus, to divide and multiply, producing cell colonies, and eventually to regenerate entire plants of a single cell origin. Protoplasts are easily obtained either from leaf mesophyll cells or from cell suspension cultures and have been extensively used in both fundamental and applied studies. Isolation and culture of protoplasts were advanced in the 1960s in the framework of the then-developed protoplast fusion, a novel technique by which unique somatic cell hybrids between sexually incompatible species could be produced. Protoplast fusion is mediated by chemical means such as polyethylene glycol (PEG) or by the use of electrical pulses, both agents causing a reversible membrane permeabilization. As discussed in the next chapter, given their ability to take up exogenous plasmid or naked DNA, cultured and subsequently regenerated protoplasts also constitute a very efficient system for both vector-independent and vector-mediated gene transfer.

Development of protoplast technology in sugar beet has been quite slow compared to other plant species and until rather recently, only in a few cases had leaf-derived protoplasts been successfully isolated and cultured to form colonies (Bhat et al., 1985; Szabados and Gaggero, 1985; Atanassov, 1986). Petiole and hypocotyl-derived protoplasts were also cultured with relative success (Pedersen et al., 1993; Majewska et al., 1994; Schlangstedt et al., 1994). In none of these cases however, was regeneration achieved. Shoot regeneration from sugar beet mesophyll protoplast callus colonies, initially described by Steen et al., (1986), was further advanced in the framework of a program for the transfer of cytoplasmic male sterility by asymmetric protoplast fusion (Krens et al., 1990) and has subsequently been reported by Lenzner et al., (1995) also.

Although culturing requirements are not particularly demanding, high plating efficiency was obtained only after protoplast immobilization in calcium alginate (Hall et al., 1993; Lenzner et al., 1995).

Despite these advances, until the mid-1990s sugar beet was still considered a recalcitrant species and successful regeneration through protoplast culture was occasional and strongly genotype dependent. A breakthrough in developing an efficient and reproducible procedure for sugar beet protoplast regeneration, came from the demonstration that, although highly specialized, leaf stomatal guard cells were totipotent, producing regenerable callus (Hall et al., 1995, 1996a, b). This important development has subsequently been exploited as a powerful, genotype independent technique in sugar beet transformation and currently presents the only practical use of protoplasts in the crop.

4.3.5 Haploid Production

Although naturally occurring haploidy had been observed for quite a long time in plant species, interest in planned production was revived in the early 1960s when haploid embryos were obtained regularly with a satisfactory frequency from *in vitro* culture of immature anthers of *Datura* (Guha and Maheshwari, 1964). Since then, haploids have been produced and exploited in many crops, mainly for applied breeding purposes. The importance of this approach lies basically in the possibilities offered for detection of recessive mutations and production of "instant" inbred lines through chromosome doubling of the regenerated haploid plants. In this manner, breeding programs can be considerably facilitated by: (i) dramatically decreasing the time needed to reach homozygosity (one step process) both in self-compatible and self-incompatible material; (ii) efficiently fixing unique recombinants and *in vitro* produced mutations; and (iii) obtaining valuable diploid genotypes from respective tetraploid material. Haploidy is achieved by either androgenesis or gynogenesis. In androgenesis, haploids are microspore-derived following anther or pollen culture, although the latter is less frequently employed. Gynogenetic haploids are obtained from unfertilized egg cells in the cultures of unpollinated ovaries or ovules.

Early studies on various methods for haploid sugar beet plant development included spontaneous induction, pollination with nonfunctional pollen, and interspecific hybridization (Bosemark, 1971), as well as crossing between different ploidy level plants (De Jong and De Bock, 1978)—all of which proved to be of very limited usefulness. Haploidization through the androgenetic pathway also did not meet with a success ensuring applicability of this technique (Banda and Tanabe, 1972;

Atanassov, 1973; Welander, 1974; Rogozinska and Goska, 1982, van Geyt et al., 1985). On the contrary, gynogenetic development appeared to be the route for obtaining haploid sugar beet plants, presenting possibilities for potential utilization in breeding. Hosemans and Bossoutrot (1983) were the first to report on such haploid production by culturing unfertilized ovules from male-sterile plants and similar results were achieved with male-fertile genotypes (Bossoutrot and Hosemans, 1985). Since then, gynogenetic haploid production has been obtained in various laboratories (D'Halluin and Keimer, 1986; Doctrinal et al., 1989; Galatowitsch and Smith, 1990; Lux et al., 1990; Paolini et al., 1994; Manojlovic et al., 1995; Svirshchevskaya and Dolezel, 2000). Ovaries can also be used as a source for haploid production, provided somatic tissue cells are removed or callusing of these tissues is chemically inhibited in the medium (van Geyt et al., 1987). In most of these cases, the haploids were produced by direct embryogenetic development, but mediation of a callus phase was occasionally also involved. As with other related techniques, success in inducing gametic cell development and eventually obtaining haploid sugar beet plants depends both on the genotype and the culturing conditions.

Initially very low, success rates subsequently improved and according to the literature, values ranging between 5-15 haploid plants per 100 cultured ovules have been obtained.

Chromosome doubling of haploids so that completely homozygous diploids (dihaploids or doubled haploids) are produced, is a procedure mainly based on the addition of colchicine in the culturing media either during the multiplication stage (Bossoutrot and Hosemans, 1985) or directly at the ovule culture stage (Hansen et al., 1994). Usually 50-60% of the regenerated plants are true diploids, meaning that still no more than 8-10 doubled haploids are obtainable in the culturing of 100 ovules.

Ovule culture can also be employed in highly heterozygous tetraploid populations endowed with desirable traits. The resultant diploid plants can be selected for superior gene combinations and used in development of diploid sugar beet hybrid varieties directly, or upon further haploidization and production of double haploids (Mavromatis et al., 1999).

The relative difficulty of haploid production through ovule culture combined with frequent presence of plants with undesirable characters due to the unmasked expression of recessive deleterious alleles, has severely restricted its large-scale use in sugar beet breeding programs. However, haploidy presents a valuable source material for more basic genetic studies.

4.3.6 Somaclonal Variation and Selection

Somaclonal variation is a term used to describe the phenomenon of variability present in plants originating from tissue culture techniques (Larkin and Scowcroft, 1981). Somaclonal variation is present in particular in plants regenerated from callus, but is also is found in *de novo* direct regenerants. Phenotypically detected variation concerns developmental, morphological, and physiological characters as well as resistance to biotic and abiotic stresses. If no observable alterations are identified, other screening procedures (i.e., molecular markers) should be employed to uncover plant variants. Somaclonal variation can be of an epigenetic or genetic nature. Epigenetic changes are almost always transient modifications induced by the tissue culture process itself and can only be transferred through asexual propagation. Contrarily, genetic changes can be heritable and might be due to numerical (polyploidy, aneuploidy) or structural (translocations, duplications, deletions) chromosomal alterations, and to base mutations of nuclear and cytoplasmic genes. Other than the genotype, factors influencing somaclonal variation include type of explant, conditions and duration of tissue culture, and selective agent if this technique is used. The major benefit of somaclonal variation is the creation of additional genetic variability that may be exploited directly or through a specific selection pressure.

Morphological and cytological somaclonal variants have been commonly observed in sugar beet tissue cultures (see, for example; Atanassov, 1986; Krens et al., 1990; Jacq et al., 1992; Zhong et al., 1993; Lenzner et al., 1995). Somaclonal variation, as already described in section 4.2.4, was detected at the nuclear and mitochondrial DNA level by isozyme, RFLP, and RAPD molecular markers (Sabir et al., 1992; Levall et al., 1994; Munthali et al., 1996; Dikalova et al., 1993; Jazdzewska et al., 2000; Sadoch et al., 2000).

However, successful exploitation of somaclonal variation has been reported only in the development of monogenic dominant sulfonylurea resistance from nonmutagenized selection of suspension cell cultures exposed to an acute dose of the herbicide and a line clone was released for breeding purposes (Saunders et al., 1992).

4.4 Genetic Engineering (G.N. Skaracis)

4.4.1 Introduction

Genetic engineering definitely comprises the most impressive category of biotechnological applications in crop improvement. Its basic advantages over the classical crossing approach for the transfer of genes are: i) considerably more precise gene manipulation and transfer of specific short DNA sequences; ii) possibility for up or down regulation of gene expression; iii) exploitation of genes regardless of taxonomic barriers with a subsequent increase in variability of the working gene pools; and iv) potential for rapid incorporation of a specific trait into a given genotype.

Since the first successful production of a genetically transformed (or transgenic) plant twenty years ago, gene transfer technology has been used for the introduction of new genetic characters in numerous commercially important plant species. As could be expected, the "first generation" genetic modifications exclusively concerned single genes involved in the expression of important traits such as resistance to biotic and abiotic factors (i.e. pest and disease resistance, herbicide resistance, and drought-temperature stress), pre- and post-harvest quality and, albeit to a lesser extent, in novel product development.

Cultivation of GM plant varieties reached a global area of about 60 million ha in 2002. Of this area, 99% was confined to four countries: USA (66%), Argentina (23%), Canada (6%), and China (4%). Crops involved were soybean (62%), maize (21%), cotton (12%) and oilseed rape (5%), and the genetic modifications pertained to herbicide resistance (75%), insect resistance (17%), and a combination of these two traits (8%).

Plant genetic modification, with the achievements realized so far and the ones expected in future, is promising to crucially supplement conventional methodology by furnishing valuable novel raw materials for mainstream breeding programs. Further, long term indirect contributions to crop improvement will come from the employment of transformation technology in the study of related important issues in the areas of genetics, biochemistry, physiology, molecular and developmental biology, etc.

Despite a considerable dedication of resources, both in the private and the public sector, successful exploitation of recombinant DNA technology in sugar beet has proven slower compared to other economically important crops (Wozniak, 1999). Advances in the last few years however, allow for quite an optimistic view on the possibilities to be offered by transgenic sugar beet, provided a favorable public perception prevails.

4.4.2 Methods of Genetic Transformation

Besides the absolute need for an efficient and repeatable whole plant regeneration procedure, success of a transformation technique depends on its ability to stably insert DNA into the genome of plant cells, and on the proper expression of the acquired trait in the transformed plants regenerated from these cells. If even one of these requirements is not met, then transgenics are either useless or impossible. Efficient regeneration protocols are now available for plant species, even those earlier considered recalcitrant.

Concomitantly, progress achieved during the last three decades in molecular biology led to today's availability of the appropriate means to ensure stable DNA insertion. In addition, a spatial, developmental, and in response to a specific inducer, regulation of transgene expression is possible.

From the numerous transformation methods developed (Potrykus, 1990; van Wordragen and Dons, 1992), two main categories are successfully employed in the production of GM plants: (i) vector mediated (or indirect) gene transfer; and (ii) direct gene transfer. For the indirect method, introduction of new genetic information into the target cells relies on the natural gene transfer capacity of the soil bacterium *Agrobacterium sp.* In direct gene transfer, several approaches are followed such as: (i) PEG or electroporation facilitated introduction of naked DNA into protoplasts; (ii) microinjection, where naked DNA is mechanically inserted into protoplasts or intact cells; and (iii) biolistics, in which insertion into the plant cell or tissue takes place through bombardment with DNA-coated particles. A recent pollen transformation system offers possibilities for a considerably more efficient engineering technology (Smith and Park, 2002).

The principal transformation methodology, also employed in sugar beet genetic modification, is summarized below.

4.4.2.1 Vector-mediated transformation

The *Agrobacterium* gene transfer system has long been the most commonly used technique for transforming plants. This is because of its simplicity, efficiency, and inexpensiveness, as well as the fact that the plants obtained contain a limited number of transgene copies and by and large DNA insertion is completed in a precise manner (De Block, 1993).

Agrobacterium tumefaciens is a plant pathogenic soil bacterium implicated in the development of crown gall disease (i.e. an uncontrolled proliferation of cells) in wounded tissues of many dicot and monocot plant species. The usefulness of this bacterium in plant transformation was realized when it was shown that crown galls resulted from the transfer and integration of bacterial genes into the plant genome (Chilton et al., 1977). The oncogenic properties of the bacterium are due to the presence of a large tumor-inducing plasmid (Ti), carrying certain kinds of sequences necessary for such transformation: (i) the oncogenic T-DNA region which is transferred to the plant cells and causes tumor development as a result of the overproduction of phytohormones IAA and cytokinin, coded for by genes in that region; and (ii) the virulence (*vir*) gene complex which is not transferred to the plant cells, but codes for the information required so that the T-DNA is excised and transferred as a single stranded molecule from the bacterium to the target cell genome. Regions to be transferred are defined by two similar border sequences flanking the T-DNA. Provided that oncogenes are removed and replaced by a gene of interest, this system can be used for the successful delivery of the gene into the target plant cells. In contemporary binary vector systems however, the transfer vehicle is a second artificially constructed plasmid endowed with the aforementioned border sequences flanking the gene under transfer. These sequences interact with the *vir* gene complex on the "disarmed" Ti plasmid and the region between them is mobilized and inserted into one of the plant's chromosomes. This plasmid vehicle also contains an origin for replication in *Agrobacterium* and a high copy number origin of replication functional in *E. coli*, allowing for an easy production and testing of the constructed plasmid prior to its transfer to the *Agrobacterium*. It also contains polylinkers (unique restriction sites to allow for vector engineering) and two selectable marker genes, one for selection in bacteria and the other to be expressed in the transformed plant cells. Genes to be transferred, as in any transformation technique, are accompanied by various regulatory elements such as promoters, terminators, and other sequences necessary for their proper transcription and expression (Gasser and Fraley, 1989; De Block, 1993).

A different *Agrobacterium* species, *A. rhizogenes,* also infects by similar mechanisms, sugar beet explants producing hairy roots whose *in vitro* culture has proven particularly useful in several studies on plant-pathogen interaction (Mugnier, 1987; Paul et al., 1987, 1990) and transgene expression (Ehlers et al., 1991; Cai et al., 1997).

Early attempts at production of genetically modified sugar beet by vector-mediated transformation, although verified at the cell level (Harpster et al., 1988; Wozniak and Owens, 1989; Kallerhöff et al., 1990), were not completed due to failure of whole-plant regeneration. The first production of transformed sugar beet, as reported in the literature, was obtained by cocultivation of shoot-base tissues with the *Agrobacterium*. The binary vector used had a kanamycin resistance (*npt-II*) selectable marker gene and either the chloramphenicol acetyltransferase (*cat*) or the a-glucuronidase (*gusA*) reporter gene. Transformation frequency was low and strongly depended on the type of explants, plant genotype, and selection conditions (Lindsey and Gallois, 1990). Significant progress in developing transformed sugar beet was realised in the work of D'Halluin et al. (1992), who were able to deliver transgenic plants expressing such an agronomically important trait as resistance to broad-spectrum herbicides gluphosinate ammonium and sulfonylurea compounds. Embryogenic friable callus, produced from various pieces of aseptically grown seedlings, was transformed by *Agrobacterium* with selectable marker genes, namely those conferring herbicide resistance and the kanamycin (*kan*) gene for antibiotic resistance. Although an adequate number of plants for field testing were produced, the protocol was both genotype dependent and time consuming. A genotype independent protocol, considered the basis for subsequent protocols developed and used by private breeding firms, was reported almost at the same time by Fry et al. (1991). But it required (as reported by Hall et al., 1996b) an extremely high number of explants per established transformed plant.

In addition to the study of factors affecting transformation efficiency (Jacq et al., 1993; Krens et al., 1996; Snyder et al., 1999), research efforts employing *A. tumefaciens*-mediated transformation in successive years were directed toward developing commercially exploitable transgenic sugar beet, mainly for characters such as herbicide and disease resistance (Nichols et al., 1992; Steen and Pedersen, 1993, 1995a,b; Tenning et al., 1995; Mannerlöf et al., 1996, 1997). Given that most of the relevant work took place in the context of private enterprises, detailed information on protocols used has not always been publicly available.

4.4.2.2 Direct DNA transformation

For almost a decade now it has been well known that *A. tumefaciens*-mediated transformation in maize, rice, and wheat is not only possible but can be very efficient. Initially however, such uttermost economically important cereal crops were considered nonamenable to this genetic modification system. It was precisely the need for species- and genotype-independent transformation methods that goaded intensive research efforts which led to eventual development of various techniques for direct naked DNA delivery to plant cells. Several kinds of DNA can be employed in such direct transformation, viz. engineered agrobacterial plasmids or any other plasmid or linearized DNA, provided that all necessary regulatory elements and selectable marker genes are included in the construct. In addition to genes of interest, cells transformed by direct DNA delivery techniques contain fragmented copies of the genes or vector at different locations in the genome and multiple integrations are more common with inserted copies usually tightly linked. So more screening has to be done to select lines with a limited number of copies and eliminate "junk DNA" by usual crossing-recombinant selection approaches (De Block, 1993).

Various methods have been employed for direct DNA transfer. In the presence of specific chemicals to fuse and uptake nucleic acids, protoplasts were the first cell types used for direct plant transformation (Krens et al., 1982). After that, several techniques, mainly PEG-mediated and electroporation were successfully used. But however efficient a protoplast transformation proved, limitations still came from the need for well established, efficient and repeatable regeneration systems, not always available in many species. These limitations were overcome by the most recently developed method for direct transfer, the biolistic or particle bombardment method, in which DNA molecules are literally shot into the intact cells by specific equipment (Klein et al., 1987). Both protoplast transformation and biolistics have been reported in sugar beet transformation.

Electroporation and sonication

In electroporation, a solution of DNA fragments containing the gene of interest is added to a protoplast suspension culture and electrical pulses used to temporarily disrupt the membrane so that DNA is absorbed by the protoplasts. This was the first method used for direct gene transfer into sugar beet protoplasts. Following a preliminary attempt in which the *cat* gene was only transiently expressed (Lindsey and Jones, 1987), the same researchers were able to establish stable transformant sugar beet cell clusters (Lindsey and Jones, 1989). A different approach for protoplast membrane permeabilization and DNA uptake has also been reported: transformation and transient expression for the *cat* gene were achieved by mild sonication, i.e., briefly exposing protoplasts to certain levels of ultrasounds in the presence of plasmid DNA (Joersbo and Brunstedt, 1990).

Neither electroporation nor sonication was pursued further for sugar beet transformation due to the absence of appropriate protoplast regeneration protocols at the time they were reported. With the efficient protoplast isolation and regeneration protocols available at present however, such direct gene transfer techniques might prove a viable means for sugar beet transformation (Wozniak, 1999).

PEG-mediated transformation

Polyethylene glycol (PEG) is a long-chain hydrophobic polymer extensively used for protoplast fusion, whose ability to cause reversible permeabilization of the plasma membranes promotes uptake of DNA molecules into protoplasts. As mentioned above, use of such a system for production of transformed sugar beet plants depended on the availability of protoplast regeneration techniques. Establishment of appropriate regeneration protocols (Krens et al., 1990; Lenzner et al., 1995) and discovery of the unique totipotency of protoplasts of stomatal guard cells (Hall et al., 1995, 1996a) led to the development of a new, efficient transformation pathway: specifically enriched from stomatal guard cells protoplast populations were used for PEG-mediated plasmid DNA uptake and both transient and stable transformation was obtained regardless of the plant genotypes involved (Hall et al., 1996b). Frequency of transformation was solely dependent on PEG concentration whereas none of the protocol parameters (i.e., PEG concentration, concentration of plasmid DNA, incubation period, and type of carrier DNA) was influential on the copy number of transgenes. Although the method was quite rapid (most transformed plants delivered in a period of 8-9 weeks) and genotype independent, the main limiting factor that has was still to be improved, was the protoplast regeneration frequency.

Biolistics-mediated transformation

The biolistics method of transformation, also called particle or microprojectile bombardment, is a process by which DNA molecules are carried on the surface of small (0.5 to 1.5 mm) gold or tungsten particles through the cell wall into the nucleus for integration. Acceleration of these microprojectiles to a supersonic velocity, necessary for cell wall penetration, is provided by helium gas discharge in an evacuated chamber. Once developed (Sanford et al., 1987; Klein et al., 1987), efforts with this method focused especially on plant species in which *Agrobacterium*-mediated and protoplast transformation was considered difficult or impossible. Indeed, as a consequence of these efforts, cereals such as rice, maize, barley, and wheat had been successfully transformed by 1992. Targets for transformation by this method are intact cells and organized tissues such as meristems, the latter providing a significant advantage, because homogeneously transformed plants can be potentially produced rapidly after the first chimeric generation. Also, a very important feature of the method is that it could be directly applied to elite breeding lines, minimizing the time needed for practical use of the transformation products.

The first report of particle bombardment-mediated transformation of sugar beet cells appeared in association with a study comparing the effect of various promoters in gene expression (Ingersoll et al., 1996). Stable levels of expression for all promoters were achieved shortly after the successful DNA transfer. Despite the fact that no whole transformed plant regeneration was involved, it was clearly demonstrated that the method was applicable to sugar beet. A successful biolistic transformation method was developed using embryogenic hypocotyl callus of a high regenerable tissue culture clone (Snyder et al., 1999). Transgenic plants contained genes encoding pathogen-defense-related proteins or the cytokinin biosynthesis gene, conferring enhanced insect resistance. Transformation efficiency was about 8% (number of transgenic plants per plate of embryogenic callus treated) compared to that of an *Agrobacterium*-mediated method, reaching on average 0.7% (number of transgenic plants per treated cotyledon). Lastly, the feasibility of using the biolistics transformation method with several advanced sugar beet breeding lines was tested by Smigocki (2001).

4.4.2.3 Recent selection approaches

A recent line of research endeavors, mainly in the framework of *A. tumefaciens*-mediated transformation, refers to the exploitation of novel selectable techniques not involving the use of antibiotic resistance or herbicide resistance (when the latter is not the targeted trait) marker genes. Such a practice could have a significant impact on diminishing public concerns about safety of genetically modified plants for man and the environment.

Several selection systems based on endowing transformed cells with a metabolic advantage instead of killing the nontransformed cells have been developed, and found to be superior in many cases to conventional approaches (Joersbo, 2001). One such system studied in sugar beet involves the phosphomannose isomerase (PMI), a protein catalyzing the reversible interconversion of mannose-6-phosphate and fructose 6-phosphate. Plant cells lacking this enzyme are not able to survive on a synthetic medium containing mannose as a carbon source. Successful transformation using a PIM gene from *E. coli* (*manA*) as selectable gene and mannose, endogenously converted to mannose-6-phosphate, as a selective agent were recently reported (Joersbo et al., 1999; Reed et al., 2001). Further, this mannose selection system proved relatively genotype independent when mannose concentration was slightly optimized during selection (Joersbo et al., 1999) and promoters causing intermediate levels of PMI expression resulted in high transformation frequencies (Joersbo et al., 2000).

A totally different approach was reported by Zhang et al. (2001): a green fluorescent protein (GFP) gene, engineered for enhanced fluorescence under blue light, was introduced into sugar beet by the *Agrobacterium* transformation method. Cocultivated shoot explants or embryogenic calli were transferred to media without selective agents and regularly examined under a blue light microscope; green fluorescent shoots were obtained at frequencies of 2-5%. The authors concluded that the GFP reporter could be vitally exploited for the development of selectable marker-free transgenic sugar beet.

4.4.3 Transgenic Traits

Sugar beet traits acquired, improved, or altered so far through genetic engineering mainly pertain to herbicide and disease resistance, resistance to environmental stress, and carbohydrate metabolism. At present, the progress status for various genetic modifications ranges from early experimental development and evaluation to a final approval for cultivation. Official field trials of transgenic sugar beet were initiated in 1990 in Europe and three years later in the USA. Until 2002, a significant proportion (around 15%) of the official trials in the EU for evaluation of GM varieties was occupied by sugar beet (EC-JRC, 2003).

4.4.3.1 Herbicide resistance

Weed competition, especially in the first two months from emergence, is a serious problem for the sugar beet crop and yield reductions of 30-50% and as high as 80% have been recorded in the absence of proper control. Complex mixtures of several herbicide active ingredients necessarily used and the timing of applications are carefully selected by the growers so that the crop is successfully protected. The cost of weed control is a significant part of the total production cost. Provided sugar beet is not affected, postemergence use of nonselective (broad spectrum) herbicide promises a considerably more efficient weed control.

Development of herbicide-resistance transgenic sugar beet refers solely to the nonselective herbicides glyphosate (Roundup) and gluphosinate ammonium (Basta, Liberty), although sugar beet has also been transformed for resistance to selective herbicides such as chlorsulforon and sulfonylurea. The overwhelming majority of officially field-tested GM sugar beet belongs to this category. Herbicide resistance was obtained with the introduction of genes that encode either insensitive variants of the enzymes and proteins targeted by the herbicides, or enzymes that convert the herbicides to inactive components. So far, both approaches have been exploited in sugar beet transformation by the *Agrobacterium* method.

Glyphosate acts by inhibiting the plant enzyme 5-enopyruvyl-3 phosphoshikimic acid synthase involved in the biosynthesis of aromatic amino acids, vitamins, and other secondary metabolites. Resistance to the herbicide was conferred by introduction of the respective *EPSPS* gene from *Agrobacterium* sp. CP4, producing a herbicide-insensitive enzyme, in contrast to the plant's respective sensitive enzyme. A second glyphosate oxidoreductase (*GOX*) gene from bacteria, encoding enzyme-degrading glyphosate into nontoxic compounds, also confers resistance and can be used alone or in combination with the former in plant transformation (Mannerlöf et al., 1997; Tenning et al., 1995; Brants et al., 1995; Brants and Harms, 1998).

Gluphosinate ammonium is a stereo-isomeric mixture of ammonium salts of l- and d- phosphinothricin (PPT). Only the l-isomer presents herbicidal activity. Gluphosinate ammonium inhibits glutamine synthetase, a plant enzyme essential for the assimilation of ammonia, resulting in its toxic accumulation in the plant. Gluphosinate ammonium tolerance is the result of a synthetic version of a gene from *Streptomyces viridochromogenes* encoding the phosphinothricin-N-acetyl transferase

(PAT) enzyme. This enzyme catalyzes the conversion of L-phosphinotricin, the active ingredient of the herbicide, to an inactive form (Loock et al., 1998; Brants and Harms, 1998).

Glyphosate and gluphosinate ammonium resistance sugar beet has been approved for release in the USA (1998). No such herbicide resistance varieties have reached commercial cultivation however, due to the unwillingness of sugar beet processors in response to public perception of the safety of such products. Gluphosinate ammonium resistance sugar beet has also been approved in Canada (2001). In Europe, despite a large number of years in official field trials (OECD, 2003) no such approvals have been granted yet due to an informal moratorium in effect these last four years, on further approvals for the cultivation of transgenic crops, and revision of the EU directive 2001/18 concerning the environmental assessment studies required (this is also true for all other kinds of transgenic sugar beet).

4.4.3.2 Rhizomania resistance

Conventional breeding has been remarkably successful in developing rhizomania-resistant material, thus allowing efficient continuation of the crop even in zones where beet growing appeared to be no longer possible. In doing so, several natural sources of quantitative and qualitative resistance found in sugar beet and in its wild relatives were exploited.

Despite such progress however, there is always the need for a further yield increase under disease conditions. Aiming at enhancing naturally occurring resistance, several approaches of genetic engineering are presently employed.

A major approach followed in developing rhizomania resistant transgenic sugar beet is that of the coat protein-mediated resistance, the most widely used procedure for obtaining transgenic virus-resistant plants (Grumet, 1994). Induced excessive production of the viral coat protein (CP) confers resistance by disrupting the infection process. Cross- protection against other related viruses has also been described (Anderson et al., 1989; Stark and Beachy, 1989).

The efficiency of this approach was evidenced initially when transgenic sugar beet protoplasts derived from cell suspensions transformed by an *Agrobacterium* vector containing the *CP* gene were infected with the rhizomania-causing agent (BNYVV). Viral multiplication was less effective compared to nontransformed protoplasts (Kallerhöff et al., 1990). The *CP* gene was also transferred and expressed, as demonstrated by molecular techniques, in hairy root cultures upon *A. rhizogenes*-mediated transformation (Ehlers et al., 1991). Coat protein-mediated rhizomania resistance at the plant level was first reported by Mannerlöf et al. (1996).

Two different constructs containing the *CP* gene were used to produce transformed sugar beet plants by the *Agrobacterium* method. Transgenic plants, after four cycles of selfing, were tested in the greenhouse and in field trials, and although *CP* accumulation was not detected by Western blotting, a reduced virus content was measured both for the greenhouse and field evaluations of the transformed plants. In the same study, cross-protection against the beet soilborne virus (BSBV) was not evident.

The *CP* gene was also introduced to stomatal guard cell protoplasts by a PEG-mediated procedure. Primary transformants with a low number of insertions were evaluated for virus content in the rootlets and in the taproot when the plantlets were infected under growth chamber and greenhouse conditions. In some of the transformants, the virus level was as low as that of the resistant control genotypes and a more homogeneous response was observed in the taproot, possibly indicating an effect of the transgene on the translocation of BNYVV in the sugar beet root. No field evaluations were reported in the study (Lathouwers et al., 1997).

A different approach for inducing pathogen-derived resistance to BNYVV involves interfering with cell-to-cell movement as regulated by a "triple gene block" (*P42, P15,* and *P13*) located on the RNAs of the virus. If the P15 protein is produced at higher than normal levels, relative to the P13 protein, cell-to-cell movement is disrupted (Bleykasten-Grosshans et al., 1997). Several genotypes were successfully transformed by a P15 coding gene (Lathouwers et al., 1997) but no further evaluation data have been reported. On the other hand, if a modified (nonfunctional in the virus) P15 protein is expressed in the plant, it will compete with the wild P15 protein and eventually deregulate viral movement (EC-JRC, 2003).

A different approach to disrupting cell-to-cell movement of the virus was attempted by the introduction of identical to P13 gene sequences, thus reducing expression of the gene (cosuppression) in transformed sugar beet plants. The same research group also employed another method for inactivation of RNA transcribed from target genes, i.e. introduction of specifically constructed hammerhead ribozymes. Ribozymes are RNA molecules with enzymatic activity. Hammerhead ribozymes possess RNA stretches that bind the complementary sequences of the target RNA which are subsequently scissored by the ribozyme at a site between the complementary sequences (Tsagri and Panopoulos, unpubl. data).

Despite the lack of further detailed information on methodology practically employed, rhizomania-resistant transgenic sugar beet has been produced by both the coat protein and the P15 approach and has participated in EU official testing alone, or in combination with other traits, since 1991 (OECD, 2003; EC-JRC, 2003).

Coat protein-mediated resistance to other viruses (i.e. BWYV, BMYV) has also been introduced to sugar beet and relevant trials are in progress (OECD, 2003).

4.4.3.3 Altered carbohydrate metabolism

In an attempt to diversify its commercial uses, transgenic sugar beet with altered carbohydrate metabolism was produced and field tested.

Fructans are low molecular weight nondigestible polymers of fructose and can be used as a natural low-caloric sweetener. Slightly longer chain fructans, because of their fat-like texture and neutral taste, can serve as a low-caloric fat replacement in specific food stuffs. Fructans have so far been produced by expensive fermentation procedures. Transgenic fructan-producing sugar beet (fructan beet) has been produced by PEG-mediated stomatal guard cell protoplast transformation with a gene from *Helianthus tuberous* encoding 1-sucrose fructosyl transferase (*1-SST*). This enzyme mediates the initial steps in fructan synthesis through the conversion of sucrose into low molecular weight fructans (Sevenier et al., 1998). Greenhouse testing revealed that sucrose stored in the taproot was almost totally converted to fructans while such conversion was very limited in the leaves. At the same time, no side effects on phenotype and taproot development were detected. Products of such an effort have been in official EU trials for some years (OECD, 2003).

The possibility of producing transgenic sugar beet with enhanced sucrose production has been reported by transformation with appropriate sucrose-phosphate synthase, sucrose synthase, and sucrose transporter genes (Hesse and Müller-Röber, 1994; Hashimoto and Shimamoto, 1999). A further experimental effort refers to sugar beet hairy root transformation by a chimeric antisense sucrose nonfermenting (SNF1) sequence derived from a potato homologue gene. This gene encodes a protein kinase that links cellular glucose levels to the derepression of a range of glucose-repressed genes. Transgenic sugar beet hairy root cultures have been identified with reduced activity of the SNF1 homologue as judged by a peptide assay (Monger et al., 1995).

4.4.3.4 Environmental stress resistance

Resistance to environmental stress (i.e., drought, salinity, temperature, etc.) involves numerous genes interacting in complex ways and for this reason successes by conventional breeding methods are relatively limited. Although the exact mechanism is not known, studies in several species have reported on the protection of transgenic plants by an increased cellular content of osmolytes and osmoprotectants.

With the objective of conferring stress (drought) resistance, a metabolic alteration toward producing the osmoprotectant carbohydrate trehalose was engineered in sugar beet by the use of a yeast trehalose phospate synthase (*TPS1*) gene. Trehalose-producing transgenic beet has been in official EU tests for a number of years (OECD, 2003).

4.4.3.5 Resistance to fungi

Defense strategies against fungal diseases tackled by genetic engineering approaches mainly consist of the introduction of genes coding either for enzymes producing antifungal metabolites or proteins and enzymes themselves acting on the fungus.

Fungal cell walls are susceptible to hydrolytic enzymes such as chitinases and β 1,3 glucanases coded for by genes to be found in a range of hosts (plants, insects, bacteria, and fungi). Ribosomal inhibitor proteins exist that selectively inactivate fungal but not plant ribosomes, resulting in fungal protein inhibition. Another approach is to render the plant insensitive to toxins produced by fungi either by genes coding for proteins extruding the fungal toxins from plant cells or by genes antagonizing the photooxidative damage caused by the fungal toxin. Lastly, "a strategic cell death" can be engineered by constructing a gene cassette consisting of an appropriate *avr-R* gene pair placed under the control of a disease inducible promoter to cause a hypersensitive reaction.

Although still at the experimental stage, all or some of these approaches have been employed in developing cercospora resistant transgenic sugar (Skaracis and Biancardi, 2000). Further, a published attempt to produce rhizoctonia-resistant sugar beet concerns the use of a pumpkin-derived chitinase gene (Hashimoto and Shimamoto, 2001). But it is obvious that unpublished work is taking place in private laboratories, given the participation of transgencic fungus-resistant sugar beet in the official EU trials (OECD, 2003).

4.4.3.6 Insect resistance

The transfer of genes encoding insecticidal crystal proteins (ICP), an approach widely used to enhance insect resistance to such major crops as maize and cotton, was recently reported in sugar beet as well. Transgenic plants produced by the *Agrobacterium*-mediated transfer of *cryIA(b)* and *cryIC* genes from *Bacillus thuringiensis* caused growth delay of cabbage armyworm larvae feeding on their leaves. In this preliminary work, the *cryIC* gene showed a slightly higher effectiveness (Kimoto and Shimamoto, 2001).

Two different approaches to developing transgenic plants resistant to the sugar beet root maggot are currently under investigation: i) identification and transfer of specific proteinase inhibitor genes to interfere with normal insect growth and ii) transformation with a cytokinin biosynthesis gene inducing insecticidal compounds (Smigocki, 2001).

Transgenic insect-resistant sugar beet has been included in official EU trials since 1997. No further information was available in 2003.

4.4.3.7 Nematode resistance

A genetic engineering approach was designed to overcome the inability of conventional interspecific hybridization methodology (low trait transmission and linkage drag) to transfer genes for resistance to the beet cyst nematode (*H. schachtii*) found in wild relative species. To this end, the resistance gene $Hs1^{pro\text{-}1}$, located on chromosome I of *B. procumbens*, was map-based cloned and a cDNA clone used for an *A. rhizogenes*-mediated transformation of sugar beet leaf segments. Hairy root cultures expressed resistance when infected with juvenile (J_2) nematodes (Cai et al., 1997). Private European breeding firms are known to be engaged in producing nematode-resistant transgenic sugar beet employing this approach.

Another approach has been postulated based on destruction of the feeding site (or syncytium). This could be achieved by local expession of the cytotoxic *barnase* gene under the control of a nematode-inducible plant promoter that is active in the syncytium. Expression of *barnase* outside the feeding site has to be neutralized by its inhibitor *barster* gene driven from a constitutive promoter that is downregulated in the syncytium. Preliminary results on the efficiency of promoters studied in a transformed hairy root culture system were encouraging (van Poucke et al., 2001).

4.4.3.8 Bolting resistance

Bolting resistance has a complicated genetic base and is strongly dependent on environmental conditions. Early spring sowings or extended cold spring cause undesirable bolting in the beet crop, resulting in considerable losses (around 0.5% sugar yield for every 1% area of bolted beets in the field), problems with harvest and factory operation, and probable weed beet development in following years. Improved bolting resistance, to allow for earlier sowings and autumn beet crops under appropriate climatic conditions, was proposed through the transfer of plant genes from beet and pumpkin, to inhibit gibberellin metabolism (van Roggen et al., 1997).

4.5 Final Considerations (G.M. Skaracis)

Production of genetically modified sugar beet, as evidenced from the results obtained so far, is a well-established approach to incorporating desirable traits and broadening the genetic base in the crop. Further advances in related technologies are ensuring that even more significant developments can be obtained in future. However, as with all genetically engineered crops, concerns regarding their consequent commercial use are many. They mainly pertain to impact on food safety, possible damage to the environment and crop diversity, as well as socioeconomic (ownership, proprietary rights, and biopiracy) and ethical issues of patenting life forms.

Such public concerns had a pronounced impact on political decisions in the EU framework and no approvals for large-scale growing and use of genetically modified crops, among them sugar beet, were granted until very recently. Near the end of 2003, mainly in response to pressures from the WTO, a decision was reached for resuming evaluation of relevant applications, thereby opening possibilities for their eventual use. The final outcome in time is not readily predictable however, as many countries still oppose use of these new products and skepticism is considerable concerning the practical adoption of the strict regulations imposed on labeling and traceability of genetic modifications in seeds, food, and feed.

References

Abe, J. and C. Tsuda. 1987. Genetic analysis for isozyme variation in the section *Vulgares* genus *Beta*. Jpn. J. Breed. 37:253-261.

Abe, J., H. Yoshikama, and C. Tsuda. 1987. Genetic analysis for annual and early-flowering habit of *Beta macrocarpa* Guss., a related species of *B. vulgaris* L.: an analysis with enzyme – coding loci as chromosome marker. J. Fac. Agric. Hokkaido Univ. 63:245-252.

Abe, J., G.P. Guan, and Y. Shimamoto. 1993. Linkage maps for nine isozyme and four marker loci in sugar beet (*Beta vulgaris* L.). Euphytica 66:117-126.

Abe, J., H. Nakashima, K. Mitsui, T. Mikami, and Y. Shimamoto. 1991. Tissue culture response of *Beta* germplasm: callus induction and plant regeneration. Plant Cell, Tiss. Org. Cult. 27:123-127.

Allen, J.F. 2003. Why chloroplasts and mitochondria contain genomes. Comp. Funct. Genom. 4:31-36.

Allen, K.D. 2002. Assaying gene content in *Arabidopsis*. Proc. Na. Acad. Sci. USA 99:9568-9572.

Anderson, E.J., D.M. Stark, R.S. Nelson, P.A. Powel, T.E. Turner, and R.N. Beachy. 1989. Transgenic plants that express the coat protein genes of tobacco mosaic virus or alfalfa mosaic virus interfere with disease development of some non-related viruses. Phytopath. 79:1284-1290.

Arumuganathan, K. and E.D. Earle. 1991. Nuclear DNA content of some important plant species. Plant Molec. Biol. Rept. 9:208-218.

Atanassov, A.I. 1973. Anther culture of tobacco and sugar beet as a tool for producing haploid plants and callus of different ploidy level. Genet. Plant Breed. 6:501-507.

Atanassov, A.I. 1976. Investigation of organogenetic ability of tissue and cell cultures of sugar beet (*Beta vulgaris* L.) with a view to use them in genetics and breeding. PhD diss., Inst. Genetics, Bulgarian Academy Science, Sofia, Bulgaria.

Atanassov, A.I. 1980. Method for continuous bud formation in tissue cultures of sugar beet (*Beta vulgaris* L.). Zeitschrift für Pflanzenzücht. 84:23-9.

Atanassov, A.I. 1986. Sugar beet. *In:* D.A. Evans, W.R. Sharp, and P.V. Ammirato (eds.). Handbook of Plant Cell Culture. MacMillan, New York, NY, vol.4, pp. 652-680.

Backert, S., B.L. Nielsen, and T. Borner. 1997. The mystery of the rings: Structure and replication of mitochondrial genomes from higher plants. Trends Plant Sci. 2:477-483.

Banda, H. and H. Tanabe. 1972. A study of anther culture in sugar beet. Bull. Sugar Beet Res. (Suppl.) 14:9-16.

Barnes, S., G. Massaro, M. Lavëbvre, M. Kuiper, and E. Verstege. 1996. A combined RFLP and AFLP genetic map for sugar beet. Proc. IIRB 59:555-560.

Barzen, E., W. Mechelke, E. Ritter, J. F. Seitzer, and F. Salamini. 1992. RFLP markers for sugar beet breeding: chromosomal linkage maps and location of major genes for rhizomania resistance, monogermy, and hypocotyl colour. Plant J. 2:601-611.

Barzen, E., W. Mechelke, E. Ritter, E. Schultekappert, and F. Salamini. 1995. An extended map of the sugar beet genome containing RFLP and RAPD loci. Theor. Appl. Genet. 90:189-193.

Barzen, E., R. Stahl, E. Fuchs, D. Borchardt, and F. Salamini. 1997. Development of coupling-repulsion-phase SCAR markers diagnostic for the sugar beet *Rr1* allele conferring resistance to rhizomania. Molec. Breed. 3:231-238.

Bellin, D., M. Werber, T. Theis, B. Schulz, B. Weisshaar, and K. Schneider. 2002. EST sequencing, annotation and macroarray transcriptome analysis identify preferentially root-expressed genes in sugar beet. Plant Biol. 4:700-710.

Bennett, M.D. and J.B. Smith. 1976. Nuclear DNA amounts in angiosperms. Phil. Trans. Roy. Soc. London, Series B 274:227-274.

Bhat, S.R., B.V. Ford-Lloyd, and J.A. Callow. 1985. Isolation of protoplasts and regeneration of callus from suspension cultures of cultivated beets. Plant Cell Rept. 4:348-350.

Bleykasten-Grosshans, C., H. Guilley, S. Bouzoubaa, K.E. Richards, and G. Jonard. 1997. Independent expression of the first two triple gene block proteins of beet necrotic yellow vein virus complements virus defective in corresponding gene but expression of the third protein inhibits viral cell-to-cell movement. Molec. Plant-Microbe Interact. 10:240-246.

Bosemark, N.O. 1971. Haploids and homozygous diploids, triploids and tetraploids in sugar beet. Hereditas 69:193-204.

Bosemark, N.O. 1993. Genetics and breeding. *In:* D.A. Cooke and R.K. Scott (eds.). The Sugar Beet Crop. Chapman & Hall, London, UK, pp. 67-119.

Bossoutrot, D. and D. Hosemans. 1985. Gynogenesis in *Beta vulgaris* L.: From *in vitro* culture of unpollinated ovules to the production of doubled haploid plants in soil. Plant Cell Rept. 4:300-303.

Boudry, P., R. Wieber, P. Saumitou-Laprade, K. Pillen, H. van Dijk, and C. Jung. 1994. Identification of RFLP markers closely linked to the bolting gene B and their significance for the study of the annual habit in beets (*Beta vulgaris* L.). Theor. Appl. Genet. 88:852-858.

Brants, I. and H. Harms. 1998. Herbicide tolerant sugar beet. Proc. IIRB 61:195-204.

Brants, I., P. Steen, S. Bisgaard, and H.C. Pedersen. 1995. Roundup Ready sugar beet. Proc. IIRB 58:557-559.

Braun, A. 1988. Cryopreservation of sugar beet germplasm. Plant Cell Tiss. Org. Cult. 14:161-168.

Burks, C. 1999. Molecular biology database list. Nucleic Acids Res. 27:1-9.

Butenko, R. G., A.I. Atanassov, and W. Urmantseva. 1972. Some features of sugar beet tissue cultures. Phytopath. 22:140-143.

Butterfass, T. 1964. Die Chloroplastenzahlen in verschiedenartigen Zellen trisomer Zuckerrüben (*Beta vulgaris* L.) Z. Bot. 52:46-77.

Cai, D., M. Kleine, S. Kifle, et al., 1997. Positional cloning of a gene for nematode resistance in sugar beet. Science 275:832-834.

Chilton, M.D., M.H. Drummond, D.J. Merlo, et al., 1977. Stable incorporation of plasmid DNA into higher plant cells: the molecular basis of crown gall-tumorigenesis. Cell 11:263-271.

Coumans-Gilles, M.F., C. Kevens, M. Coumans, E. Ceulemans, and T. Gaspar. 1981. Vegetative multiplication of sugar beet through *in vitro* culture of inflorescence pieces. Plant Cell, Tiss. Org. Cult. 1:93-101.

D'Halluin, K. and B. Keimer. 1986. p.307-309. *In:* W. Horn, C.J. Jensen, W. Odenbach, and O. Schieder (eds.). Genetic Manipulation in Plant Breeding. Walter de Gruyter and Co., Berlin, Germany.

D'Halluin, K., M. Bossut, E. Bonne, B. Mazur, J. Leemans, and J. Botterman. 1992. Transformation of sugar beet (*Beta vulgaris* L.) and evaluation of herbicide resistance in transgenic plants. Bio/Tech. 10:309-314.

De Block, M. 1993. The cell biology of plant transformation: Current state, problems, propects and the implications for plant breeding. Euphytica 71:1-14.

De Greef, W. 1978. Callus initiation and growth of the sugar beet (*Beta vulgaris* L.). Bull. Soc. Roy Bot. Belg. 111:69-76.

De Greef, W. and M. Jacobs. 1979. *In vitro* culture of sugar beet: description of a cell line with a high regeneration capacity. Plant Sci. Lett. 17:55-61.

De Jong, J.H. and T.S.M. De Bock. 1978. Use of haploids of *Beta vulgaris* L. for the study of orcein and giemsa stained chromosomes. Euphytica 27:41-47.

De las Rivas, J., J.J. Lozano, A.R. Ortiz. 2002. Comparative analysis of chloroplast genomes: Functional annotation, genome-based phylogeny, and deduced evolutionary patterns. Genome Res. 12:567-583.

De los Reyes, B.G. and J.M. McGrath. 2003. Cultivar-specific seedling vigor and expression of a putative oxalate oxidase germin-like protein in sugar beet (*Beta vulgaris* L.). Theor. Appl. Genet. 107:54-61.

De Riek, D.J., E. Calsyn, I. Evaraert, V.E. Bockstaele, and D.M. Loose. 2001. AFLP based alternatives for the assessment of distinctness, uniformity and stability of sugar beet varieties. Theor. Appl. Genet. 103:1254-1265.

Desel, C., R. Jansen, D.D. Gue, and T. Schmidt. 2002. Painting of parental chromatin in *Beta* hybrids by multi-colour fluorescent in situ hybridization. Ann. Bot. London 89:171-181.

Desplanque, B., F. Viard, J. Bernard, et al. 2000. The linkage disequilibrium between chloroplast DNA and mitochondrial DNA haplotypes in *Beta vulgaris* ssp. *maritima* (L.): The usefulness of both genomes for population genetic studies. Molec. Ecol. 9:141-154.

Detrez, C., T. Tetu, R.S. Sangwan, and B.S. Sangwan-Norreel. 1988. Direct organogenesis from petiole and thin cell layer explants in sugar beet cultured *in vitro*. J. Exper. Bot. 39:917-926.

Dikalova, A.E., N.A. Dudareva, M. Kubalakova, and R.I. Salganik 1993. Rearrangements in sugar beet mitochondrial DNA induced by cell suspension, callus cultures and regeneration. Theor. Appl. Genet. 86:699-704.

Doctrinal, M., R.S. Sangwan, and B.S. Sangwan-Norreel. 1989. *In vitro* gynogenesis in *Beta vulgaris* L.: Effects of plant growth regulators, temperature, genotypes and season. Plant Cell, Tiss. Org. Cult. 17:1-12.

Doley, W.P., and J.W. Saunders. 1989. Hormone-free medium will support callus production and subsequent shoot regeneration from whole plant leaf explants in some sugarbeet (*Beta vulgaris* L.) populations. Plant Cell Rept. 8:222-225.

Driessen, S., M. Pohl, and D. Bartsch. 2001. RAPD-PCR analysis of the genetic origin of sea beet (*Beta vulgaris* ssp. *maritima*) at Germany's Baltic Sea coast. Basic Appl. Ecol. 2:341-349.

EC-JRC (The European Commission Joint Research Centre), 2003. Biotechnology and GMOs (www.vbiotech.jrc.it).

Ehlers, U., U. Commandeur, R. Frank, J. Landsmann, R. Koenig, and W. Burgermeister. 1991. Cloning of the coat protein gene from beet necrotic yellow vein virus and its expression in sugar beet hairy roots. Theor. Appl. Genet. 81:777-781.

El-Mezawy, A., F. Dreyer, G. Jacobs, and C. Jung. 2002. High resolution mapping of the bolting gene B of sugar beet. Theor. Appl. Genet. 105:100-105.

Flavell, R.B., M.D. Bennet, and J.B. Smith. 1974. Genome size and the proportion of repeated nucleotide sequence DNA in plants. Biochem. Genet. 12:257-269.

Ford-Lloyd, B.V., M. Munthali, and H.J. Newburg. 1993. The use of RAPD for the identification of sugar beet varieties. J. Sugar Beet Res. 30:291-298.

Francis, S.A. and M.J.C. Asher. 2000. Exploiting novel sources of disease resistance in *Beta* germplasm using molecular markers. J. Sugar Beet Res. 37:89-95.

Francis, S.A., M. Redfearn, D.M. Chwarzczynska, M.J.C. Asher, and R.T. Lewellen. 1998. Use of molecular markers in breeding for disease resistance in sugar beet. Aspects Appl. Biol. 52:297-285.

Freytag, A.H., S.C. Anand, A.P. Rao-Ardelli, and L.D. Owens. 1988. An improved medium for adventitious shoot formation and callus induction in *Beta vulgaris* L. *in vitro*. Plant Cell Rept 7:30-34.

Fry, J.E., A.R. Barnason, and M. Hinchee. 1991. Genotype-independent transformation of sugar beet using *Agrobacterium tumefaciens*. Molecular Biology and Plant Development. Proc. 3rd Intl. Cong. ISPMB, Tucson, AR, pp. 101-102.

Fulton, T.M., R. van der Hoeven, N.T. Eannetta, and S.D. Tanksley. 2002. Identification, analysis, and utilization of conserved ortholog set markers for comparative genomics in higher plants. Plant Cell 14:1457-1467.

Galatowitsch, M.W. and G.A. Smith. 1990. Regeneration from unfertilized ovule callus of sugarbeet (*Beta vulgaris* L.). Can. J. Plant Sci. 70:83-89.

Gamborg, O.L., R.A. Miller, and K. Ojima. 1968. Nutrient requirements of suspension cultures of soybean root cells. Exper. Cell Res. 50:151-158.

Gasser, C.S. and R.T. Fraley 1989. Genetically engineering plants for crop improvement. Science 244:1293-1299.

Gindullis, F., C. Desel, I. Galasso, and T. Schmidt. 2001. The large-scale organization of the centromeric region in *Beta* species. Genome Res. 11:253-265.

Giorio, G., M. Gallitelli, and F. Carriero. 1997. Molecular markers linked to rhizomania resistance in sugar beet, *Beta vulgaris*, from two different sources map to the same linkage group. Plant Breed. 116:401-408.

Goff, S.A., D. Ricke, T.H. Lan, et al. 2002. A Draft Sequence of the Rice Genome (Oryza Sativa L. ssp. Japonica) Science 296:92-100.

Goska, M. and M. Szota. 1992. Micropropagation of sugar beet (*Beta vulgaris* L.) trisomics in *in vitro* culture. Genetica Polonica 33:115-118.

Grieve, T.M., K.M.A. Gartland, and M.C. Elliot. 1997. Micropropagation of commercially important sugar beet cultivars. Plant Growth Regul. 21:15-18.

Grumet, R. 1994. Development of virus resistant plants via genetic engineering. Plant Breed. Rev. 12:47-79.

Guha, S. and S.C. Maheshwari. 1964. *In vitro* production of embryos from anthers of *Datura*. Nature 204:497-498.

Haberlandt, G. 1902. Kulturersusce mit isolierten Pflanzenzellen. Sitzungser Math-Naturwiss. Kl. Kaiser Akad. Wiss. 111:69-92.

Hagihara, E., N. Itchoda, T. Kubo, Y. Habu, S. Iida, and T. Mikami. 1999. AFLP and RAPD markers tightly linked to fertility restorer genes for the Owen cytoplasmic male sterility in sugarbeet. Proc. Jpn. Soc. Sugar Beet Tech. 41:73-77.

Hall, R.D., C. Pedersen, and F.A. Krens. 1993. Improvement of protoplast culture protocols for *Beta vulgaris* L. (sugar beet). Plant Cell Rept. 12:339-342.

Hall, R.D., H.A. Verhoeven, and F.A. Krens. 1995. Computer-assisted identification of protoplasts responsible for rare division events reveals guard cell totipotency. Plant Physiol. 107:1379-1386.

Hall, R.D., Riksen-Bruinsma, T., Weyens, G., Lefebvre, M., Dunwell, J.M., and Krens, F.A. 1996a. Stomatal guard cells are totipotent. Plant Physiol. 112:889-892.

Hall, R.D., T. Riksen-Bruinsma, G. Weyens, et al. 1996b. A high efficiency technique for the generation of transgenic sugar beets from stomatal guard cells. Nature Biotech. 14:1133-1138.

Hallden, C., T. Sall, K. Olsson, N.O. Nilsson, and A. Hjerdin. 1997. The use of bulked segregant analysis to accumulate RAPD markers near a locus for beet cyst nematode resistance in *Beta vulgaris*. Plant Breed. 116:18-22.

Hallden C., D. Ahrén, A. Hjerdin, T. Säll, and N.O. Nilsson. 1998. No conserved homoeologous regions found in the sugar beet genome. J. Sugar Beet Res. 35:1-13.

Hallden, C., A. Hjerdin, I. Rading, et al. 1996. A high density RFLP linkage map of sugar beet. Genome 39:634-645.

Hansen, A.L., C. Plever, H.C. Pedersen, B. Keimer, and S.B. Andersen 1994. Efficient *in vitro* chromosome doubling during *Beta vulgaris* ovule culture. Plant Breed. 112:89-95.

Harpster, M.H., J.A. Townsend, J.D.G. Jones, J. Bedbrook, and P. Dunsmuir. 1988. Relative strengths of the 35S cauliflower mosaic virus, 1', 2' and nopaline synthase promoters in transformed tobacco, sugar beet and oilseed rape callus tissue. Molec. Gen. Genet. 212:182-190.

Hashimoto, R. and Y. Shimamoto. 1999. Growth of transgenic sugar beet plants with sucrose-phosphate synthase (SPS). Proc. Jpn. Soc. Sugar Beet Tech. 41:85-89.

Hashimoto, R. and Y. Shimamoto. 2001. Transgenic sugar beet plants harboring a pumpkin chitinase gene demonstrating improved resistance to *Rhizoctonia solani*. Proc. Jpn. Soc. Sugar Beet Tech. 43:24-28.

Herwig R., B. Schulz, B. Weisshaar, et al. 2002. Construction of a 'unigene' cDNA clone set by oligonucleotide fingerprinting allows access to 25,000 potential sugar beet genes. Plant J. 32:845-857.

Hesse, H. and B. Müller-Röber. 1994. DNA sequences and plasmids for the preparation of sugar beet with changed sucrose concentrations. Hoechst-Schering, AgrEvo GMBH. Intl. patent application number PCT/EP94/01671, World Intellectual Organization, Publ. No. WO 94/28146.

Hieter, P. and M. Boguski. 1997. Functional genomics: It's all how you read it. Science 278:601-602.

Hjerdin-Panagopoulos, A., T. Kraft, I. Rading, S. Tuvesson, and N. O. Nilsson. 2002. Three QTL regions for restoration of Owen CMS in sugar beet. Crop Sci. 42:540-544.

Hofer, J. and N. Ellis. 2002. Conservation and diversification of gene function in plant development. Curr. Opin. Plant Biol. 5:56-61.

Hooker, M.P. and M.W. Nabors. 1977. Callus initiation, growth, and organogenesis in sugarbeet *(Beta uulgaris* L.). Z. Pflanzenphysiol. 84:237-246.

Hosemans, D. and D. Bossoutrot. 1983. Induction of haploid plants *in vitro* culture of unpollinated beet ovules (*Beta vulgaris* L.). Z. Pflanzenzücht. 91:74-77.

Hunger, S, G. Di Gaspero, S. Mohring, et al. 2003. Isolation and linkage analysis of expressed disease-resistance gene analogues of sugar beet (*Beta vulgaris* L.). Genome 46:70-82.

Hussey, G. and A. Hepher. 1978. Clonal propagation of sugar beet plants and the formation of polyploids by tissue cultures. Ann. Bot. 42:447-478.

Ingersoll, J.C., T.M. Heutte, and L.D. Owens. 1996. Effect of promoter leader sequences on transient expression of reporter gene chimeras biolistically transferred into sugar beet (*Beta vulgaris*) suspension cells. Plant Cell Rept. 15:836-840.

Ingle, J., J.N. Timmis, and J. Sinclair. 1975. The relationship between satellite desoxyribonucleic acid, ribosomal ribonucleic acid redundancy, and genome size in plants. Plant Physiol. 55:496-501.

Ivic S.D. and A.C. Smigocki, 2001. Evaluation of the Biolistic Transformation Method for Commercially Important Sugar Beet Breeding Lines. Proc. ASSBT 31:207-213.

Jacq, B., T. Tetu, R.S. Sangwan, A. De Laat, and B.S. Sangwan-Norreel. 1992. Plant regeneration from sugarbeet *(Beta vulgaris* L.) hypocotyls cultured *in vitro* and flow cytometric nuclear DNA analysis of regenerants. Plant Cell Rept. 11:329-333.

Jacq, B., T. Tetu, R.S. Sangwan, A. De Laat, and B.S. Sangwan-Norreel. 1993. Efficient production of uniform plants from cotyledons of sugar beet (*Beta vulgaris* L.). Plant Breed. 110:185-191.

Janssen, G.J.M., M. Nihlgaard, and T.Kraft. 2003. Mapping of resistance genes to powdery mildew (*Erysiphe betae*) in sugar beet. Intl. Sugar J. 105:448-451.

Jazdzewska, E., Z. Sadoch, A. Niklas, and S.A. Majewska. 2000. Plant regeneration from sugar beet leaf protoplasts: analysis of shoots by DNA fingerprinting and restriction fragment length polymorphism. Canadian J. Bot. 78:10-18.

Joersbo, M. 2001. Advances in the selection of transgenic plants using non-abiotic marker genes. Physiol. Plant. 111:269-272.

Joersbo, M. and J. Brunstedt. 1990. Direct gene transfer to plant protoplasts by mild sonication. Plant Cell Rept. 9:207-210.

Joersbo, M., S.G. Pedersen, and F.T. Okkels. 1999. Parameters interacting with mannose selection employed for the production of transgenic sugar beet. Physiol. Plant. 105:109-115.

Joersbo, M., J. Mikkelsen, and J. Brunstedt. 2000. Relationship between promoter strength and transformation frequencies using mannose selection for the production of transgenic sugar beet. Molec. Breed. 6:207-213.

Jung, C. and G. Herrmann. 1991. A DNA probe for rapid screening of sugar beet (*Beta vulgaris* L.) carrying extra chromosomes for wild beets of the *Procumbentes* section. Plant Breed. 107:275-279.

Jung, C., K. Pillen, L. Frese, S. Fahr, and A. E. Melchinger. 1993. Phylogenetic relationships between cultivated and wild species of the genus *Beta* revealed by DNA "fingerprinting". Theor. Appl. Genet. 86:449-457.

Jung, C., R. Koch, F. Fisher, A. Brandes, G. Wricke, and R.G. Herrmann. 1992. DNA markers closely linked to nematode resistance genes in sugar beet (*Beta vulgaris* L.) mapped using chromosome additions, and translocations originating from wild beets of the *Procumbentes* section. Molec. Gen. Genet. 232:271-278.

Kallerhöff, J., P. Perez, S. Bouzoubaa, S. Ben Tahar, and J. Perret. 1990. Beet necrotic yellow vein virus coat protein-mediated protection in sugarbeet (*Beta vulgaris* L.) protoplasts. Plant Cell Rept. 9:224-228.

Kimoto, Y. and Y. Shimamoto. 2001. Differences in toxicity to larvae of cabbage armyworm between transgenic sugar beet lines with *crylA*(b) and *crylC*. Proc. Jpn. Soc. Sugar Beet Tech. 43:20-23.

Klein, T.M., E.D. Wolf, R. Wu, and J.C. Sanford. 1987. High-velocity microprojectiles for delivering nucleic acids into living cells. Nature 327:70-73.

Kleine, M., H. Voss, D. Cai, and C. Jung. 1998. Evaluation of nematode-resistant sugar beet (*Beta vulgaris* L.) lines by molecular analysis. Theor. Appl. Genet. 97:896-904.

Kloos D.U., H. Oltmanns, C. Dock, D. Stahl, and R. Hehl. 2002. Isolation and molecular analysis of six taproot expressed genes from sugar beet J. Exper. Bot. 53:1533-1534.

Knight, H., and M.R. Knight. 2001. Abiotic stress signalling pathways: specificity and cross-talk. Trends Plant Sc. 6:262-267.

Knoop, V. and Brennicke. 2002. A molecular biology of the plant mitochondrion. Crit. Rev. Plant Sci. 21:111-126.

Koch, G. and C. Jung. 2000. Genetic localization of cercospora resistance genes. *In: Cercospora beticola* Sacc. Biology, Agronomic Influence and Control Measures in Sugar Beet. Advances in Agronomy, IIRB, Bruxelles, Belgium, vol. 2, pp. 197-209.

Kraft, T., B. Fridlund, A. Hjerdin, T. Sall, S. Tuvesson, and C. Hallden. 1997. Estimating genetic variation in sugar beets and wild beets using pools of individuals. Genome 40:527-533.

Krens, F.A., L. Molendijk, G.J. Wullems, and R.A. Schilperoort. 1982. *In vitro* transformation of plant protoplasts with Ti plasmid DNA. Nature 296:72.

Krens, F.A., D. Jamar, G.J.A. Rouwendal, and R.D. Hall. 1990. Transfer of cytoplasm from new *Beta* CMS sources to sugar beet by asymmetric fusion. I. Shoot regeneration from mesophyll protoplasts and characterization of regenerated plants. Theor. Appl. Genet. 79:390-396.

Krens, F.A., A. Trifonova, P. Keizer, and R.D. Hall. 1996. The effect of exogenously-applied phytohormones on gene transfer efficiency in sugar beet (*Beta vulgaris* L.). Plant Sci. 116:97-106.

Krikorian, A.D. and D.I. Berquam. 1969. Plant cell and tissue cultures. Bot. Rev. 35:59-67.

Kubalakova, M. 1990. Somatic embryogenesis and cytoplasmic sterility in *Beta vulgaris* L. var. *saccharifera*. Biol. Plant. 32:414-419.

Kubis, S., J.S. Heslop-Harrison, and T. Schmidt. 1997. A family of differentially amplified repetitive DNA sequences in the genus *Beta* reveals genetic variation in *Beta vulgaris* subspecies and cultivars. J. Molec. Evol. 44:310-320.

Kubo, T., S. Nishizawa, and T. Mikami. 1999. Alterations in organization and transcription of the mitochondrial genome of cytoplasmic male sterile sugar beet (*Beta vulgaris* L). Molec. Gen. Genet. 2:283-290.

Kubo, T., Y. Yanai, T. Kinoshita, and T. Mikami. 1995. The chloroplast TRNP-TRNW-PETG gene-cluster in the mitochondrial genomes of *Beta vulgaris, B. trigyna* and *B. webbiana* — evolutionary aspects. Curr. Genet. 27:285-289.

Kubo T., S. Nishizawa, A. Sugawara, N. Itchoda, A. Estiati, and T. Mikami. 2000. The complete nucleotide sequence of the mitochondrial genome of sugar beet (*Beta vulgaris* L.) reveals a novel gene for $tRNA^{Cys}$(GCA). Nucleic Acids Res. 28:2571-2576.

Kulshreshtha, S. and R.H.A. Coutts. 1997. Direct somatic embryogenesis and plant regeneration from mature sugarbeet (*Beta vulgaris* L.) zygotic cotyledons. Plant Growth Regul. 22:87-92.

Laporte, V., D. Merdinoglu, P. Saumitou-Laprade, G. Butterlin, P. Vernet, and J. Cuguen. 1998. Identification and mapping of RAPD and RFLD markers linked to a fertility restorer gene for a new source of cytoplasmic male sterility in *Beta vulgaris* ssp. *maritima*. Theor. Appl. Genet. 96:989-996.

Larkin, P.J. and W. R. Scowcroft. 1981. Somaclonal variation -- a novel source of variability from cell cultures for plant improvement. Theor. Appl. Genet. 60:197-214.

Lathouwers, J.E., C. Bleykasten, S. Bouzoubaa, et al. 1997. Sugar beet transformation for rhizomania resistance: introduction and expression of different viral sequences. Proc. IIRB 61:491-495.

Leister, D. 2003. Chloroplast research in the age of genomics. Trends Genet. 19:47-56.

Lenzner, S., K. Zoglauer, and O. Schieder. 1995. Plant regeneration from protoplasts of sugar beet (*Beta vulgaris*). Physiol. Plant. 94:342-350.

Levall, M.W., K. Bengtsson, N.O. Nilsson, A. Hjerdin, and C. Hallden. 1994. Molecular characterization of UV-treated sugar beet somaclones using RFLP markers. Physiol. Plant. 90:216-220.

Levin, D.A. 2001. 50 years of plant speciation. Taxon 50:69-91.

Levin, D.A. 2003. The cytoplasmic factor in plant speciation. Syst. Bot. 28:5-11.

Lewin B. 1997. Genes Vol. VI. Oxford Univ. Press, New York, NY

Lindsey, K. and M.G.K. Jones. 1987. Transient gene expression in electroporated protoplasts and intact cells of sugar beet. Plant Molec. Biol. 10:43-52.

Lindsey, K. and M.G.K. Jones. 1989. Stable transformation of sugar beet protoplasts by electroporation. Plant Cell Rept. 8:71-74.

Lindsey, K. and P. Gallois. 1990. Transformation of sugar beet (*Beta vulgaris*) by *Agrobacterium tumefaciens*. J. Exper. Bot. 41:529-536.

Linsmaier, E.M. and F. Skoog. 1965. Organic growth factor requirements of tobacco tissue cultures. Physiol. Plant 18:100-227.

Loock, A., J.R. Stander, J. Kraus, and R. Jansen. 1998. Performance of transgenic glufosinate ammonium (Liberty) tolerant sugar beet hybrids. Proc. IIRB 61:339-343.

Lux, H., L. Herman, and C. Wetzel. 1990. Production of haploid sugar beet (*Beta vulgaris* L.) by culturing unpollinated ovules. Plant Breed. 104:177-183.

Mackenzie, S. and L. McIntosh. 1999. Higher plant mitochondria. Plant Cell 11:571-585.

Majewska, S.A., H. Nakashima, and K. Mori. 1994. Isolation and culture of suspension-derived protoplasts of *Beta vulgaris* L. Biol. Plant. 36:9-13.

Mannerlöf, M., B. L. Lennerfors, and P. Tenning. 1996. Reduced titer of BNYVV in transgenic sugar beet expressing the BNYVV coat protein. Euphytica 90:293-299.

Mannerlöf, M., S. Tuvesson, P. Steen, and P. Tenning. 1997. Transgenic sugar beet tolerant to glyphosate. Euphytica 94:83-91.

Manojlovic, V., Z. Nikolic, and S. Radovic. 1995. Sugarbeet haploids obtained from unpollinated ovules. Proc. IIRB 58:147-150.

Margara, J. 1970. Neoformation de bourgeons *in vitro* chez la betterave sucriere *Beta vulgaris* L. C.R. Hebd. Seanc. Acad. Sci., Paris 270:698-701.

Margara, J. 1977. La multiplication vegetative de la betterave (*Beta vulgaris* L.) en culture *in vitro*. C.R. Acad. Sci. D285:1041-1044.

Martin, G.B., J.G.K. Williams, and S.D. Tanksley. 1991. Rapid identification of markers linked to a *Pseudomonas* resistance gene in tomato by using random primers and near-isogenic lines. Proc. Natl. Acad. Sci. 88:2336-2340.

Mavromatis, A., G.N. Skaracis, D. Roupakias, F. Dubois, B. Sangwan-Norreel, and R. Sangvan. 1999. Production of diploids after *in vitro* culture of unfertilized ovules originated from tetraploid sugarbeet genotypes. Proc. Cost 824 WG2 Meeting "Gametic Embryogenesis in Dicots". Krakow, Poland, pp. 19-21.

McGrath, J.M. 2003. Alphabet soup for beets: status of ESTs, BACs, RILs and other genomic sundries. Proc. ASSBT/IIRB, 1:261-266.

McGrath, J.M., C. A. Derrico, and Y. Yu. 1999. Genetic diversity in selected, historical US sugarbeet germplasm and *Beta vulgaris* ssp. *maritima*. Theor. Appl. Genet. 98:968-976.

McGrath, J.M., S. Myers, and B. de los Reyes. 2001. Gene expression analyses to identify and target genes for agronomic traits, a case of improving emergence in sugar beet. Plant & Animal Genome IX Conference. www.intl-pag.org/pag/9/abstracts/W55_03.html.

McGrath, J.M., C.A. Derrico, M. Morales, L.O. Copeland, and D.R. Christenson. 2000. Germination of sugar beet (*Beta vulgaris* L.) seed submerged in hydrogen peroxide and water as a means to discriminate cultivar and seedlot vigor. Seed Sci. Tech. 28:607-620.

Merdinoglu, D., F. Pelsy, G. Butterlin, and P. Soreau. 1995. Use of RAPD to identify genetic markers for agronomically interesting characters in beet. Proc. IIRB 58:45-51.

Meyers, B.C., A. Kozik, A. Griego, H. Kuang, and R.W. Michelmore. 2003. Genome-wide analysis of NBS-LRR–encoding genes in *Arabidopsis*. Plant Cell 15:809-834.

Michelmore, R.W., I. Paran, and R.V. Kesseli. 1991. Identification of markers linked to disease-resistance genes by bulked segregant analysis: A rapid method to detect markers in specific genomic regions by using segregating population. Proc. Natl. Acad. Sci. USA 88:9828-9832.

Miedema, P. 1982a. A tissue culture technique for vegetative propagation and low temperature preservation of *Beta vulgaris* L. Euphytica 31:635-643.

Miedema, P. 1982b. Vegetative propagation of *Beta vulgaris* by leaf cuttings with axillary buds. Euphytica 31:771-772

Mita, G., M. Dani, P. Casciari, et al. 1991. Assessment of the degree of genetic variation in beet based on RFLP analysis and the taxonomy of *Beta*. Euphytica 55:1-6.

Moghaddam, B.E., M. Mesbah, and N. Yavari. 2000. The effect of in planta TIBA and proline treatment on somatic embryogenesis of sugar beet (*Beta vulgaris* L.). Euphytica 112:151-156.

Monger, W., P.J. Barrington, A.M. Acaster, N.G. Halford, C. Ainsworth, and H. Thomas. 1995. A transgenic approach towards investigating carbon metabolism in sugar beet. Proc. IIRB 58:193-196.

Morgante, M. and A.M. Olivieri. 1993. PCR-amplified microsatellites as markers in plant genetics. Plant J. 3:175-182.

Mugnier, J. 1987. Infection by *Polymyxa betae* and *Plasmodiophora brassicae* of roots containing root-inducing transferred DNA of *Agrobacterium rhizogenes*. Phytopathology 77:539-542.

Munthali, M.T., H.J. Newbury, and B.V. Ford-Lloyd. 1996. The detection of somaclonal variants of beet using RAPD. Plant Cell Rept. 15:474-478.

Murashige, T. 1974. Plant propagation through tissue culture. Annu. Rev. Plant Physiol. 25:135-166.

Murashige, T. and F. Skoog. 1962. A revised medium for rapid growth and bioassay with tobacco tissue cultures. Physiol. Plant 15:473-497.

Nagamine, T., G.A. Todd, K.P. McGann, H.J. Newburg, and B.V. Ford-Lloyd. 1989. Use of restriction fragment length polymorphism to fingerprint beet at the genotype and species levels. Theor. Appl. Genet. 78:847-851.

Nichols, J.B., C.C. Dalton, G.A. Todd, and N.W. Broughton. 1992. Genetic engineering of sugar beet. Zuckerindustrie. 117:797-800.

Nilsson, N.O., C. Hallden, M. Hansen, A. Hjerdin, and T. Sall. 1997. Comparing the distribution of RAPD and RFLP markers on a high density linkage map of sugar beet. Genome 40:644-651.

Nilsson, N.O., M. Hansen, A.H. Panagopoulos. 1999. QTL analysis of *Cercospora* leaf spot resistance in sugar beet. Plant Breed. 118:327-334.

OECD (Organisation for Economic Co-operation and Development). 2003. OECD's Database for Field Trials. (www.webdomino.oecd.org).

Owens, L.D. and D.R. Eberts. 1992. Sugarbeet leaf disc culture: an improved procedure for inducing morphogenesis. Plant Cell, Tiss. Org. Cult. 31:195-201.

Paolini, M., P. Jenabzadeh, and E. Ambrogio. 1994. Gynogenesis and chromosome doubling in sugarbeet breeding. Sementi Elette 40:5-12.

Paul, H., C. Zijlstra, J.E. Leeuwangh, F.A. Krens, and H.J. Huizing. 1987. Reproduction of the beet cyst nematode *Heterodera schachtii* Schm. on transformed root cultures of *Beta vulgaris* L. Plant Cell Rept. 6:379-381.

Paul, H., J.E.M. van Deelen, B. Henkel, T. De Bock, W. Lange, and F.A. Krens. 1990. Expression *in vitro* of resistance to *Heterodera schachtii* in hairy roots of an alien monotelosomic addition plant of *Beta vulgaris*, transformed by *Agrobacterium rhizogenes*. Euphytica 48:153-157.

Pedersen, C., R.D. Hall, and F.A. Krens. 1993. Petioles as the tissue source for isolation and culture of *Beta vulgaris* and *B. maritima* protoplasts. Plant Sci. 95:89-97.

Pelsy, F. and D. Merdinoglu. 1996. Identification and mapping of random amplified polymorphic DNA markers linked to a rhizomania resistance gene in sugar beet (*Beta vulgaris* L.) by bulked segregant analysis. Plant Breed. 115:371-377.

Pillen, K., G. Steinrücken, R. G. Herrmann, and C. Jung. 1993. An extended linkage map of sugar beet (*Beta vulgaris* L.) including nine putative lethal genes and the restorer gene X. Plant Breed. 111:265-272.

Pillen, K., G. Steinrücken, G. Wricke, R.G. Herrmann, and C. Jung. 1992. A linkage map of sugar beet (*Beta vulgaris* L.). Theor. Appl. Genet. 84:129-135.

Potrykus, I. 1990. Gene transfer to plants: assessment and perspectives. Physiol. Plant. 79:125-134.

Pruitt R.E., J.L. Bowman, and U. Grossniklaus. 2003. Plant genetics: a decade of integration. Nature Genet. 33 (suppl.):294-304.

Qiu, Y.L. and J.D. Palmer. 1999. Phylogeny of early land plants: insights from genes and genomes. Trends Plant Sci. 4:26-30.

Rae, S.J., C. Aldam, I. Dominguez, M. Hoebrechts, S.R. Barnes, and K.J. Edwards. 2000. Development and incorporation of microsatellites markers into linkage map of sugar beet (*Beta vulgaris* spp.). Theor. Appl. Genet. 100:1240-1248.

Rafalski, J.A. 2002. Novel genetic mapping tools in plants: SNPs and LD-based approaches. Plant Sci. 162:329-333.

Ranalli, P. and G. Mandolino. 1999. Il contributo delle biotecnologie al miglioramento genetico della barbabietola da zucchero. *In:* B. Casarini, E. Biancardi, and P. Ranalli (eds.) La barbabietola da zucchero in ambiente mediterraneo. Edagricole, Bologna, Italy, pp. 183-244.

Reamon-Ramos, S.M. and G. Wricke. 1992. A full set of monosomic addition lines in *Beta vulgaris* from *Beta webbiana*: Morphology and isozyme markers. Theor. Appl. Genet. 84:411-418.

Reed, J., Privalle, L., Powell, M.L., et al. 2001. Phosphomannose isomerase: an efficient selectable marker for plant transformation. *In Vitro* Cell. Devel. Biol-Plant. 37:127-132.

Ritchie, G.A., K.C. Short, and M.R. Davey. 1989. *In vitro* shoot regeneration from callus, leaf axils and petioles of sugar beet *(Beta vulgaris* L.). J. Exper. Bot. 40:277-283.

Rogozinska, J. and M. Goska. 1978. Induction of differentiation and plant formation in isolated sugar beet leaves. Bull. Acad. Polon. Sci. 25:343-345.

Rogozinska, J.H. and M. Goska. 1982. Attempts to induce haploids in anther culture of sugar fodder and wild species of beet. Acta Soc. Bot. Polon. 51:91-105.

Roussy, I., F. Dubois, R.S. Sangwan, and B.S. Sangwan-Noreel. 1996. In planta 2,3,5 triiodobenzoic acid treatment promotes high frequency and routine *in vitro* regeneration of sugar beet *(Beta vulgaris* L.) plants. Plant Cell Rept. 16:142-146.

Sabir, A., H.J. Newbury, G. Todd, J. Catty, and B.V. Ford-Lloyd. 1992. Determination of genetic stability using isozymes and RFLPs in beet plants regenerated *in vitro*. Theor. Appl. Genet. 84:113-117.

Sadoch, Z., S.A. Majewska, E. Jazdzewska, and A. Niklas. 2000. Changes in sugar beet mitochondrial DNA induced during callus stage. Plant Breed. 119:107-110.

Salentijn, E.M.J., M.J.B. Arens-De Reuver, W. Lange, T. De Bock, W.J. Stiekema, and R. Klein-Lankhorst. 1995. Isolation and characterization of RAPD-based markers linked to the beet cyst nematode resistance locus (Hs1pat-1) on chromosome 1 of *B. patellaris*. Theor. Appl. Genet. 90:885-891.

Salentijn, E.M.J., N.N. Sandal, W. Lange, et al. 1992. Isolation of DNA markers linked to a beet cyst nematode resistance locus in *Beta patellaris* and *Beta procumbens*. Molec. Gen. Genet. 235:432-440.

Sand, C. 1995. Direct and indirect organogenesis in *Beta vulgaris* L. Lucrari Stiintifice 20:32-34.

Sanford, J.C., T.M. Klein, E.D. Wolf, and N. Allen. 1987. Delivery of substances into cells and tissues using a particle bombardment process. Part. Sci. Tech. 5:27-37.

Santoni, S. and A. Berville. 1992. Characterization of the nuclear ribosomal DNA units and phylogeny of *Beta* L. wild forms and cultivated beets. Theor. Appl. Genet. 83:533-542.

Saunders, J.W. 1982. A flexible *in vitro* shoot culture propagation system for sugarbeet that includes rapid floral induction of ramets. Crop Sci. 22:1102-1104.

Saunders, J.W. and Mahoney, M. 1982. Benzyladenine induces foliar adventitious shoot formation on young plants of 2 sugar beet (*Beta vulgaris* L.) cultivars. Euphytica 31:801-804.

Saunders, J.W. and K. Shin. 1986. Germplasm and physiologic effects of induction of high-frequency hormone autonomous callus and subsequent shoot regeneration in sugar beet. Crop Sci. 26:1240-1245.

Saunders, J.W. and Tsai, C.J. 1999. Production of somatic embryos and shoots from sugar beet callus: effects of abscisic acid, other growth regulators, nitrogen source, sucrose concentration and genotype. *In Vitro* Cell. Devel. Biol. Plant 35:18-24.

Saunders, J.W., I.D. Kltykendali, and A.C. Smigocki 2001. First report of adventitious shoots from intact sugarbeet plants and its implications. Proc. ASSBT 31:184-188.

Saunders, J.W., G. Acquaah, K. Renner, and W. Doley. 1992. Monogenic dominant sulfonylurea resistance in sugar beet from somatic cell selection. Crop Sci. 32:1357-1360.

Schafer-Pregl, R., D. Borchardt, E. Barzen, et al. 1999. Localization of QTLs for tolerance to *Cercospora beticola* on sugar beet linkage groups. Theor. Appl. Genet. 99:829-836.

Schlangstedt, M., K. Zoglauer, S. Lenzner, B. Hermans, and M. Jacobs. 1994. Im-provement of sugar beet *(Beta vulgaris* L.*)* protoplast culture: leaf petioles as a protoplast source. J. Plant Physiol. 143:227-233.

Schmidt, D., A. Polley, S. Reich, C. Kehling , M.W. Ganal. 2003. Identification of SNP markers in the sugar beet (*Beta vulgaris*) genome. Plant & Animal Genome XI Conf. http://www.intl-pag.org/11/abstracts/P3e_P246_XI.html.

Schmidt, T. and J.S. Heslop-Harrison 1996. The physical and genomic organization of microsatellites in sugar beet. Proc. Natl. Acad. Sci. USA 93:8761-8765.

Schmidt, T. and J.S. Heslop-Harrison. 1998. Genomes, genes and junk: the large scale organization of plant chromosomes. Trends Plant Sci. 3:195-199.

Schmidt, T., H. Junghans, and M. Metzlaff. 1990. Construction of *Beta procumbens*-specific DNA probes and their application for the screening of *B.vulgaris* x *B.procumbens* (2n = 19) addition lines. Theor. Appl. Genet. 79:177-181.

Schmidt, T., T. Schwarzacher, and J.S. Heslop-Harrison. 1994. Physical mapping of ribosomal RNA genes by fluorescent in situ hybridization and structural analysis of 5S ribosomal RNA genes and intergenic spacer sequences in sugar beet (*Beta vulgaris*). Theor. Appl. Genet. 88:629-636.

Schmidt, T., S. Kubis, and J.S. Heslop-Harrison. 1995. Analysis and chromosomal localization of retrotransposons in sugar beet (*Beta vulgaris* L.): LINES and Ty1-*copia*-like elements as major components of the genome. Chrom. Res. 3:335-345.

Schmidt, T., K. Boblenz, M. Metzlaff, D. Kaemmer, K. Weising, and G. Kahl. 1993. DNA fingerprinting in sugar beet (*Beta vulgaris*)–identification of double-haploid breeding lines. Theor. Appl. Genet. 85:653-657.

Schmitz-Linneweber C., R.M. Maier, J. Alcaraz, A. Cottet, R.G. Herrmann, and R. Mache. 2001. The plastid chromosome of spinach (*Spinacia oleracea*): complete nucleotide sequence and gene organization Plant Molec. Biol. 45:307-315.

Schnable, P.S. and R.P. Wise. 1998. The molecular basis of cytoplasmic male sterility and fertility restoration. Trends Plant Sci. 3:175-180.

Schneider, K., B. Weisshaar, D.C. Borchardt, and F. Salamini. 2001. SNP frequency and allelic haplotype structure of *Beta vulgaris* expressed genes. Molec. Breed. 8:63-74.

Schneider, K., R. Schäfer-Pregl, D.C. Borchardt, and F. Salamini. 2002. Mapping QTLs for sucrose content, yield and quality in a sugar beet population fingerprinted by EST-related markers. Theor. Appl. Genet. 104:1107-1113.

Schneider, K., D.C. Borchardt, R. Schäfer-Pregl, et al. 1999. PCR-based cloning and segregation analysis of functional gene homologues in *Beta vulgaris*. Molec. Gen. Genet. 262:515-524.

Scholten, O., T. De Bock, R. Klein-Lankhorst, and W. Lange. 1999. Inheritance of resistance to beet necrotic yellow vein virus in *Beta vulgaris* conferred by a second gene for resistance. Theor. Appl. Genet. 99:740-746.

Scholten, O., R. Jansen, P. Keizer, T. De Bock, and W. Lange. 1996. Major gene resistance to beet nectrotic yellow vein virus (BNYVV) in *Beta vulgaris*. Euphytica 91:331-339.

Scholten, O., R. Klein-Lankhorst, D. Esselink, T. De Bock, and W. Lange. 1997. Identification and mapping of random amplified polymorphic DNA (RAPD) markers linked to resistance against beet necrotic yellow vein virus (BNYVV) in *Beta* accessions. Theor. Appl. Genet. 94:123-130.

Schondelmaier, J. and C. Jung. 1997. Chromosomal assignment of the nine linkage groups of sugar beet (*Beta vulgaris* L.) using primary trisomics. Theor. Appl. Genet. 95:590-596.

Schondelmaier, J., G. Steinrücken, and C. Jung. 1996. Integration of AFLP markers into a linkage map of sugar beet (*Beta vulgaris* L.). Plant Breed. 115:231-237.

Schondelmaier, J., T. Schmidt, J.S. Heslop-Harrison, and C. Jung. 1997. Genetic and chromosomal localization of the 5S rDNA locus in sugar beet (*Beta vulgaris* L). Genome 40:171-175.

Schondelmaier, J., R. Heller, K. Pillen, G. Steinrücken, and C. Jung. 1995. A linkage map of sugar beet. Proc. IIRB. 58:37-45.

Schumacher, K., J. Schondelmaier, E. Barzen, et al. 1997. Combining different linkage maps in sugar beet (*Beta vulgaris* L.) to make one map. Plant Breed. 116:23-38.

Schwarzacher, T. 2003. Meiosis, recombination and chromosomes: a review of gene isolation and fluorescent *in situ* hybridization data in plants. J. Exper. Bot. 54:11-23.

Setiawan, A., G. Koch, S.R. Barnes, and C. Jung. 2000. Mapping quantitative trait loci (QTLs) for resistance to cercospora leaf spot disease (*Cercospora beticola* Sacc.) in sugar beet (*Beta vulgaris* L.). Theor. Appl. Genet. 100:1176-1182.

Sevenier, R., R.D. Hall, I.M. van der Meer, H.J.C. Hakkert, A. van Tunen, and A.J. Koops. 1998. High level fructan accumulation in a transgenic sugar beet. Nature Biotechn. 16:843-846.

Shen, Y., H. J. Newburg, and B.V. Ford-Lloyd. 1996. The taxonomy characterisation of annual *Beta* germplasm in genetic resources collection using RAPD markers. Euphytica 91:205-212.

Shimamoto, Y., and T. Mikami. 1993. Callus induction and plant regeneration of *Beta* germplasm. J. Sugar Beet Res. 30:317-319.

Skaracis, G.N. and E. Biancardi. 2000. Breeding for cercospora resistance in sugar beet. *In: Cercospora beticola* Sacc.: Biology, Agronomic Influence and Control Measures in Sugar beet. Advances in Agronomy IIRB, Bruxelles, Belgium, vol. 2, pp. 177-195.

Smigocki, A.C. 2001. Engineering sugarbeets with multiple proteinase inhibitor genes for enhanced tolerance to the sugarbeet root maggot. Sugarbeet Development Foundation Research Report, Denver, CO, pp. 17-24.

Smith, R.H. and S.H. Park. 2002. Tissue culture for crop improvement. *In:* M.S. Kang (ed.) Quantitative Genetics, Genomics and Plant Breeding. CABI Intl. Wallingford, UK, pp. 189-196.

Snyder, G.W., J.C. Ingersoll, A.C. Smigocki, and L.D. Owens. 1999. Introduction of pathogen defense genes and a cytokinin biosynthesis gene into sugar beet (*Beta vulgaris* L.) by *Agrobacterium* or particle bombardment. Plant Cell Rept. 18:829-834.

Staginnus, C., B. Huettel, C. Desel, T. Schmidt, and G. Kahl. 2001. A PCR-based assay to detect En/Spm-like transposon sequences in plants. Chrom. Res. 9:591-605.

Stark, D.M. and R.N. Beachy. 1989. Protection against polyvirus infection in transgenic plants: evidence for broad spectrum resistance. Bio-Tech. 7:1257-1262.

Steen, P. and H.C. Pedersen. 1993. Gene transfer for herbicide resistance. J. Sugar Beet Res. 30:267-274.

Steen, P. and H.C. Pedersen. 1995a. Yield and quality characters in transgenic herbicide tolerant sugar beet (*Beta vulgaris* L.). Proc. IIRB 58:185-188.

Steen, P. and H.C. Pedersen. 1995b. Strategies in creating transgenic herbicide tolerant sugar beet (*Beta vulgaris* L.) varieties. Proc. IIRB 58:189-192.

Steen, P., B. Keimer, K. D'Halluin, and H.C. Pedersen. 1986. Variability in plants of sugar beet *(Beta vulgaris* L.) regenerated from callus, cell suspension and protoplasts. *In:* W. Horn, C.J. Jensen, W. Odenbach and O. Schieder (eds.). Genetic Manipulation in Plant Breeding. Walter de Gruyter and Co., Berlin, Germany, pp. 633-635.

Svirshchevskaya, A.M. and J. Dolezel. 2000. Production and performance of cynogenetic sugarbeet lines. J. Sugar Beet Res. 37:117-133.

Szabados, L. and C. Gaggero. 1985. Callus formation from protoplasts of a sugarbeet cell suspension culture. Plant Cell Rept. 4:195-198.

Tabor, H.K., N.J. Risch, and R.M. Myers. 2002 Candidate-gene approaches for studying complex genetic traits: Practical considerations. Nature Rev. Genet. 3:391-397.

Tanksley, S.D., N.D. Young, A.H. Paterson, and M.W. Bonierbale. 1989. RFLP mapping in plant breeding: New tools for an old science. Bio-Tech. 7:257-263.

Tenning, P., E.W. Weich, U.B. Kjarsgaard, M.A. Lelu, and M. Nihlgard. 1992. Somatic embryogenesis from zygotic embryos of sugar beet *(Beta vulgaris* L.). Plant Sci. 81:103-109.

Tenning, P., J. Bensefelt, P. Fouillard, M. Mannerlof, and S. Tuvesson. 1995. Glyphosate tolerance in transgenic sugar beet. Proc. IIRB 58:183-184.

Tetu, T., R.S. Sangwan, and B.S. Sangwan-Norrell. 1987. Hormonal control of organogenesis and somatic embryogenesis in *Beta vulgaris* callus. J. Exper. Bot. 38:506-517.

The *Arabidopsis* Genome Initiative. 2000. Analysis of the genome sequence of the flowering plant *Arabidopsis thaliana*. Nature (London) 408:796-815.

Thorpe, T.A. 2000. History of plant cell culture. *In:* R.H. Smith (ed.) Plant Tissue Culture Techniques and Experiments. Acad. Press, New York, NY, pp. 1-32.

Toldi, O., G. Gyulai, J. Kiss, I. Tarns, and E. Balazs. 1996. Antiauxin enhanced microshoot initiation and plant regeneration from epicotyl-originated thin layer explants of sugar beet *(Beta vulgaris* L.). Plant Cell Rept. 15:851-854.

Tsai, C.J. and J.W. Saunders. 1995. Somatic embryos from callus of sugar beet biotechnology clone REL-1. J. Sugar Beet Res. 32:215-227.

Tsai, C.J. and J.W. Saunders. 1999. Encapsulation, germination and conversion of somatic embryos in sugarbeet. J. Sugar Beet Res. 36:11-32.

Unseld, M., J.R. Marienfeld, P. Brandt, and A. Brennicke. 1997. The mitochondrial genome of *Arabidopsis thaliana* contains 57 genes in 366,924 nucleotides. Nature Genet. 15:57-61.

Uphoff, H. and G. Wricke. 1995. A genetic map of sugar beet (*Beta vulgaris*) based on RAPD markers. Plant Breed. 114:355-357.

Van der Hoeven, R., C. Ronning, J. Giovannoni, G. Martin, and S. Tanksley. 2002. Deductions about the number, organization, and evolution of genes in the tomato genome based on analysis of a large expressed sequence tag collection and selective genomic sequencing. Plant Cell 14:1441-1456.

Van Geyt, J. and M. Jacobs. 1985. Suspension culture of sugar beet (*Beta vulgaris* L.). Induction and habituation of dedifferentiated and self-regenerating cell lines. Plant Cell Rept. 4:66-69.

Van Geyt, J., K. D'Halluin, and M.Jacobs. 1985. Induction of nuclear and cell divisions in microspores of sugarbeet (*Beta vulgaris* L.). Z. Pflanzenzücht. 95:325-335.

Van Geyt, J., G.J.Jr. Spreckman, K. D'Halluin, and M.Jacobs. 1987. *In vitro* induction of haploid plants from unpollinated ovules and ovaries in sugar beet. Theor. Appl. Genet. :920-925.

Van Poucke, K., M. Karimi, and G. Gheysen. 2001. Analysis of nematode-responsive promoters in sugar beet hairy roots. Meded Rijksuniv Gent Fak. Landbouwkd Toegep Biol Wet. 66:591-598.

Van Roggen, P.M., S.G. Thomas, P. Hedden, et al. 1997. Genetic engineering for bolting resistance in sugar beet. Proc. IIRB 60:571-574.

Van Wordragen, M.F. and H.J.M. Dons. 1992. *Agrobacterium tumefaciens*-mediated transformation of recalcitrant crops. Plant Molec. Biol. Rept. 10:12-36.

Vandenbussche, B. and M.P. De Proft 1996. Cryopreservation of *in vitro* beet shoot-tips using the encapsulation-dehydration technique: development of a basic protocol. Cryo-Lett. 17:137-140.

Vandenbussche, B., G. Weyens, and M. De Proft. 2000. Cryopreservation of *in vitro* sugar beet (*Beta vulgaris* L.) shoot tips by a vitrification technique. Plant Cell Rept. 19:1064-1068.

Venter, J.C., S. Levy, T. Stockwell, K. Remington, and A. Halpern. 2003. Massive parallelism, randomness and genomic advances. Nature Genet. 33 (suppl): 219-227.

Vos, P., R. Hogers, M. Bleeker, et al. 1995. AFLP: a new technique for DNA fingerprinting. Nucleic Acids Res 23:4407-4414.

Wagner, H., W.E. Weber, and G. Wricke. 1992. Estimating linkage relationship of isozyme markers and morphological markers in sugar beet (*Beta vulgaris* L.) including families with distorted segregation. Plant Breed. 108:89-96.

Walbot V. and R. Goldberg. 1979. Plant genome organization and its relationship to classical plant genetics. *In:* T. Hall and J. Davies (eds.) Nucleic Acids in Plants. CRC Press, Boca Raton, FL, pp. 3-40.

Weber, W.E., D. Borchardt, and G. Koch. 1999. Combined linkage maps and QTLs in sugar beet (*Beta vulgaris* L.) from different populations. Plant Breed. 118:193-204.

Weber, W.E., D. Borchardt, and G. Koch. 2000. Marker analysis for quantitative traits in sugar beet. Plant Breed. 119:97-106.

Weiland, J.J. and M.H. Yu. 2001. Molecular genetic tagging of resistance in sugarbeet to root-knot nematode (*Meloidogyne* species). J. Sugar Beet Res. 38:103.

Weiland, J.J. and M.H. Yu. 2003. A cleaved amplified polymorphic sequence (CAPS) marker associated with root-knot nematode resistance in sugarbeet (*Beta vulgaris* L.). Crop Sci. 43:1814-1818.

Welander, T. 1974. Callus and root formation in explants of *Beta vulgaris*. Physiol. Plant. 32:305-307.

Williams, J.G.K., A.R. Kubelic, K.J. Livak, J.A. Rafalsky, and S.V. Tingey. 1990. DNA polymorphisms amplified by arbitrary primers are useful as genetic markers. Nucleic Acids Res. 18:6531-6535.

Wozniak, C.A. 1999. Transgenic sugar beet: progress and development. Appl. Plant Biotech. 10:301-324.

Wozniak, C.A. and L.D. Owens. 1989. Transformation of sugarbeet cell suspension cultures mediated by *Agrobacterium tumefaciens*. J. Cell. Biochem. Suppl. 13D:272.

Yu, J., S.N. Hu, J. Wang, et al. 2001. A draft sequence of the rice (*Oryza sativa* ssp. *indica*) genome. Chinese Sci. Bull. 46:1937-1942.

Yu, M.H. 1989. Callus induction and differentiation from leaf explants of different species of the genus *Beta*. Crop Sci. 29:205-209.

Yu, M.H. 1995. Identification of a *Beta maritima* source of resistance to root-knot nematode for sugarbeet. Crop Sci. 35:1288-1290.

Yu, M.H., L.M. Pakish, and H. Zhou. 2001. An isozyme marker for resistance to root-knot nematode in sugar beet. Crop Sci. 41:1051-1053.

Zhang, C.L., D.F. Chen, A.C. McCormac, N.W. Scott, M.C. Elliott, and A. Slater. 2001. Use of the GFP reporter as a vital marker for *Agrobacterium*-mediated transformation of sugar beet (*Beta vulgaris* L.). Molec. Biotech. 17:109-117.

Zhong, Z., H.G. Smith, and T.H. Thomas. 1993. Micropropagation of wild beet (*Beta maritima*) from inflorescence pieces. Plant Growth Regul. 12:53-57.

5

Seed Production

Significant yield increases in sugar beet are, in part, the result of improved seed technology and germination (Bee and Jaggard, 1996). Among the factors influencing seed quality are: i) genetic characteristics; ii) cultural practices used in producing the seed crop; iii) diseases affecting seed-bearing plants; iv) climate during flowering, pollination and ripening of the seed (Scott and Longden, 1970); v) drying, processing, and packaging techniques; and vi) storage conditions and age of the seed. Geographical areas that consistently produce high yields of quality seed have been identified. The most noteworthy of these are Emilia-Romagna in Italy, southwest France, and Oregon in the USA. Seed production is concentrated in these regions to meet the quality demands of growers.

5.1 Agronomic Practices (E. Biancardi)

There are two basic systems of seed production: direct and indirect seeding. In the first, genotypes to be reproduced are seeded, overwintered, and harvested in the same place. With the indirect seeding method, seed is planted in a nursery and the resultant stecklings transplanted in another location for seed production (Bornscheuer et al., 1993).

Direct seeding is used in the USA (Oregon), southwest France, England, Yugoslavia, Hungary, and Greece. Its foremost advantage is that no transplanting is necessary, thereby eliminating the costs involved. Since the plants complete their life cycle in one place, they have a deeper and broader root system than the transplanted stecklings. As a consequence, lodging is increased, less irrigation is required, vegetative development is enhanced, and they tend to flower slightly earlier than the transplanted stecklings.

The disadvantages of direct seeding include susceptibility to frost and the risk of contamination by weed beet seeds in the soil. With direct seeding, the beets are seeded more sparsely than in a steckling nursery and therefore are less protected from the cold and may experience lower winter survival rates. Temperatures below –12°C may cause severe losses, particularly when dealing with monogerm germplasm (Campbell, 1968). Frost damage can be reduced by overseeding, followed by thinning in early spring.

Cover crops, seeded between the rows and eliminated after winter, are frequently used to protect sugar beet plants from frost. Barley is suitable for this purpose (Jansen, 1963). Additional frost-protection systems are discussed later. A further disadvantage of direct seeding, compared with indirect, is the problem of unwanted beet seed in seed production fields. Unwanted seed may also be present in steckling nursery sites, but such nurseries require less space and hence allow more flexibility in field selection. In fact, a hectare of stecklings will produce enough plants for a 10-hectare seed field.

Cultural practices used to produce seed are similar to those used in breeding nurseries except that a precision drill is used to plant seeds 6-14 cm apart in rows spaced 60-75 cm apart. Rarely, the seed of the two parents of a hybrid can be mixed (the multigerm pollinator seed can be eliminated after harvest by grading), or they may be sown in separate rows. In the first case, the pollinators and CMS parent are mixed in a 1:4 ratio: in the latter case, a row of pollinator is generally sown between every three or four CMS rows. These proportions may vary, depending on the pollen-producing capability of the pollinator (Smith, 1987).

Indirect seeding is more laborious than the direct seeding system, but generally produces higher quality seed. It is widely used in Italy and southwest France. The steckling nursery is planted during the last half of August, preferably in light well-drained soil. Fields must be disease-free so that the health of the plants is not jeopardized and no pathogen spread in subsequent transplanting. Hence, tests for soil diseases are compulsory in many countries.

As with direct seeding, there is risk of contamination by seed left in the soil from previous crops. Seed of annual weed beets that bolt late in the season can be a serious problem in seed-production fields. Unwanted plants between the rows can be eliminated, unlike those that grow in the rows. In the latter case, it is often necessary to destroy the nursery. To preclude this problem, the cropping history of the field should be obtained and at least 10 years (Bornscheuer et al., 1993) left between beet crops.

As an extra precaution, prospective nursery sites can be irrigated in July; if any beet plants appear, there is ample time to select another field.

The precise germination percent of the genotypes must be determined prior to planting because stand affects steckling size. Smaller roots can be used for mechanical transplanting than for hand transplanting. Regular stands reduce waste of plants smaller or larger than optimal. An ideal stand is between 1,000,000 and 1,200,000 plants per hectare (CAC, 1996). Spacing of monogerm seeds in the row generally ranges from 2 to 3 cm. For multigerm seed, spacing depends on the number of embryos per seed cluster.

Spacing between rows ranges from 20 to 25 cm depending on the region, harvesting system-and climate. In areas with cold winters, dense stands may compensate for anticipated losses due to frost and provide better protection by the canopy. The stand of stecklings at harvest also depends on each genotype's sensitivity to cold. Frost damage to multigerm pollinators is less frequent. CMS lines and especially O-type lines are more sensitive; heavy losses may occur when temperatures fall below −10°C. Special water-permeable plastic fabric covers that provide thermal protection for the stecklings may prevent frost damage to sensitive genotypes. Within a genotype, the heaviest losses occur in irregular stands which cause some roots to enlarge and become more sensitive to frost.

Nursery fertilization roughly follows the amounts recommended for the sugar crop, paying special attention to avoid excess nitrogen that may cause excessive leaf and root development. Great care is taken to protect against parasites that can spread rapidly after transplanting. The most dangerous diseases are cercospora l.s., which attacks the nursery in October (in hot climates), *Phoma*, *Alternaria*, powdery mildew, *Botrytis*, *Pseudomonas,* and *Peronospora*. Insects (flea beetles, aphids, cutworms, etc.) can cause severe damage and therefore must be controlled with chemicals. Dicloropropene fumigation 20-30 days before seeding may be necessary to control nematodes in some areas. Weed control is carried out with the same methods and herbicides used on the sugar crop. However, the greater commercial value of the nursery means increased protection is required, including manual labor, when necessary.

Stecklings are harvested in February-March and transplanted immediately (**Fig. 5.1**). In cold regions it is better to harvest before winter and store the stecklings in piles with leaves to outside and roots inside. Dimensions of the roots at the end of winter depend on weather conditions, but uniformity is of greater importance than size. Generally speaking, roots measuring 3-4 cm in diameter are more productive and suffer the least losses due to transplanting, but they cause difficulty during

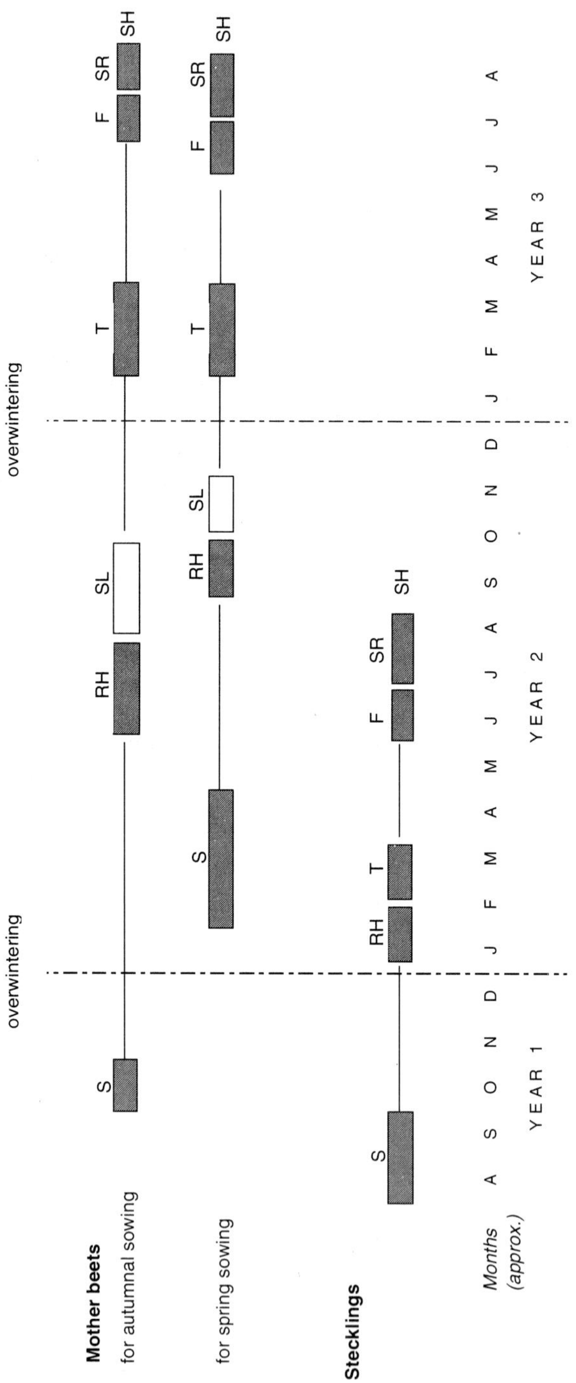

Fig. 5.1 Reproduction of germplasm using mother roots and stecklings (S - sowing; RH - root harvest; SL - root selection; RS - root storage; T - transplanting; F - flowering; SR - seed ripening; SH - seed harvest).

mechanical planting operations. Smaller roots are less costly to transport and better suited to mechanical transplanting. They are, however, more susceptible to drought and frequently produce plants which do not flower.

Leaves are cut mechanically before uprooting so as to leave petioles 4-5 cm in length. The taproot is trimmed terminally to stimulate subsequent abundant production of lateral roots. Lastly, the stecklings are cleaned and submerged in chemical solutions that provide protection from the most dangerous fungal diseases, such as *Phoma*, *Peronospora,* and downy mildew.

As with direct seeding, the parents of the hybrid varieties are transplanted in distinct rows or mixed in appropriate proportions. Transplanting in different rows is preferable since it allows inspection of the CMS parent at the beginning of flowering and the elimination of fertile, anomalous, and off-type plants. Furthermore, it allows for trimming the stalks to synchronize flowering of the two parents. Stecklings are normally spaced 40-50 cm in rows 70-80 cm apart, which results in about 36,000 plants per hectare. Clay soils are preferable, as long as they are well drained and properly prepared. Spatial isolation among crossing fields must not be less than determined distances (300, 600, and 1,000 meters), to avoid pollen contamination.

Regardless of whether the stecklings are transplanted manually or mechanically, only the petioles must protrude from the soil. It is important that the surrounding soil be carefully compressed so it is in contact with the roots, enabling the plants to take root quickly. As in the steckling nursery, nitrogen fertilization must be carried out with great care since excess nitrogen causes delayed ripening and reduces germination. Nitrogen is generally distributed twice: once during transplanting, together with phosphorus and potassium, and again before seed stalk initiation. The amount of nitrogen depends on residual fertility, which must be determined by soil analyses.

Weeds are eliminated by hoeing or tillage and with herbicides. Any wild beet readily is recognized as it will be at a different stage of development. With indirect seeding, wild beets are not the serious problem described for nurseries and seed production fields established by direct seeding. For fields with transplanted plants, lower contamination risks mean shorter field rotations can be used 4 years versus the 10 required for nurseries and direct seeding fields.

Diseases which compromise yield and seed quality must be controlled. Particular attention should be paid to *Phoma*, *Alternaria*, *Uromyces*, *Ramularia*, *Cercospora*, *Erysiphe*, *Botrytis*, *Peronospora*, *Verticillium*, and *Pseudomonas*. The presence of black and green peach aphids before and after the appearance of the stalk must be prevented because they might transmit viruses that depress crop development. Any type of chemical treatment should be avoided during flowering. Trimming the stalks of one parent to delay flowering may be necessary if flowering of the parents is not synchronized.

Irrigation after transplanting is often needed and should be repeated as necessary during ripening. High yields of quality seed are obtained with drop irrigation, which does not moisten leaves and stalks and therefore minimizes development of fungi on both plants and seeds.

Development of the seed stalk and initiation of new flowers continue up to harvest in July-August. Therefore, a single plant may have flowers forming, functional flowers, flowers that are closed, and seeds at different stages of ripening. In terms of quantity and quality, the most valuable seed production is that which ripens earliest, while the late-ripening is smaller and has poor germination. There is a close correlation between seed or embryo weight and germination potential. The highest quality seed develops on the main stalk (Alcaraz, 1981).

Pollinators are frequently eliminated at the end of June when they have completed their pollinating function. They are usually machine cut and sometimes harvested to collect the multigerm seed. Harvesting begins when most of the seed has turned a light tobacco color and threshes easily. Earlier harvests will reduce seed losses, but some of the seed will be partially ripe and reduce germination. Late harvesting increases losses as the best seed has shattered.

Harvesting is generally carried out with machines so operated as to preclude shaking the plants insofar as possible, causing seed to fall. For this same reason, it is preferable to work early morning when higher relative humidity reduces the fragility of the stalks and losses caused by seed shattering. When the seeds are dried in the fields, the stalks are laid out in windrows and left for some days. Rainy weather in this period is very damaging because: (i) it delays drying; (ii) promotes development of fungal parasites on the seeds; (iii) increases seed losses; and (iv) lowers seed germination.

Seeds are dried in the fields wherever the climate allows, as in Italy. Where such is not possible, the harvested stalks are transported to the processing facility to be dried as quickly as possible to prevent mould development on the pericarp and consequent lowering of germination.

5.2 Pollen Isolation (E. Biancardi)

Sugar beet is normally allogamous and self-sterile. Over medium and long distances, wind pollination is predominant (Artschwager, 1927; Stewart and Tingey, 1927). Except for some strictly self-fertile genotypes, pollination control measures are necessary both during advancement of breeding populations and production of commercial seed. Isolation systems include: i) paper or cloth bags for one or more branches of the stalk; ii) cloth or plastic coverings for one or more plants; iii) glass and metal structures for a few plants; and iv) distance isolation for larger groups and for commercial seed production (Down and Lavis, 1930; Knapp, 1958). Using bags or small isolators often leads to insufficient seed production and low quality due to increase in temperature and moisture that occurs inside the bags or isolators (Raleigh, 1936). Space isolation, which resolves many of these problems, uses distance to lower the foreign pollen concentration in the air (Archimowitsch, 1949). Stewart and Campbell, (1952) state that pollen concentration diminishes as the squared distance from the source, but air movement and meteorological conditions cause significant variations. According to Knapp (1958), vegetal barriers (hemp or rye) improve isolation conditions but the results are often uncertain due to lodging of the rye or slow vertical growth of hemp during the beet's early flowering stage (Biancardi, unpubl. data).

In commercial seed production, fields must be appropriately spaced from one another to prevent or minimize contamination. The pollen circulating around each field includes that desired for fertilization, and foreign or contaminating background pollen from other sources (Chamberlain, 1967). The concentration of foreign pollen depends on: i) distance and dimensions of the sources of contamination; ii) prevalence of flowering beets in the area; iii) meteorological conditions; and iv) other factors dealt with below.

Damage caused by pollen contamination in commercial seed fields depends not only on the percentage of undesired crosses, but also on the origin of the contaminating pollen. Although rare, crosses with wild beets, e.g. sea beet and *B. macrocarpa,* are very damaging. More commonly contamination is caused by annual beet originating from sugar or seed crops. Since the annual trait is dominant, the bolting tendency is readily transmitted to the progeny by the pollen.

Damage caused by pollen from bolting sugar beets is limited by the fact that seed plants flower a few weeks earlier than the annual beets. Moreover, seeds produced by foreign pollen are generally smaller than normal and many can be eliminated by grading. Crosses between beets for different uses (leaf, garden, fodder) are very damaging because immediately recognizable on the commecial field. This type of contamination of sugar beet can drastically reduce productivity and processing quality.

Lastly, crosses can occur between fields where different, but similar varieties are grown. Morphological and productivity differences between commercial pollinators are generally small, so the contamination is difficult to detect unless there are ploidy differences between the pollinators. In this case, if the field is to produce triploid hybrids (3n) with the cross between a tetraploid pollinator and a 2n CMS line, the presence of foreign pollen originating from 2n pollinators increases the frequency of 2n hybrids. Risks of contamination by tetraploid pollen are lower in fields producing 2n hybrids, as the foreign tetraploid pollen is less competitive (Scott and Longden, 1970).

There are other factors to consider, such as the period during which foreign pollen is present, the receptiveness of the stigma of commercial CMS parents, the competitive effect of domestic pollen on the seed-bearer's stigma, incompatibility phenomena etc. The diverse nature of pollen contamination makes it necessary to adopt distances in the seed production fields that are directly proportional to the degree of risk posed by the contaminating pollen. In many countries the minimum distance between seed crops required by law leads to acceptable contamination levels, as long as early flowering annual beet is eliminated before flowering. Increasing the distance would reduce the risk of contamination only slightly (Chamberlain, 1967), while increasing the cost of transportation of the seed to processing facilities.

During reproduction of breeding materials precautions must be exercised, especially when CMS or tetraploid genotypes are being reproduced. In these cases, the greatest problems are not caused by pollen from bolting beet or beet for other purposes, whose hybrids can be easily identified and eliminated, but by pollen which transmits recessive characters that are difficult to identify in the progenies.

The height and branching of the stalk, abundance of pollen and pollen dispersal over long periods when atmospheric conditions are favorable, facilitate wind pollination (van Roggen, 1997). Nevertheless, the flower retains some ancestral characteristics typical of plants with insect pollination, such as joined flowers, nectar secretion, and the characteristic smell (Archimowitsch, 1927). Pollen is produced inside the anthers (five for each flower). Since a flower produces up to 85,000 pollen granules and there can be as many as 20,000 flowers per plant (Knapp, 1958), a total production of several billion pollen granules per plant is possible. Seed beet generally begins to flower about the middle of May and continues until harvest (from mid-July onward). Stalk branch growth is continuous as is the initiation of new flowers. Therefore, at harvest one can find ripe seeds, newly formed flowers, and all stages in between on the same plant.

Based on observations by Oksijuk (1927), confirmed by Artschwager and Starrett (1933), flowers begin to open when the sun rises. The sepals gradually move into a position perpendicular to the flower axis while the anthers simultaneously begin to open lengthways. At first, the pollen granules stick together but with reduction in moisture (Scott, 1970), separate and are pushed out of the anthers. While the flowers are opening, the stigma lobes (usually three) are closed and arranged in a tubular pattern. The lobes gradually open on the second or third day after anthesis, and generally are fully open by the third day. On the outside of the stigma there are very fine papillas, which increase the surface area and the probability of capturing pollen grains (Artschwager and Starrett, 1933).

Beet pollen tends to be spherical. Its surface has a characteristic relief and small circular marks from which the pollen tube can emerge. Granule diameter averages 16-20 μm and appears to depend on the genotype (Artschwager and Starrett, 1933). For tetraploid plants the diameter is 5-10μm greater (Knapp, 1958). A diameter of about 20 μm was most commonly recorded in a survey of 586 plant species throughout Europe (Stanley and Linskens, 1974). We can therefore assume that this size is most suitable for wind pollination and, hence for long-distance movement.

Pollen vitality depends on climatic conditions. At low temperatures and humidity, germination may occur after 50 days. Under field conditions, if humidity is low, the ability to germinate lasts no more than a day. Intense sunlight can reduce the time pollen is viable to three hours or less (Artschwager and Starrett, 1933; Knapp, 1958).

Information about spatial isolation of beet is not definitive. This may be due to difficulties involved in conducting experiments in which the effects to be evaluated are strongly influenced by meteorological factors. Pollen dispersal varies during the course of the day. Around midday it reaches maximum, almost in correspondence with minimum humidity and maximum air temperature (Scott, 1970). The same author found that in England, pollen concentrations in the air were reduced by rain or high humidity. In many cases, however, rain at night increased pollen production the next day. The highest release recorded over a three-year period in the aforementioned environment was observed between the 1st and 15th of July. The same experiments demonstrated the importance of year effects on pollen concentration in the air. Indeed, pollen concentrations can vary by as much as 100% from one year to another. Variability probably depends on differences in seasonal factors that control development of stalks and flowers.

In 1970, Scott and Longden showed that pollen dispersal times also depend on ploidy level. This difference should be recognized in the production of genetic monogerm hybrids with 2n versus tetraploid pollinators. The tetraploid plants produce less pollen than 2n plants throughout the day as well as during the season. Furthermore, there is about an hour delay between maximum pollen emission by tetraploid plants, compared to 2n plants. Because of its larger size, pollen granules originating from tetraploid plants have a faster sedimentation speed than those from 2n plants; it is therefore less mobile (Stanley and Linskens, 1974). Hence in fields with tetraploid pollinators, there is a greater probability of contamination by pollen from 2n plants.

To determine optimum spatial isolation, the concentration of beet pollen at various distances and directions from the source was examined. Petri dishes coated with glycerine (Archimowitsch, 1949) or automatic devices described by Hirst (1952) were used to capture and count the pollen granules. Flowering CMS plants could also be employed, by observing the frequency of hybrids in the offspring. For these experiments, monogenic dominant traits are used as indicators, for example the red root skin of garden beet.

In 1949, around a pollen source of about 500 red beets, Archimowitsch placed groups of sugar beet at increasing distances from the centre and along eight rays corresponding to the four cardinal points and the intermediate directions. Results varied greatly over three years, due to the weather; in particular wind direction. In 1938, German, using an aeroplane and petri dishes coated with glycerine, collected pollen up to a height of 2,500 meters. In 1938, Meyer and Artschwager, using similar methods, flew over about 400 hectares of seed beet between El Paso, Texas and Las Cruces, New Mexico. Large quantities of pollen were found up to an altitude of 1,500 meters. With strong rising currents the granules were carried to heights of 10,000 meters and more and were thus able to travel long distances (German, 1939). Pine pollen (double the diameter and much heavier than beet pollen) has traveled 1,750 km from its point of emission (Stanley and Linskens, 1974).

The problem of spatial isolation and transportation of pollen was confronted from a unique point of view in the early 1960s (Dark, 1971; Tyldesley, 1977). The author used knowledge gleaned from other disciplines, such as the atmospheric spread and fallout of volcanic or radioactive dusts, with which pollen can be compared, at least in terms of aerodynamic properties. Moreover, distribution of human allergenic reactions triggered by certain types of pollen was studied in depth. Further research into this subject was stimulated by the risk of contamination of wild populations posed by transgenic crops (Luby und McNicol, 1995; Schmidt, 1995; Bartsch et al., 1996, 2001; Brand, 1997; Saeglitz et al., 2000).

Gregory (1945) and Charmberlain (1967) developed a system for estimating the ratio of foreign to domestic pollen in a beet seed production field. In this system, several variables which were difficult to quantify, such as wind direction and speed, were excluded. Further simplifications and assumptions were also required: i) all sources of foreign pollen are uniformly distributed around the field and produce the same quantity of pollen over the course of a season; ii) the probability of foreign pollination is proportional to the number of pollen granules collected in the source field; iii) vitality of the pollen remains constant; iv) the wind does not have prevalent directions during the course of the season; and v) meteorological conditions are uniform. The number of pollen granules emitted per plant and their ability to travel long distances make determination of safe limits for seed reproduction difficult. The cross-polination tolerance is higher for commercial seed, as long as the pollen is not from wild beet, early flowering beet, or beet for other purposes.

The presence of foreign pollen in reproduction fields depends on many, highly variable factors; hence the safety of spatial isolation is always considered in terms of probability. The matter has been dealt with in several ways starting with numerous simplified assumptions. Several considerations of general validity can be made: i) the greater the spatial isolation, the safer the field; ii) the same effect is achieved with an abundant quantity of domestic pollen, obtainable with highly productive pollinators or with a greater plant density; iii) pollen contamination can come from afar due to the number of pollen granules produced by each plant, the distance they can be transported, and their relatively long vitality.

To conclude, higher safety levels can be obtained not so much by increasing distances, but rather by eliminating other flowering beets and sources of wild contamination. To produce seed of different types of cultivated beets (garden, fodder, leaf, sugar), it is necessary to dedicate separate adequately spaced areas to each type of beet.

5.3 Biodiversity and Gene Flow (D. Bartsch)

Gene flow is by definition the active or passive dispersal of genes via seed, pollen, or clonal parts of a plant into the environment. Biodiversity is the multitude of different organisms and their biological relationships in ecosystems. Biodiversity can be seen and studied at different levels of biological organization: genes, species, populations and biocenoesis. In the context of gene flow, biodiversity can be influenced both within a species (intraspecific) as well as between species (interspecific). Intraspecific gene flow limits natural selection through homogenization of populations (Lenormand, 2002) and can therefore counteract loss of diversity caused by genetic drift or bottleneck effects. Lack of genetic diversity can increase the susceptibility of a species to pathogens and increase the risk of harvest decline in agriculture (Marshal, 1977; Wood and Lenne, 1997). Interspecific biodiversity is important for plant productivity as well (Tilman et al., 2001). For example, Zhu et al., (2000) demonstrated that a mixture of rice genotypes was effective in limiting the infection of a pathogen.

Two plant compartments are designed for hybridization and gene flow during sexual reproduction: the ovule as the maternal part and the pollen as paternal part. Only pollen and the resultant seeds play an important role in natural long-distance distribution, but seed companies may unintentionally distribute seed lots with an adventitious presence of genetically modified plants (GMP) over country borders. When seed is the main distribution organ, it does not matter whether the genetically modified crop is the male or female parent. Crop-to-weed gene flow as a consequence of hybridization has had important practical and economic consequences since it promotes the evolution of more aggressive weeds (Ellstrand and Schierenbeck, 2000). Less attention has been focused on crop-wild plant gene flow, although Ellstrand et al., (1999) pointed out that 12 of the world's 13 most important crops hybridize with wild relatives somewhere in their cultivation area (**Tab. 5.1**). Hybridization and escape of genes of cultivated varieties have led to disadvantages for wild plant populations in some documented cases. Hybridization with domesticated species has been implicated in the extinction of certain wild crop relatives (Ellstrand and Elam, 1993). Here, one of the most important goals is conservation of genetic resources, in this case the genetic diversity of wild relatives of crop plants.

Traditional breeding has focused primarily on biomass improvement factors (Gepts, 2002). Compared to wild relatives, cultivated varieties are in general genetically less diverse and therefore less adapted to natural environments. Nevertheless, varieties are not completely incapable of thriving in the wild (Sukopp and Sukopp, 1993). Minor adaptation of varieties to conditions outside agricultural areas may change if fitness-enhancing traits are transformed into GMP. Introduction of GMP has raised

Table 5.1 12 of the 13 most important crops worldwide that hybridize with wild relatives somewhere in their cultivation area. Sugar beet was added to the list given by Ellstrand et al., (1999) as no.14, given the importance of this species in Europe.

Crop	*Relative(s)*	*Crop*	*Relative(s)*
1. WHEAT	Wild *T. turgidum* subspecies, some *Aegilops* species	8. MILLETS	*Eleusine coracana* ssp. *africana,* Wild *Pennisetum* species
2. RICE	Wild *Oryza* species	9. BEANS	Wild *Phaseolus* species
3. MAIZE	Wild *Zea mays* subspecies	10. OILSEED RAPE	Some wild *Brassica* species
4. SOYBEAN	*Glycine gracilis, G. soya*	11. PEANUT	No report
5. BARLEY	*Hordeum spontaneum*	12. SUNFLOWER	Wild *Helianthus annuus*
6. COTTON	Wild *Gossypium* species	13. SUGAR CANE	Wild *Saccharum* species
7. SORGHUM	Wild *Sorghum* species	14. SUGAR BEET	Some wild *Beta* species

questions about its potential impact on gene flow and biodiversity (Tiedje et al., 1989; Gepts and Papa, 2003). Until now, reports of a fitness advantage for hybrids between crops and wild relatives are rare (oilseed rape: Stewart et al., 1997; sunflower: Snow et al., 2003). Since crop to wild or weed plant gene flow is likely to occur in outcrossing species, the current discussion on impacts with regard to sugar beet are summarized below.

5.3.1 Inter and Intra-specific Diversity in Genus *Beta*

An illustrative example of the interaction of gene flow and biodiversity is the genus *Beta*: beets have been cultivated for more than 2,000 years in the eastern Mediterranean region. *Beta vulgaris* comprises an extraordinarily variable group, in which cultivated and wild forms are often difficult to distinguish genetically (Bartsch and Ellstrand, 1999). This is mainly due to the extensive use of sea beet (*Beta vulgaris* ssp. *maritima*) in conventional breeding programs, especially of sugar beet (*Beta vulgaris* ssp. *vulgaris*). Sea beet is largely a coastal taxon, with a wide distribution from Cape Verde and the Canary Islands in the west, northward along Europe's Atlantic coast to the North and Baltic Seas. It also extends eastward through the Mediterranean region into Asia where it occurs in Asia Minor, central and outer Asiatic steppes, and desert areas as far as western India. Sea beet varies from self-compatible annuals to self-incompatible perennials with a life span between one and more than 11 years (Letschert 1993; Hautekeete et al., 2002). Cultivated *B. vulgaris*, including Swiss chard, red garden beet, and sugar beet, are biennial. The latter is partially self-incompatible due to the extensive use of male-sterility genes in sugar beet breeding. All cultivated and wild subspecies of *B. vulgaris* are mostly wind pollinated, although some insect pollination has been noted. The intraspecific diversity of *Beta vulgaris* is well documented since it is an important contributor to sugar beet breeding (Frese et al., 2001; Bartsch et al., 2003).

Besides wild and cultivated beet, a weed beet form also grows in sugar beet fields, where, due to its genetically based annuality, it bolts late in the season (**Fig. 5.2**). Producing lignescent inflorescence and large amounts of seed instead of the commercially desirable fleshy roots, weed beets can be a serious problem for sugar beet farmers (Soukup et al., 2002; Desplanque et al., 2002). Due to the similarity of their seedling morphology and physiology to sugar beet, conventional methods do not control annual weed beets. *Beta vulgaris* is a good example for a wild/cultivated varieties complex, in which gene flow and hybridization can be observed (Van Dijk and Desplanque, 1999; Mücher et al., 2000).

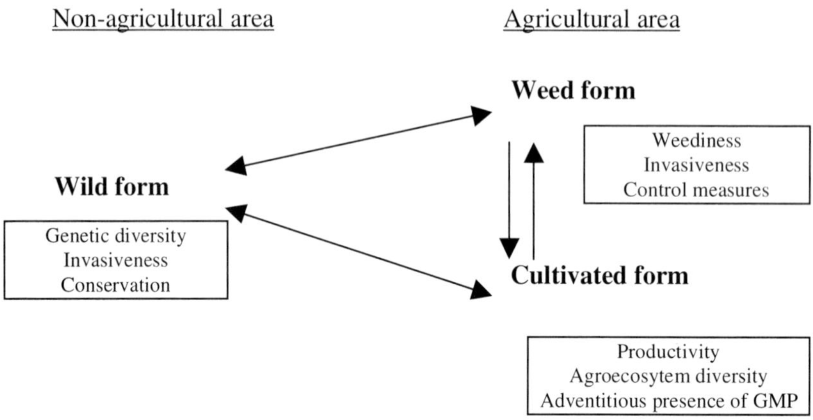

Fig. 5.2 Levels of potential impact (rectangles) of gene flow (arrows) in the crop/wild/weed plant complex of *Beta vulgaris*.

5.3.2 Impact of Genetically Modified Beet on Biodiversity

Gene flow from GM sugar beet to wild and weedy relatives is very likely to occur. Hybrids formed by spontaneous crosses between cultivated beets and south European wild types are typically annual, because this life cycle trait is genetically dominant and common in southern wild populations. In Germany, a research program was undertaken to address biosafety issues related to the interaction of genetically modified sugar beet with wild and weed beet. The transgenes studied had a single copy of the beet necrotic yellow vein virus (BNYVV) coat protein (*cp*), phosphinothricin-acetyl-transferase (*bar*) and neomycin-phospho-transferase (*nptII*) genes. My research group found the virus-tolerant phenotype of hybrids with wild sea beet did not significantly differ from naturally tolerant genotypes (Pohl-Orf et al., 2000). In addition, there is no evidence that the virus has any ecological role since it has not been found in sea beet populations. My colleague and I concluded that ecological implications due to the introduction and spread of virus resistant hybrids would be minimal in this specific case (Bartsch and Schuphan, 2002). However, long-term monitoring by collection of basic data on geographic distribution and genetic diversity of wild plant populations is necessary for detection of any effect, if wild beet genetic resources need to be addressed. So my research team, studied the ecological impact of a century of gene flow from traditionally bred cultivated beets into the wild sea beet populations of northeastern Italy (Bartsch et al., 1999). We demonstrated that gene flow from a crop to a wild relative does not necessarily result in decrease of genetic diversity of the wild plant, albeit cultivated beets are less diverse and outnumber wild relatives by a factor 10,000 to 1. In French seed production and cultivation areas of sugar beet also gene flow seem to differ qualitatively (Desplanque et al., 1999; Viard et al., 2002). These data support the view commonly held among scientists in biosafety research that gene flow *per se* should not be regarded as an adverse environmental effect of GMP (e.g. Crawley et al., 2001).

However, the concept of ecological risk assessment and risk management of GMO, as defined in EU Directive 2001/18, relies on precommercialization biosafety research and postcommercialization monitoring, which may be "case specific" or "general surveillance". Risk is defined as a function of both effect (hazard) and exposure (Sharples, 1991). Scientific consensus holds that risk assessment must first target the specific effect associated with the new technology (Saeglitz and Bartsch, 2002), and secondly the exposure (e.g. frequency of transgenes and their expression products). Because laboratory and field biosafety studies can target only a limited number of species and parameters, broad and long-term monitoring programs may be necessary to detect potential side effects (e.g. population decline of non-target arthropod species based on lethal and sublethal effects, biometrical changes, changes in species composition).

Genetically modified sugar beet will not be exempt from the biological rules of gene flow, but there is no evidence that genetically modified plants are an environmental risk *per se*. Over a period of 15 years, 81 international biosafety research projects have been supported by the EU. Within the same period, national programs (e.g. in Germany, see www.biosicherheit.de) added significant knowledge about the environmental effects of GMO. Based on the literature available in biosafety databases (e.g. www.icgeb.trieste.it/~bsafesrv/), we found no evidence that the use of GMO conflicts with sustainable agriculture and nature conservation per se (Bartsch and Schuphan, 2002). New studies suggest that intelligent application of genetically modified sugar beet may even increase interspecific biodiversity in agroecosystems (Strandberg and Bruus Petersen, 2002, Dewar et al., 2003), but generalizations based on these studies should be made with caution.

5.3.3 Monitoring of GMO Gene Flow

Biosafety research cannot solve every open and basic question of general ecology (Kareiva et al., 1997). After the best pragmatic use of the case-by-case and step-by-step approach, a well-designed monitoring program is necessary following commercialization. A combination of case-specific monitoring and general environmental surveillance will help us to confirm the risk assumptions made prior to commercialization (Schmitz et al., 2003). Intelligent monitoring should use causal analytic approaches based on comparison of genetically modified and isogenic plants. However, as in other nonexperimental field studies, researchers will be confronted with the difficulty of identifying a single effect. This is due, for instance, to the high complexity of the ecosystem, "natural" fluctuations, and regional differences in biocenoses. Finally, any measurable effect of transgenes on non-target herbivores must be evaluated.

The beet example is a well-documented case of how unwanted economic effects arise through uncontrolled gene flow in conventional breeding (Soukup et al., 2002). The primary question related to genetically modified crops is whether the transgenes will be fixed in recipient plant populations (Arnaud et al., 2003). The secondary more ecologically important question is whether the transgenes cause effects that differ fundamentally from conventional gene flow consequences. To answer the later question, we need baseline data; otherwise we shall not be able to assess any effect observed within genetically modified crop cultivation.

5.4 Seed Quality and Conservation (M. De Biaggi)

Selecting an area that consistently produces quality seed is an important initial step in the production of commercial seed. Sugar beet seed with high germination potential is crucial for growers. Fauchere (2001) found root yields were more than 4% higher with a population of 100.000 plants ha^{-1} than with 70,000 plants ha^{-1} in French variety trials between 1988 and 1999. The sugar concentration decreased 0.36% for every 10% reduction in population below 100,000 plants ha^{-1}. Therefore, poor seed quality may prevent a variety from realizing its inherent yield potential.

Raw seed is processed and selected to optimize seed quality. The quality of raw seed determines the sequence of these stages. ISTA methods are usually employed to determine the germination potential of seed in the various steps of processing. Moreover, companies often modified these procedures to develop laboratory tests that predict field emergence more accurately (van Swaaij et al., 2001).

Seed companies producing pelleted seed often use so-called "stress conditions", that is, 20°C and 70 ml water instead of 40 ml. If the 4-day germination count exceeds 70-75%, the lot is considered adequate for commercial purposes. Other producers use their specific methods to measure this character. The pelleted seed distributed to growers is a mixture of seed sizes with high vigor. Small seeds and relatively heavy seed caps appear to be associated with poor germination (Morris et al., 1985). Hence breeders need to consider seed characteristics when selecting F_1 CMS seed parents.

Seed companies often need to maintain the quality of naked seed lots for an extended time, depending on market demand for a variety. Storing pelleted seed poses an additional problem because insecticides in the covering material can have a negative effect on germination.

Breeders need to store enough basic seed to reproduce a variety for a number of years. They also need to maintain a sample of each genetic source over the years. There are different solutions, but the common goal is to avoid quality deterioration and loss of vigor. Therefore, breeding companies must invest in facilities with temperature and relative humidity controls, either in existing warehouses or in specially designed cold-storage facilities. To ensure the availability of seed in case of crop failure, it is common practice to plan for an overproduction of at least 50% for each variety in each year.

It is necessary to maintain seed storage rooms at about 15°C and below 55/60% relative humidity. These conditions are satisfactory for conservation of commercial seed lots (De Biaggi, unpubl. data). Breeders often store basic seed of parental lines in rooms with low temperature (10°C) and dry air (35-40% relative humidity). Pack and Owen (1950) report that after 22 years, the germination of seeds placed in a commercial cold-storage house was only 10% less than the original prior to storage.

Long-term storage of seed samples for preservation of genetic diversity requires more extreme conditions. Yndgaard (1984) described two methods for long-term seed storage: i) in standard glass bottles of the Bordeaux type sealed airtight in cold store at –20°C, and ii) in aluminum and plastic bags sealed airtight and stored at –20°C.

References

Alcaraz, G. 1981. etude de certaines aspects de l'éterogénéité de la semence de la betterave sucrière. PhD diss., Univ. Curie, Paris.

Archimowitsch, A. 1927. The problem of isolation of seed plants in sugar beet. Bull. Belaya Cercov. 1:7-36.

Archimowitsch, A. 1949. Control of pollination in sugar beet. Bot. Rev. 15:613-629.

Arnaud, J.F., F. Viard, M. Delescluse, and J. Cuguen. 2003. Evidence for gene flow via seed dispersal from crop to wild relatives in *Beta vulgaris* (*Chenopodiaceae*): consequences for the release of genetically modified crop species with weedy lineages. Proc. Roy. Soc. London, Biol. Sci. 270:1565-1571.

Artschwager, E. 1927. Development of flowers and seed in the sugar beet. J. Agric. Res. 34:1-25.

Artschwager, E. and R.C. Starrett. 1933. The time factor in fertilization and embryo development in the sugar beet. J. Agric. Res. 47:823-843.

Bartsch, D. and N.C. Ellstrand. 1999. Genetic evidence for the origin of Californian wild beets (genus *Beta*). Theor. Appl. Genet. 99:1120-1130.

Bartsch, D., and I. Schuphan. 2002. Lessons we can learn from ecological biosafety research. J. Biotech 98:71-77.

Bartsch, D., J. Cuguen, E. Biancardi, and J. Sweet. 2003. Environmental implications of gene flow from sugar beet to wild beet-current status and future research needs. Envir. Biosafety Res. 2:105-115.

Bartsch, D., M. Schmidt, M. Pohl-Orf, C. Haag, and I. Schuphan. 1996. Competitiveness of transgenic sugar beet resistant to beet necrotic yellow vein virus and potential impact on wild beet populations. Molec. Ecol. 5:199-205.

Bartsch, D., M. Lehnen, J. Clegg, M. Pohl-Orf, I. Schuphan, and N.C. Ellstrand. 1999. Impact of gene flow from cultivated beet on genetic diversity of wild sea beet populations. Molec. Ecol. 8:1733-1741.

Bartsch, D., U. Brand, C. Morak, M. Pohl-Orf, I. Schuphan, and N.C. Ellstrand. 2001. Biosafety of hybrids between transgenic virus-resistant sugar beet and Swiss chard. Ecol. Appl. 11:142-147.

Bee, P. and K. Jaggard. 1996. Plant populations for today's conditions. Brit. Sugar Beet Rev. 3:28-30.

Bornscheuer, E., K. Meierholz, and K.H. Wunderlich. 1993. Seed production and quality. *In:* D.A. Cooke and R.K. Scott (eds.). The Sugar Beet Crop. Chapman & Hall, London, UK, pp. 121-153.

Boschetti, W., G. Negrini, M. Colombo, and E. Biancardi. 1997. Scomposizione della varianza e stabilità varietale in prove agronomiche di barbabietola da zucchero. Sementi Elette 6:27-32.

Brand, U. 1997. Untersuchungen zur Diversität in italienischen Wildpopulationen von *Beta vulgaris* L. ssp. *maritima*. MS thesis, Univ. Aachen, Germany.

CAC (Cooperativa Agricola Cesenate). 1996. Tech. Publ., Cesena, Italy.

Campbell, S.C. 1968. Sugar beet seed production in Oregon. Proc. IIRB 31:165-174.

Chamberlain, A.C. 1967. Cross-pollination between fields of sugar beet. Quart. J. Roy. Meteor. Soc. 93:509-515.

Crawley, M.J., S.L. Brown, R.S. Hails, D. Kohn, and M. Rees. 2001. Transgenic crops in natural habitats. Nature (London) 408:682-683.

Dark, S.O.S. 1971. Experiments on the cross-pollination of sugar beet in the field. J. Nat. Inst. Agric. Bot. 12:242-266.

Desplanque, B., N.C. Hautekeete, and H. van Dijk, H. 2002. Transgenic weed beets: possible probable, avoidable? J. Appl. Ecol. (London) 39:561-571.

Desplanque, B., P. Boudry, K. Broomberg, P. Saumitou-Laprade, J. Cuguen, and H. van Dijk. 1999. Genetic diversity and gene flow between wild, cultivated and weedy forms of *Beta vulgaris* L. (*Chenopodiaceae*), assessed by RFLP and microsatellite markers. Theor. Appl. Genet. 98:1194-1201.

Dewar, A.M., H.M.J. May, I.P. Woiwood, et al., 2003. A novel approach to the use of genetically modified herbicide tolerant crops for environmental benefit. Proc. Roy. Soc. London. 270:335-340.

Down, E.E. and C.A. Lavis. 1930. Studies on methods for control of pollination in sugar beet. J. Amer. Soc. Agron. 22:1-9.

Ellstrand, N.C. and D.R. Elam. 1993. Population genetic consequences of small population size, implications for plant conservation. Ann. Rev. Ecol. Syst. 24:217-242.

Ellstrand, N.C. and K.A. Schierenbeck. 2000. Hybridization as a stimulus for the evolution of invasiveness in plants? Proc. Natl. Acad. Sci. USA 97:7043-7050.

Ellstrand, N.C., H.C. Prentice, and J.F. Hancock. 1999. Gene flow and introgression from domesticated plants into their wild relatives. Ann. Rev. Ecol. Syst. 30:539-563.

Fauchere, J. 2001. Influence de la qualite des graines sur les varietes experimentees. Proc. IIRB 64:257-268.

Frese, L., B. Desprez, and D. Ziegler. 2001. Potential of genetic resources and breeding strategies for base-broadening in *Beta*. p. *In:* H.D. Cooper, C. Spillane, and T. Hodgkin (eds.). Broadening the Genetic Base of Crop Production. IPGRI/FAO, Rome, Italy, pp. 295-309.

Gepts, P. 2002. A comparison of crop domestication, classical breeding, and genetic engineering. Crop Sci. 42:1780-1790.

Gepts, P. and R. Papa. 2003. Possible effects of (trans)gene flow from crops on the genetic diversity from landraces and wild relatives. Envir. Biosafety Res. 2:89-103.

German, I.V. 1939. Transference of sugar beet pollen by air currents in vertical direction. C.R. Acad. Sci. USSR 26:77-85.

Gregory, P.M. 1945. The dispersion of airborne spores. Trans. Brit. Mycol. 28:26.

Hautekeete, N., Y. Piquot, and H. van Dijk. 2002. Life span in *Beta vulgaris maritima*: the impacts of disturbance and of age at first reproduction. J. Ecol. 90:508-516.

Hirst, J.M. 1952. An automatic volumetric spore trap. Ann. Appl. Biol. 39:257-265.

Jansen, V. 1963. Anbau von Zuckerrübensamen in Dänemark im besonderen Hinblik auf die Überwinterung von Stecklingen im Feldbestand. Zucker 16:596-600.

Kareiva, P., I.M. Parker, and M. Pascual. 1997. Can we use experiments and models in predicting the invasiveness of genetically engineered organisms? Ecology 77:1670-1675.

Knapp, E. 1958. *Beta* Rüben. *In:* R. Roemer and W. Rudorf (eds.). Handbuch der Pflanzenzüchtung. Paul Parey, Berlin, Germany, vol. 3, pp. 196-284.

Lenormand, T. 2002. Gene flow and the limits to natural selection. Trends Ecol. Evol. 17:183-189.

Letschert, J.P.W. 1993. *Beta* section *Beta:* biogeographical patterns of variation and taxonomy. Wageningen Agric. Univ. Publ. 93:1-153.

Luby, J.J. and R.J. McNicol. 1995. Gene flow from cultivated to wild raspberries in Scotland: developing a basis for risk assessment for testing and development of transgenic cultivars. Theor. Appl. Genet. 90:1133-1137.

Marshal, D.R. 1977. The advantages and hazards of genetic homogeneity. Ann. New York Acad. Sci. 287:1-20.

Meyer, F.C. and E. Artschwager. 1938. Airplane collection of sugar beet pollen. Science 25:507-508.

Morris, P.C., D.G. Grierson, and W.J. Whittington. 1985. The influence of fruit structure on germination of sugar beet. Seed Sci. Tech. 13:41-51.

Mücher, T., P. Hesse, M. Pohl-Orf, N.C. Ellstrand, and D. Bartsch. 2000. Characterization of weed beets in Germany and Italy. J. Sugar Beet Res. 37:19-38.

Oksijuk, P. 1927. Entwicklungsgeschichte der Zuckerrübe (*Beta vulgaris*). Bull. Jard. Bot. Kiev 5-6:145-164.

Pack, D.A. and F.W. Owen. 1950. Viability of sugar beet held in cold storage for 22 years. Proc. ASSBT 6:127-129.

Pohl-Orf, M., C. Morak, U. Wehres, et al., 2000. The environmental impact of gene flow from sugar beet to wild beet—an ecological comparison of transgenic and natural virus tolerance genes. *In:* C. Fairbairn, G. Scoles, and A. McHughen (eds.) Proc. 6th Intl. Symp. Biosafety of Genetically Modified Organisms, July 2000, Univ. Press, Saskatoon, Canada, pp. 51-55.

Raleigh, S.M. 1936. Environmental factors affecting seed setting in sugar beets. J. Amer. Soc. Agron. 28:34-51.

Saeglitz, C. and D. Bartsch. 2002. Plant gene flow consequences. AgBiotechNet 4, ABN 084.

Saeglitz, C., M. Pohl, and D. Bartsch. 2000. Monitoring gene escape from transgenic sugar beet using cytoplasmic male sterile bait plants. Molec. Ecol. 9:2035-2040.

Schmidt, M. 1995. Populationsökologische Untersuchungen und Wildformen der Species *Beta vulgaris* L.-Risikoabschätzung der Folgen eines inverkehrbringens transgener Zuckerrüben. MS thesis, Univ. Aachen, Germany.

Schmitz, G., D. Bartsch, and P. Pretscher. 2003. Selection of relevant non-target herbivores for monitoring the environmental effects of Bt maize pollen. Envir. Biosafety Res. (in press)

Scott, R.K. 1970. The effect of weather on the concentration of pollen within sugar beet seed crops. Ann. Appl. Biol. 66:119-127.

Scott, R.K. and P.C. Longden. 1970. Pollen release by diploid and tetraploid sugar beet plants. Ann. Appl. Biol. 66:129-135.

Sharples, F.E. 1991. Ecological aspects of hazard identification for environmental uses of genetically engineered organisms. *In:* M.A. Levin and H.S. Strauss (eds.). Risk Assessment in Genetic Engineering. McGraw-Hill, New York, NY, pp. 18-31.

Smith, G.A. 1987. Sugar beet. *In:* W.R. Fehr (ed.). Principles of Cultivar Development. MacMillan Publ. Co., New York, NY, vol. 2, pp. 577-625.

Snow, A.A., D. Pilson, L.H. Rieseberg, et al., 2003. A *Bt* transgene reduces herbivory and enhances fecundity in wild sunflowers. Ecol. Appl. (in press)

Soukup, J., J. Holec, P. Vejl, S. Skupinova, and P. Sedlak. 2002. Diversity and distribution of weed beet in the Czech Republic. J. Plant Dis. Prot., Spec. Iss. XVIII, pp. 67-74.

Stanley, R.G. and H.F. Linskens. 1974. Pollen: biology, biochemistry, Management. Springer Verlag, Berlin, Germany.

Stewart, C.N., J.N. All, P.L. Raymer, and S. Ramachandran. 1997. Increased fitness of transgenic insecticidal rapeseed under insect selection pressure. Molec. Ecol. 6:773-779.

Stewart, D. and S.C. Campbell. 1952. The dispersion of pollen in sugar beet seed plot. Proc. ASSBT 7:459-469.

Stewart, G. and D.C. Tingey. 1927. A method for controlling pollination of sugar beet. J. Amer. Soc. Agron. 19:126-128.

Strandberg, B. and M. Bruss Pedersen. 2002. Biodiversity in glyphosate tolerant fodder beet fields. Timing of herbicide application. Natil. Envir. Res. Inst. Silkeborg, Denmark.

Sukopp, H. and U. Sukopp. 1993. Ecological long-term effects of cultigens becoming feral and of naturalization of non-native species. Experientia 49:210-218.

Tiedje, J.M.R., R.L. Colwell, Y.I. Grossman, et al., 1989. The planned introduction of genetically engineered organisms, ecological considerations and recommendations. Ecology 70:298-315.

Tilman, D., P.B. Reich, J. Knops, D. Wedin, T. Mielke, and C. Lehman. 2001. Diversity and productivity in a long-term grassland experiment. Science 294:843-845.

Tyldesley, J.B. 1977. Out-crossing in sugar beet due to airborne pollen. Agric. Meteor. 19:463-469.

Van Dijk, H. and B. Desplanque. 1999. European *Beta*: crops and their wild and weedy relatives. *In:* L.W.D. Van Raamsdonk and J.C.M. Den Nijs (eds.) Plant Evolution in Man-made Habitats. Hugo de Vries Laboratory, Amsterdam, Netherlands, pp. 257-270.

Van Roggen, P.M. 1997. The sex life of a sugar beet. Brit. Sugar Beet Rev. 1:28-30.

Van Swaaij, A.C.P.M., W. Heijbroek, and J.L. Basting. 2001. Testing and improving seed vigour in sugar beet. Proc. IIRB 64:237-546.

Viard, F., J. Bernard, and B. Desplanque. 2002. Crop-weed interactions in the *Beta vulgaris* complex at a local scale: allelic diversity and gene flow within sugar beet fields. Theor. Appl. Genet. 104:688-687.

Wood, D. and J.M. Lenne. 1997. The conservation of agrobiodiversity on-farm: Questioning the emerging paradigm. Biodiv. Conserv. 6:106-120

Yndgaard, F. 1984. A procedure for packing long-term storage seed. Intl. Board Plant Genet. Res. Newsl. No. 54.

Zhu, Y.Y., H.R. Chen, J.H. Fan, et al., 2000. Genetic diversity and disease control in rice. Nature 406:718-722.

6

Prospects in Sugar Beet Breeding

Increase in sugar yield through breeding and improved beet production methods is resulting in reduced sugar beet growing areas worldwide, especially in Western Europe, a development likely to become even more evident in future. Parallelly, sugar production costs are under pressure to decrease, thereby strengthening competition between sugar cane and sugar beet.

Future activities in sugar beet breeding have to focus on improvements in sugar-producing capacity and profitability of sugar beets versus sugar cane. Breeding will change from "phenotypic" to "genotypic" selection, implying the need for analysis of the difference, i.e., how the environment is influencing the expression of genes.

6.1 Breeding Targets (G. Steinrücken)

Sugar itself has to compete with other sweet products, such as nutritive sweeteners (e.g. glucose syrup, fructose syrup, isoglucose) and artificial sweeteners (e.g. saccharin, cycloamate, aspartam). Despite a growing world market of sugars and sweeteners in future due to an increasing human population worldwide, there will be a strengthened competition between sugar and other sweeteners forced by consumer needs and production costs.

In all future scenarios, production of sugar from sugar beets needs to become competitive *vis-a-vis* other producers of sweetness and targeted breeding is one way to minimize sugar production costs. In an open world market, sugar from sugar beet has to compete with sugar from sugar cane. Experience has shown that productivity of beet is as good, sometimes even better than that of cane under comparable climatic conditions, but labor costs are much higher in most sugar beet growing areas than in cane areas.

Competitiveness of sugar beets will also be influenced by political circumstances: an open world market versus a protected and regulated European market. On-going discussions on political regulations such as the "farm bill", sugar import from the 48 poorest countries into the EU market, sugar import from Mexico into the US market, and not at all the least, modification of the EU quota system will have negative impact on the global sugar beet acreage. In a protected European market, sugar consumption would stagnate due to limited population growth and changed consumer behavior. Increasing productivity of new sugar beet hybrids would consequently result in reduced acreage and market size for sugar beet seed, as already experienced over the last 10 years. Reduced acreage would automatically result in selection of the most productive growing areas and thus accelerate reduction in sugar beet acreage.

However, opportunities for increasing sugar beet acreage are also in reach, depending on other political decisions, e.g. use of beet for ethanol production, the ability to develop new markets, and acceptance of a genetically modified sugar beet crop.

Regardless of possible scenarios, future breeding activities will have to focus on:

- Minimizing production costs by reducing necessary inputs such as fertilizers and water, pesticides and herbicides, thereby increasing the ecological value for a sustainable agriculture,
- Maximizing output in sugar yield per acreage and sugar per ton beet to increase competitiveness with cane and other sweeteners.

6.1.1 Fertilizers and Water

The need for fertilizers and water could possibly be reduced were sugar beets capable of similar growth under less favorable conditions (low input adaptation). Natural genetic variation for drought and salt resistance is available within the *Beta* germplasm. Introducing or selecting genes for higher drought resistance could considerably strengthen yield stability, although there is likely to be an initial yield penalty. *Beta maritima* grows on very light sandy soils in which nutrition is minimal indicating that selection for plants is giving little fertilizer to produce the same biomass might be possible.

Drought resistance could be an advantage when possible future global climatic changes warm up the atmosphere and water becomes an even more limiting factor. Already today calculations made by Broom's Barn in the UK (Pidgeon, 2000) indicate that sugar productivity could have benefited significantly from improved drought resistance. Even stress induced by increased sunlight leads to the production of reactive oxygen species that cause photooxidative cell damage. *Arabidopsis* plants overexpressing the *chyB*-gene that encodes beta-carotene hydroxylase-an enzyme in the zeaxanthin biosynthetic pathway-showed higher resistance to high light intensity and high temperature (Davidson et al., 2002).

However, an overall reduction in acreage would lead to concentration of beet cultivation in those fields most suitable and well supplied with water. Developments in fertilizer application technologies, especially use of "precision farming" and GPS (global positioning system) would further support the efficient application and use of fertilizers by all crops and therefore, at least temporarily, possibly reduce the benefits of crops with lower fertilizer requirements.

Development of systems in which beet plants are capable of nitrogen fixation is not likely to happen within the next 20–30 years despite high expectations since the early 1980s. Researchers only recently identified a major gene (*SYMRK*) involved in nitrogen fixation in *Lotus,* which is required for a symbiotic signal transduction pathway leading from the perception of microbial signal molecules to rapid symbiosis related gene activation (Stracke et al., 2002). Experience from other crops shows that much more knowledge about the complex interaction between higher plants and microorganisms is needed before the nitrogen-fixing ability can be introduced into species that naturally lack the characteristic.

6.1.2 Pesticides

Pesticides are used when pests and/or diseases cause or may cause yield losses. Applications of pesticides could be reduced if more resistance genes versus different plant diseases and pests were introduced into the sugar beet. We already have several successful examples, such as resistance against rhizomania, cercospora, curly top, root aphids, and rhizoctonia. In all markets, resistance traits are likely to give the greatest potential of cost savings in future. Also associated ecological benefits will be of great importance.

Further improvements in resistance various diseases can generally be expected through recombination of different resistance genes as well as by the targeted selection of minor genes involved. Application of molecular markers in this context is of special importance. Since most of the genes involved in resistance express a nondominant phenotypic behavior, creation of homozygous hybrids will improve the level of resistance of course, but with a potential risk of inbreeding depression.

6.1.3 Herbicides

Weed control by herbicides is still a prerequisite for successful sugar beet production. Depending on the pressure of weed seed in the soil, costs for herbicide applications contribute a considerable part to input costs. Use of herbicides could be facilitated and reduced by growing herbicide-resistant crops where herbicide resistance is introduced through gene transfer by transformation. Herbicide-resistant sugar beets are already available and we can expect that new herbicides and related resistance systems will be developed in future, which will even allow use of various herbicide resistance systems in a crop rotation.

6.1.4 Sugar Yield

Increasing sugar yield could reenforce alternative uses of sugar and sugar byproducts, e.g. for ethanol production, bioplastics, ascorbic acid, or antifreeze compounds.

The average annual total yield increase in sugar beet is about 1.5%, twice as much as for cane. Experience in sugar beets has shown that increase in sugar yield due to breeding has been about 1% per year. There is no obvious theoretical reason why this trend cannot be maintained for many years, eventually doubling sugar yield until the end of this century. Depending on setting the right priorities, improved sugar yield could be mainly based on increasing sugar content versus root yield. During the first decades of this century the necessary genetic variation will in most cases be found within the current sugar beet germplasm. But new genetic variation has to be integrated from related or unrelated genetic resources to allow further selection progress. In other crops, e.g. rice, it has been shown that not only can resistance genes be usefully integrated from wild species but even new alleles for yield.

Molecular modeling of photosynthesis, inner-plant sugar transport, and sugar storage could further add to achievements for the most important agronomic traits.

6.1.5 Processing Costs

Profitability of sugar beet processing could be further improved: beets with higher sugar content are already reducing costs for transportation per unit sugar. Decreasing levels of impurities and increasing extractability of sugar (ca. 1% per year over the past 50 years) would further improve sugar extraction. Limitations for increasing sugar content levels and sugar extractability might be there, but they have get to be defined.

Extending the processing period could be a major factor in reducing fixed costs in beet processing. Models have been developed showing how sugar stability in stored beets could be improved by modification of certain enzymes involved in the sugar physiology (Rosenkranz et al., 2001).

Specific resistance genes against several root-rotting related diseases, e.g. *Aphanomyces*, *Rhizoctonia*, *Erwinia*, *Fusarium*, *Actinomyces*, and *Ditylenchus* would result in less sugar degradation by pathogens and could further improve the general storability of beets.

Beet growing could be extended over winter even under more northern European conditions. Such beets would need frost-resistance genes and complete bolting resistance whereby flowering could be induced artificially, e.g. by chemical treatments that allow seed production. Isolation and characterization of genes involved in vernalization, bolting, and flowering of plants would generate better understanding of the physiological mechanisms and could result in development of solutions such as those indicated above (Simpson et al., 1999; Levy et al., 2002).

6.1.6 Low Tare

"Low tare" beets are those delivered to factories with the minimal soil adherence possible. Soil adherents could be further lowered by 5 to 30% through breeding compared to current averages. However, it is doubtful whether breeding could compete here with further technical developments for cleaning beets directly on the field during and after harvest.

6.1.7 New Products

The production of new traits (output traits) resulting in new products has been a dream of biotechnologists for years. Specifically, fructose and biodegradable plastic (Nonato et al., 2001) have been discussed and studied in research projects. Production of existing byproducts could be enhanced, e.g. betaine to allow its efficient application as a heat transfer liquid. However, to date no other product than that already produced today by sugar beets is visible and none is likely to appear in the near future. Concomitantly there is always the risk that other crops, e.g. corn and potato, will be developed into fields currently covered by beet through modification to produce sugar instead of starch, and thereby increase competition.

Besides production of new traits a crop could become modified to fit into a completely different application: a plant gene that takes up toxic heavy metals from soils and stores them in the leaves has been identified (Ghoshal et al., 1998; Thomine et al., 2000). Engineered into beet plants the crop could be used to detoxify soils while simultaneously producing sugar.

6.2 Breeding Methods (G. Steinrücken)

Breeding is a cyclic process of creating genetic variation and narrowing variation down again through selection of the superior new combination of genes. Breeding methods have been subjected to major changes during the last years:

- Gene isolation and transformation technologies have been established as new tools to create genetic variation.
- Chromosome and gene mapping technologies as well as marker systems have been developed to identify specific individuals (genotypes) or genes in a genome.
- Methods have been established to study expression of genes and the dynamics of protein metabolism and function (proteomics).

Breeding progress in future has to take advantage of further changes and improvements in these technologies and possibly other, not yet invented ones.

6.2.1 Molecular Markers

The main focus here has to be more effective use of existing genetic variation wherein genetic markers play an important role. New marker systems have been developed over the last 20 years from the simple, visible color gene influencing hypocotyl color, to isoenzymes, RFLPs, AFLPs, microsatellites, to SNPs. The driving force for this development has been to find a marker system for which the markers are abundant, highly polymorphic, and amenable for automatization (allowing high-throughput screening). Today, SNPs are considered the most promising system, the only drawback being the rather low level of polymorphism, but the large number of potential markers easily compensates this. SNP markers can be developed from any kind of sequence, both noncoding and coding regions. In many applications, the latter is highly interesting, since it allows development of a marker directly on the gene that the breeders want to select for.

When the most important genes responsible for phenotypic variation have been identified, this will allow a fundamental shift in plant breeding from selecting and combining phenotypic traits to combining markers representing DNA variation at the gene level. Plant breeding will, with the help of molecular markers, convert to plant modeling in which not traits are selected or combined with each other, but certain sequences of DNA (nucleotides, alleles, or genes), identified as different between single plants. Crosses between selected individuals will result in a positive recombination of genetic information precisely where recombination is most wanted.

This new technology would first be applied to more simply inherited traits, such as disease resistances wherein major quantitative trait loci (QTLs) would be combined even with minor ones to enforce sustainable resistance.

A key factor for success is correct identification of the genes responsible for a particular phenotypic variation, i.e., trait dissection. This is a complicated, time-consuming, and costly process. However, research in other species can be used. Once the genes involved in a trait have been identified in one species, it is possible to find homologous genes in sugar beet also. In general, most genes play more or less the same role in all plants and the sequences are quite conserved among related species. The gene sequence in one species can therefore be used to develop a marker for the homologous gene in another species.

Genomic research will be shared among different crops and only in this way can the huge investments be compensated over time. Most of the principal pathways of photosynthesis and sugar metabolism are known from various crops, e.g. *Daucus, Citrus, Lycopersicon* or *Arabidopsis* (Luo et al., 1993; Koch 1996; Shakya and Sturm, 1998; Tymowska and Kreis, 1998; Sturm et al., 1999; Pego and Smeekens, 2000; Chikano et al., 2001; Eastmond et al., 2002; Komatsu et al., 2002; Miron et al., 2002), and the enzymes involved identified, including the nucleotide sequences of the corresponding genes. This information can be used to identify equivalent genes within sugar beet and their involvement in the most important quantitative characters as yield, sugar content, and extractability.

6.2.2 Genetic Resources

With further improved breeding tools such as those described above, use of genetic resources of related species will enter into a new dimension. If the positive alleles responsible for interesting traits were more readily identified and separated from all unwanted genes, existing genetic variation in wild *Beta* germplasm could be much better exploited than is currently the case. Today's limitation to simply inherited, monogenic traits such as disease resistances would no longer be valid. Linkage drag would be much less of a problem and even more complex inherited, polygenic traits such as yield could benefit from incorporation of new genetic variation, even making use of the integration of new alleles for known genes.

6.2.3 New Genes

Besides targeted use of markers in breeding, crop independent genomic research can supply breeding programs with new genes to be introduced via classical transformation techniques **(Tab. 6.1)**. Sequencing the genome

Table 6.1 Achievement of breeding target, with and without gene transfer.

Target	*Achievement without gene transfer*	*Achievement with gene transfer*
Root yield	1% per year	10-50 % (for one innovation)
Stress adaptation	Recombination of existing genetic variation	Different solution for different stress conditions, possible, first results for othe crops available (e.g. Lehr et al., 1999; Bajaj et al., 1999)
Low dirt tare	Recombination of existing genetic variation	No solutions available
"Output traits"	No solutions available	Not yet available, but theoretically possible
Sugar content	1% per year, physiological limits expected	Modification of sugar metabolism (e.g. Tang and sturm, 1999; Tang et al., 1999; Chiou and Bush, 1996; Chengappa et al., 1999; Häusler et al., 2002)
Other sugars and sweeteners	No solution	Modification of sugar metabolism (e.g. Börnke et al., 2002)
Impurities	1% per year, physiological limits expected	Modification of sugar metabolism
Bolting resistance	Highest expression reached	Modification of vernalization metabolism (e.g. Simpson et al., 1999; Levy et al., 2002)
Rhizomania resistance	Recombination of existing genetic variation	Different approaches already realized (e.g. Lauber et al., 2001)
Cercospora resistance	Recombination of existing genetic variation	Not yet available, but theoretically possible (Nielsen et al., 1993; Kuykendall et al., 2002)
Erysiphe resistance	Recombination of existing genetic variation	Not yet available, but theoretically possible
Ramularia resistance	Recombination of existing genetic variation	Not yet available, but theoretically possible
Phoma resistance	Recombination of existing genetic variation	Not yet available, but theoretically possible
Peronospora resistance	Recombination of existing genetic variation	Not yet available, but theoretically possible
Alternaria resistance	Recombination of existing genetic variation	Not yet available, but theoretically possible

(Table 6.1 Contd.)

(Table 6.1 Contd.)

Target	*Achievement without gene transfer*	*Achievement with gene transfer*
Verticillium resistance	Recombination of existing genetic variation	Not yet available, but theoretically possible
Pseudomonas resistance	Recombination of existing genetic variation	Not yet available, but theoretically possible
Erwinia resistance	No solutions available	Not yet available, but theoretically possible
Aphanomyces resistance	Recombination of existing genetic variation	Not yet available, but theoretically possible
Rhizoctonia resistance	Recombination of existing genetic variation	Not yet available, but theoretically possible
Fusarium resistance	Recombination of existing genetic variation	Not yet available, but theoretically possible
Actinomyces resistance	Recombination of existing genetic variation	Not yet available, but theoretically possible
Curly-top resistance	Recombination of existing genetic variation	Not yet available, but theoretically possible
BVY resistance	Recombination of existing genetic variation	Not yet available, but theoretically possible
BMVY resistance	Recombination of existing genetic variation	Not yet available, but theoretically possible
Root aphid resistance	Recombination of existing genetic variation	Not yet available, but theoretically possible
Heterodera resistance	Recombination of existing genetic variation	Not yet available, but theoretically possible
Meloidogyne resistance	Recombination of existing genetic variation	Not yet available, but theoretically possible
Ditylenchus resistance	No solutions available	Not yet available, but theoretically possible
Herbicide resistance	No solutions available	Solutions already available

of a certain crop is possible today but is still a nightmare; hundreds of scientists needed years to sequence the human genome—but it is only the start. Tiny differences in genes between individuals cause millions of trait variations. For their detection we need to map the genetic code of thousands or millions of individuals, which is impossible with today's technology but will be possible with the technology of tomorrow.

Introducing new genes into the genome is not the only way to modify the genome to obtain a superior plant. Transformation of genomes in organelles needs to be better explored. Knocking out a gene can, in some cases, improve the performance of the plant; in other cases it is advantageous to increase the number of copies of a native gene. Introducing a native gene into a new position in the genome can drastically alter gene expression, thereby creating a better plant.

The still problematic public acceptance of GM plants and GM food triggers other developments as well, such as "Exorcist" wherein integrated foreign DNA is spliced out again and destroyed before the crop is flowering or harvested (Keenan and Stemmer 2002).

More research has to be done to understand the complex relationships of genes within a certain genome and to facilitate integration of new genes into a particular genomic environment. During evolution various mechanisms developed in plants to protect them against attacks of pathogens and also integration of foreign DNA into the genome—a big help in their resistance responses against viruses, but posing a big problem in engineering genes into plants. Once a gene is successfully integrated into the genome, control of gene expression—switching on the gene when it is needed without altering the expression of other genes—is difficult. Therefore, fine tuning of engineered gene expression and establishment of optimal expression levels in the target site (cell or organ) requires further intensive research.

Beside new genes, small changes to a plant's DNA by arranging genes in a new order, transposable elements, or modifying gene sequences (inversion, translocation), have been a natural way in evolution to create new genetic variation without the need for integrating foreign DNA. New technologies could even exploit those ways in a more efficient way.

6.2.4 Genes and Environment

More intensive use of molecular markers will eventually result in varieties, that are no longer as heterogeneous as they are today. Precisely modeled and inbred plants will be used as parents to hybrids, which will then combine complementary gene sequences or alleles in the most favorable combination.

All new breeding tools will facilitate the targeted recombination of genes and alleles and part of the breeding work will move from the field to the lab. However, the final confirmation of a newly developed line or variety still has to be done in real biological environments, e.g. disease screenings or field trials.

More knowledge about the environment will be necessary to understand the complicated interactions between single genotypes, plants or hybrids, and certain environmental factors. The use of globally available information temperature, precipitation, and soil types will help to describe more precisely what constitutes a particular environment. To learn how different environments result in different market segments or how an environment is influencing the growth of a certain plant or genotype -- all this is an ongoing process and will remain so for a long time.

6.3 Challenges (G. Steinrücken)

Research in general but sugar beet breeding in particular will generate new information at an immense pace. The first challenge therefore is to make positive use of this information and to convert it into practical applications.

Estimating the impact of single genes and gene sequences as part of a multigene trait will not only be more precise and accurate, but even give many more data points for what is estimated today as a single trait, e.g. yield as the final result of all interactions between all genes involved and all environmental factors. By using molecular markers single plants will be characterized by hundreds of thousands of data points, including conventional information on agronomic traits. Two plants might be crossed differing only by one data point related to the expression of a desired trait. Besides the presence or absence of a single gene or gene sequence, the interaction between genes is resulting not only in simple additive effects; billions of interactions between millions of genes are possible, organized in spatially or chemically isolated functional modules. Complex network models are already able to describe metabolic networks for model plants in a hierarchical manner (Ravasz et al., 2002). With more accurate information, the complexity for making decisions will increase but new software development will be instrumental in coping with the complexity.

Considering breeding and sugar beet growing is a mental reflection on all interactions between genes and environmental factors. Despite the information available on certain genes and certain environmental conditions, we have to consider that most of the information necessary for understanding their relationship is still missing. Examples can readily be found (**Plate 6.1**) showing the difference between:

Plate 6.1 Nature structured.

- Nature as we would like to see it in an organized and structured way and
- Nature as it is—looking unorganized and chaotic because we still do not understand the underlying system.

That is why many scientists try to use the principles of the complexity theory—how components work together rather than simply studying individual components.

The second big challenge is to find ways and means, that will allow us to utilize the information available in complex systems, taking into consideration that the data is not complete but nevertheless trying to create the "most probable reality".

The third challenge is maintenance of a successful business — getting the right perspective, setting the right priority, realizing breeding progress in the shortest possible time-frame, and generating thereby profitability in a market where competition will put increasing pressure on the sugar beet business.

Acknowledgments: The author thanks his colleagues Bengt Bentzer, Philippe Chatin, Maria Nilgård, and Thomas Kraft for their ideas and critical reading of this manuscript.

References

Bajaj, S., J. Tagolli, L.F. Liu, T.H.D. Ho, and R. Wu. 1999. Transgenic approaches to increase dehydration-stress tolerance in plants. Molec. Breed. 5:493-503.

Börnke, F., M. Hajirezaei, D. Heinike, M. Melzer, K. Herbers, and U. Sonnewald. 2002. High-level production of the non-cariogenic sucrose isomer in transgenic tobacco plants strongly impairs development. Planta 214:356-364.

Chengappa, S., M. Guilleroux, W. Philipps, and R. Shields. 1999. Transgenic tomato plants with decreased sucrose synthase are unaltered in starch and sugar accumulation in the fruit. Plant Molec. Biol. 40:213-221.

Chikano, H., M. Ogawa, Y. Ikeda, N. Koizumi, T. Kusano, and H. Sano. 2001. Two novel genes encoding SNF1-related protein kinases from *Arabidopsis thaliana*: Differential accumulation of AtSR1 and AtSR2 transcripts in response to cytokinins and sugars, and phosphorylation of sucrose synthase by AtSR2. Molec. Gen. Genet. 264:674-681.

Chiou, T.J. and D.R. Bush. 1996. Molecular cloning, immunochemical localization to the vacuole, and expression in transgenic yeast and tobacco of putative sugar transporter from sugar beet. Plant Physiol. 110:511-520.

Davidson, P.A., C.N. Hunter, and P. Horton. 2002. Overexpression of beta-carotene hydroxylase enhances stress tolerance in *Arabidopsis*. Nature 418:203-206.

Eastmond, P.J., A.J.H. van Dijken, M. Spielman, et al., 2002. Trehalose-6-phosphate synthase 1, which catalyses the first step in trehalose synthesis, is essential for *Arabidopsis* embryo maturation. Plant J. 29:225-235.

Ghoshal, K., Z. Li, and S.T. Jacob. 1998. Overexpression of the large subunit of the protein Ku suppresses metallothionein-I induction by heavy metals. Proc. Natl. Acad. Sci. 95:10390-10395.

Häusler, R.E., H.J. Hirsch, F. Kreuzaler, and C. Petrerhänsel. 2002. Overexpression of C4-cycle enzymes in transgenic C3 plants: a biotechnological approach to improve C3-photosynthesis. J. Exper. Bot. 53:591-607.

Keenan, R.J. and W.P.C Stemmer. 2002. Nontransgenic crops from transgenic plants. Nature Biotech. 20:215-216.

Klann, E.M., B. Hall, and A.B. Bennett. 1996. Antisense acid invertase (TIV1) gene alters soluble sugar composition and size in transgenic tomato fruit. Plant Physiol. 112:1321-1330.

Koch, K.E. 1996. Carbohydrate-modulated gene expression in plants. Ann. Rev. Plant Physiol. Plant Molec. Biol. 47:509-540.

Komatsu, A., T. Moriguchi, K. Koyama, M. Omura, and T. Akihama. 2002. Analysis of sucrose synthase genes in citrus suggests different roles and phylogenetic relationships. J. Exper. Bot. 53:61-71.

Kuykendall, L.D., T.M. Stockett, and J.W. Saunders. 2002. Transformation of sugar beet with a Cercosporin export gene. Proc. Meet. Amer. Phytopath. Soc., July 27-31 2002, Milwaukee, WI, p. 43.

Lauber, E., L. Janssens, G. Weyens, et al., 2001. Rapid screening for dominant negative mutations in the beet necrotic yellow vein virus triple gene block proteins P13 and P15 using a viral replicon. Transg. Res. 10:293-302.

Lehr, A., M. Kirsch, R. Viereck, J. Schiemann, and T. Rausch. 1999. CDNA and genomic cloning of sugar beet V-type H+ -ATPase subunit A and c isoforms: evidence for coordinate expression during plant development and coordinate induction in response to high salinity. Plant Molec. Biol. 39:463-475.

Levy, Y.Y., S. Mesnage, J.S. Mylne, A.R. Gendall, and C. Dean. 2002. Multiple roles of *Arabidopsis* VRN1 in vernalization and flowering time control. Science 297:243-246.

Luo, M., P. Piffanelli, L. Rastelli, and R. Cella. 1993. Molecular cloning and analysis of a CDA coding for the bifunctional dihydrofolate reductase-thymidylate synthase of *Daucus carota*. Plant Molec. Biol. 22:427-435.

Miron, D., M. Petreikov, N. Carmi, et al., 2002. Sucrose uptake, invertase localization and gene expression in developing fruit of *Lycopersicon esculentum* and the sucrose-accumulating *Lycopersicon hirsutum*. Physiol. Plant. 115:35-47.

Nielsen, K., J.D. Mikkelsen, K.M. Kragh, and K. Bojsen. 1993. An acidic class III chitinase in sugar beet: Induction by *Cercospora beticola*, characterization, and expression in transgenic tobacco plants. Molec Plant Mic. Intl. 6:495-506.

Nonato, R.V., P.E. Mantelatto, and C.E.V. Rossell. 2001. Integrated production of biodegradable plastic, sugar and ethanol. Appl. Microbiol. Biot. 57:1-5.

Pego, J.V. and S.C.M. Smeekens. 2000. Plant fructokinases: A sweet family get-together. Trends Plant Sci. 5:531-536.

Pidgeon, J.D. 2000. Drought stress, climate change and research strategy for sugar beet crop. Inst. Arable Crops Res. p. 36-39.

Ravasz, E., A.L. Somera, D.A. Mongru, Z.N. Oltvai, and A.L. Barabási. 2002. Hierarchical organization of modularity in metabolic networks. Science 297:1551-1555.

Rosenkranz, H., R. Vogel, S. Greiner, and T. Rausch. 2001. In wounded sugar beet (*Beta vulgaris* L.) tap-root, hexose accumulation correlates with the induction of a vacuolar invertase isoform. J. Exper. Bot. 52:2381-2385.

Shakya, R. and A. Sturm. 1998. Characterization of source- and sink-specific sucrose/h+ symporters from carrots. Plant Physiol. 118:1473-1480.

Simpson, G.C., A.R. Gendall, and C. Dean. 1999. When to switch to flowering. Ann. Rev. Cell. Dev. Biol. 99:519-550.

Stracke, S., C. Kistner, S. Yoshida, et al., 2002. A plant receptor-like kinase required for both bacterial and fungal symbiosis. Nature 417:959-962.

Sturm, A., S. Lienhard, S. Schatt, and M. Hardegger. 1999. Tissue-specific expression of two genes for sucrose synthase in carrots. Plant Molec. Biol. 39:349-360.

Tang, G.Q. and A. Sturm. 1999. Antisense repression of sucrose synthase in *carrot (Daucus carota* L.) affects growth rather than sucrose partitioning. Plant. Molec. Biol. 41:465-479.

Tang, G.Q., M. Lüscher, and A. Sturm. 1999. Antisense repression of vacuolar and cell wall invertase in transgenic carrot alters early plant development and sucrose partitioning. Plant Cell 11:177-189.

Thomine, S., R. Wang, J.M. Ward, N.M. Crawford, and J.I. Schroeder. 2000. Cadmium and iron transport by members of a plant metal transporter family in *Arabidopsis* with homology to Nramp genes. Proc. Natl. Acad. Sci. USA 397:4991-4996.

Tymowska, L.Z. and M. Kreis. 1998. Expression of the *Arabidopsis thaliana* invertase gene family. Planta 207:259-265.

Additional Information Sources

Monographs

Casarini, B., E. Biancardi, and P. Ranalli. 1999. La Barbabietola da Zucchero in Ambiente Mediterraneo (in Italian). Edagricole, Bologna, Italy.

Cooke, D.A. and R.K. Scott. (eds). 1993. The Sugar Beet Crop. Chapmann & Hall, London, UK.

Johnson R.T., J.T. Alexander, G.E. Bush, and G.R. Hawkes. 1971. Advances in Sugar Beet Production. Iowa State Univ. Press, Ames, IA.

Märländer, B. 1991. Zuckerrüben (in German). Bernhardt-Patzold, Stadthagen, Germany.

Winner, C. 1981. Zuckerrübenbau (in German). DLG Verlag, Munich, Germany.

Proceedings and International Journals

Betteravier Française (in French)
25 Rue da Madrid, 75008 Paris, France
Site: www.lebetteravier.com
E-mail: lebetteravier@wanadoo.fr

British Sugar Beet Review
Oundle Road, Peterborough, Cambs PE2 9QU, UK
E-mail: beetreview@britishsugar.co.uk

IIRB: Proceedings of Annual Congresses (in English, French or German; summaries in languages cited)

IIRB, Avenue de Tervuren 195, B-1150 Brussels, Belgium.
Site: www.iirb.org. E-mail: mail@iirb.ory

IRBAB: Beet Information Bulletin
Moolenstraat 45, B-3300 Tienen, Belgium
Site sme.belgium.eu.net.irbab-kbivb
E-mail: www.bib@irbab.be

Journal of Sugar Beet Research
ASSBT, 800 Grant Street, Suite 300, Denver, CO 80203
Site: www.bsdf-assbt.com

Zuckerindustrie (in German or English, summaries in German, English, French, and Spanish)
Postfach 380250 D-14122 Berlin, Germany.
Site: www.bartens.com
E-mail: zuckerindustrie@bartens.com

History

Holbrich, U. 1989. Zucker-Museum (in German). Druckhaus Hentrich, Berlin, Germany.

Institut für Zuckerrübenforschung (IFZ). 1984. Geschichte der Zuckerrübe (in German). 200 Jahre anbau und Züchtung. Albert Bartens, Berlin, Germany.

Institut International de Recherches Betteravières (IIRB). 1982. 50 Annèes de Recherches Betteravières (in French; also available in English and German). IIRB, Brussels, Belgium.

Lippmann E.O. 1925. Geschichte der Rübe (*Beta*) als Kulturpflanze (in German). Verlag von Julius Springer, Berlin, Germany.

Lippmann E.O. 1929. Geschichte des Zuckers (in German). Verlag von Julius Springer, Berlin, Germany.

Müller, H.H. 2002. Franz Carl Achard (in German). Verlag Albert Bartens, Berlin, Germany.

Winner, C. 1993. History of the crop. *In:* D.A. Cooke, and R.K. Scott (eds.). The Sugar Beet Crop. Chapman & Hall, London, UK, pp. 1-35.

Plant Breeding and Genetics

Fehr, W.R. 1987. Principles of Cultivar Development, vol. 1. Macmillan Publishing Company, London, UK.

Hayward, M.D., N.O. Bosemark, and I. Romagosa. 1993. Plant Breeding. Chapman & Hall, London, UK.

Hallauer, A.R. and J.B. Miranda. 1988. Quantitative Genetics in Maize Breeding. Iowa State Univ. Press, Ames, IA (2nd ed.).

Simmonds, N.V. 1987. Principles of Crop Improvement. John Wiley & Sons Inc., New York, NY.

Breeding and Genetics of Sugar Beet

Bandlow, G. 1955. Die Genetik der *Beta vulgaris* Rüben (in German). Der Züchter 4-5:104-122.

Barocka, K.H. 1985. Zucker und Futterrüben (*Beta vulgaris* L.). *In:* W. Hoffmann, A. Mudra, and W. Plarre (eds.). Lehrbuch der Züchtung landwirtschaftlicher Kulturpflanzen (in German). Verlag Paul Parey, Berlin, Germany, vol. 2, pp. 245-287.

Biancardi, E. 1999. Miglioramento genetico. *In:* B. Casarini, E. Biancardi, and P. Ranalli (eds). La Barbabietola da Zucchero in Ambiente Mediterraneo (in Italian). Edagricole, Bologna, Italy, pp. 57-145.

Bosemark, N.O. 1993. Genetics and breeding. *In:* D.A. Cooke and R.K. Scott (eds). The Sugar Beet Crop. Chapman & Hall, London, UK, pp. 67-119.

Campbell, L.G. 2002. Sugarbeet breeding and improvement. *In:* M.S. Kang (ed.). Crop Improvement in the Twenty-first Century. Food Product Press, New York, NY, pp. 193-221.

Coons, G.H. 1936. Improvement of the sugar beet. Yearbook of Agriculture, pp. 625-657.

Coons, G.H., F.W. Owen, and D. Steward. 1955. Improvement of the sugar beet in the United States. Advances in Agronomy 7:89-139.

Hecker, R.J. and R.M. Helmerich. 1985. Sugar beet breeding in the United States. *In:* G.E. Russel (ed.). Progress in Pant Beeding. Butterworths, London, UK, pp. 37-71.

Knapp, E. 1958. *Beta* Rüben. *In: R.* Roemer and W. Rudorf (eds.). Handbuch der Pflanzenzüchtung (in German) Paul Parey, Berlin, Germany, vol. 3, pp. 196-284.

Lasa, J.M. and I. Romagosa. 1992. Mejora Genetica de la Ramolacha Azucarera (in Spanish) AIMCRA, Valladolid, Spain.

Lewellen, R.T. 1992. Use of plant introductions to improve populations and hybrids of sugar beet. *In:* Use of Plant Introductions in Cultivar Development, part 2. Crop Science Society of America, Madison, WI, pp. 117-135.

McFarlane, J.S. 1971. Variety development. *In:* R.T. Jonhson, J.T. Alexander, G.E. Bush, and G.R. Hawkes (eds.). Advances in Sugar Beet Production. Iowa State Univ. Press, Ames, IA, pp. 402-435.

Nakamura, C., G.N. Skaracis, and I. Romagosa. 1992. Cytogenetics and breeding in sugar beet. *In:* T. Tsuchiya and P.K. Gupta (eds.). Chromosome Engineering in Plants: Genetics, Breeding, Evolution, part B. Elsevier, Amsterdam, pp. 295-313.

Smith, A.L. 1983. Influence of External Factors on Growth and Development of Sugar Beet (*Beta vulgaris* L.). Pudoc, Wageningen, Netherlands.

Smith, G.A. 1980. Sugarbeet. *In:* W.R. Fehr and H.H. Hadley (eds.). Hybridization of Crop Plants. ASA, CSSA, and SSSA. Madison, WI, pp. 601-616.

Smith, G.A. 1987. Sugar Beet. *In* W.R. Fehr (ed.). Principles of Cultivar Development. Macmillan Publishing Company, New York, NY, vol. 2, pp. 577-625.

Sneep, J. and J.T. Hendriksen (ed.). 1979. Plant Breeding Perspectives. Pudoc, Wageningen, Netherlands.

Biotechnology

Wolzniak, C.A. 1999. Transgenic sugarbeet: progress and development. *In:* V.L. Chopra, V.S. Malik, and S.R. Baht (eds.). Applied Biotechnology. Science Publisher Inc, Enfield, NH, pp. 395-413.

Agronomy and Physiology

Cavazza L., M.T. Amaducci, G. Venturi, and C. Pesci. 1983. Ecologie de la Betterave à Sucre dans la Région Méditerraneenne (in French, also available in English and German). IIRB, Brussels, Belgium.

Draycott, A.P. 1972. Sugar Beet Nutrition. Applied Science Publishers Ltd., London. UK.

Elliott, M.C. and G.D. Weston. 1993. Biology and physiology of the sugar beet plant. *In:* D.A. Cooke and R.K. Scott (eds). The Sugar Beet Crop. Chapman & Hall, London, UK, pp. 37-66.

IIRB. 2002. Variety trials in sugar beet: Methodology and Design. Advances in Sugar Beet Research. IIRB, Brussels, Belgium, vol. 4.

Märländer B. 1991. Zuckerrüben (in German) Bernhardt-Patzold, Stadthagen, Germany.

Martin, S.S. 1983. Sugarbeet. *In:* I.D. Teare and M.M. Peet (eds.). Crop Water Relations. Wiley, New York, NY, pp. 423-443.

Scott, R.K. and K.W. Jaggard. 1993. Crop physiology and agronomy. *In:* D.A. Cooke and R.K. Scott (eds.). The Sugar Beet Crop. Chapman & Hall, London, UK, pp. 179-237.

Smit A.L. 1983. Influence of External Factors on Growth and Development of Sugar Beet (*Beta vulgaris* L.), (in English with Dutch summaries) Pudoc, Wageningen, Netherlands.

Wolf I. 1995. Sorte und Sortenwahl bei Zuckerrüben und deren Wechselwirkung zu Umwelt und Qualitätsbezahlung (in German). Cuvillier Verlag, Göttingen, Germany.

Wyse, R.I. 1982. The sugarbeet chemistry: the sugarbeet and sucrose formation. *In:* R.A. McGinnis (ed.). Sugar-beet Technology. Beet Sugar Development Foundation, Fort Collins, CO, pp. 17-24 (3rd ed.).

Pathology

Asher, M.J.C. 1993. Rhizomania. *In:* D.A. Cooke and R.K. Scott (eds.). The Sugar Beet Crop. Chapman & Hall, London, UK, pp. 311-346.

Benada, J., J. Šedivw, and J. Špaèek. 1987. Atlas of Diseases and Pests in Beet (in English, French, and German). Elsevier, Amsterdam, Netherlands.

Cooke, D.A. 1993. Pests. *In:* D.A. Cooke and R.K. Scott (eds.). The Sugar Beet Crop. Chapman & Hall, London, UK, pp. 429-483.

Duffus, J.E. and E.G. Ruppel. 1993. Diseases. *In:* D.A. Cooke and R.K. Scott (eds.). The Sugar Beet Crop. Chapman & Hall, London, UK, pp. 347-427.

IIRB. 2000. *Cercospora beticola* Sacc. Biology, agronomic influence and control measures in sugar beet. Advances in Sugar Beet Research. IIRB, Brussels, Belgium, vol.2.

Lejealle, F. and J. d'Aguilar. 1982. Pests, Disease, and Disorders of Sugar Beet (also available in French, German, Italian, and Spanish) Deleplanque & Cie, Maison Laffitte, France.

Lüdecke, H. and Winner C. 1959. Farbtafelatlas der Krankenheiten und Schädigungen der Zuckerrübe (in German). DLG-Verlag, Frankfurt am Main, Germany.

Mukhopadhyay, A.N. 1987. Handbook on Diseases of Sugar Beet. CRC Press Inc., Boca Raton, FL.

Scholten, O.E. 1997. Characterization and inheritance of resistance to sugar beet to beet necrotic yellow vein virus in *Beta*. Ph.D. diss. (in English with Dutch summaries) Wageningen Agricultural University, Wageningen, Netherlands.

Tuitert, G. 1994. Epidemiology of Rhizomania Disease in Sugar Beet. IRS, Bergen op Zoom, Netherlands.

Whitney, E.D. and J.E. Duffus. 1986. Compendium of Beet Disease and Insects. APS Press, St. Paul, MN.

Seed production

Bornscheuer, E., K. Meierholz and K.H. Wunderlich. 1993. Seed production and quality. *In:* D.A. Cooke and R.K. Scott (eds.). The Sugar Beet Crop. Chapman & Hall, London, UK, pp. 121-153.

Processing

Clarke, M.A. and M.A. Godshall. 1988. Chemistry and Processing of Sugarbeet and Sugarcane. Elsevier Science Publisher B.V., Amsterdam, Netherlands.

McGinnis, R.A. 1982. Sugar-beet Technology. Beet Sugar Development Foundation, Fort Collins, CO (3rd ed.).

Vukov K. 1977. Physics and Chemistry of Sugar-beet in Sugar Manufacture. Elsevier Sci. Publ. Co., Amsterdam, Netherlands.

Processing Quality

Campbell, L.G. 2002. Sugarbeet quality improvement. *In:* A.S. Basra (ed.) Quality Improvement in Field Crop. Food Products Press, New York, NY, pp. 395-413.

Harvey, C.W. and J.W. Dutton. 1993. Root quality and processing. *In:* D.A. Cooke and R.K. Scott (eds.). The Sugar Beet Crop. Chapmann & Hall, London, UK, pp. 571-617.

Oltmann, W., M. Burba, and G. Bolz. 1984. Die Qualität der Zuckerrübe, Bedeutung, Beurteilungskriterien und züchterische Massnahmen zu ihre Verbesserung (in German, with English summary) Paul Parey, Berlin, Germany.

Dictionaries

Haensch, G. and G. Habercamp de Anton. 1987. Dictionary of Agriculture. (German, English, French, Spanish, Italian, and Russian) BLV Verlaggesellschaft, Munich, Germany.

IIRB. 1999. Sugar Beet Dictionary (English, French, and German) IIRB, Brussels, Belgium.

Universities of Geneva, Georgetown, Heidelberg, Mainz, Paris, and Trieste. 1970. Glossary of Genetics (English, French, Spanish, Italian, German, and Russian) Elsevier Sci. Publ. Co., Amsterdam, Netherlands.

Index

About the Editors

Enrico Biancardi (Graduate of Bologna University) has been involved in sugar beet breeding activities at the Istituto Sperimentale per le Colture Industriali, Sezione di Rovigo (Italy) since 1977. He has released several breeding lines with resistance to rhizomania and cercospora leaf spot that were subsequently used for commercial purposes. His current research activities include the localization and characterization of wild beet populations and the physiology of the sugar beet root system.

Larry G. Campbell (Ph.D. Kansas State University) has served as a sugar beet breeder-geneticist with the USDA-Agricultural Research Service in Fargo, North Dakota (USA) since 1978. His research has focused on improving sugar beet root maggot resistance, post-harvest storage traits, impurity components, and the expansion of the genetic diversity of sugar beet through the introgression of exotic germplasm into cultivated sugar beet.

George N. Skaracis (M.Sc. and Ph.D Colorado State University) has been a sugar beet breeder with Hellenic Sugar Industry SA., Thessalonica (Greece), since 1976 and has developed and registered for commercial use a large number of varieties, emphasizing on disease resistance and quality. He has served in many national and international organizations and committees relevant to his research interests. Earlier in 2004, he was elected as Professor of Plant Breeding and Biometry at the Agricultural University of Athens.

Marco De Biaggi (Graduate of Padua University) began his sugar beet breeding career in 1974 at the Istituto Sperimentale per le Colture Industriali (ISCI), Sezione di Rovigo. In 1980, he assumed breeding responsibilities at S.E.S. Italy and Agra-Agrosem seed companies. He developed "Rizor", the first monogenic rhizomania resistant-variety in 1985. Recently, he rejoined the ISCI-Rovigo staff. Currently, he is involved in the development of additional rhizomania-resistant lines in collaboration with Lion Seeds Ltd.